Efficient Energy-Saving Control and Optimization for Multi-Unit Systems

Fulai Yao · Yaming Yao

Efficient Energy-Saving Control and Optimization for Multi-Unit Systems

A Guide for Electrical Engineers

 Springer

Fulai Yao
School of Artificial Intelligence
Hebei University of Technology
Tianjin, China

Yaming Yao
Beijing IAO Technology Development
Company
Beijing, China

ISBN 978-981-97-4491-6   ISBN 978-981-97-4492-3 (eBook)
https://doi.org/10.1007/978-981-97-4492-3

The project was supported by the authors of this project.

This Springer imprint is published by the registered company Springer Nature Singapore Pte Ltd.
The registered company address is: 152 Beach Road, #21-01/04 Gateway East, Singapore 189721, Singapore

If disposing of this product, please recycle the paper.

Preface

Excessive carbon emissions lead to global warming and threaten human survival. "Carbon peaking" and "carbon neutrality" with the goal of reducing carbon dioxide emissions have become the consensus of all mankind. In order to reduce the use of fossil energy, we need green and efficient energy production for power generation systems; for energy consumption systems, energy conservation and efficient operation are required. We also need efficient carbon utilization and carbon sequestration. These three aspects all involve energy efficiency. Therefore, the International Energy Agency (IEA) believes that "energy efficiency is the first fuel" and will make the greatest contribution to the climate goals of the Paris Agreement.

In various fields of the national economy, there are a large number of electric motors, generators, steam turbines, engines, aircraft, ships, automobiles, trains, boilers, air conditioners, water pumps, fans, and home appliances. As long as they use energy, there are energy efficiency issues.

During the COVID-19 epidemic, we conducted a random operation energy efficiency survey on 501 secondary water supply pumping stations in buildings in an international metropolis (population of more than 20 million). The average power waste is 54%, and the operation energy efficiency was surprisingly low.

Electrical control and automation control, firstly, replace labor, improve output and precision, and create a safe and comfortable environment; secondly, they can reduce energy consumption and improve energy efficiency while completing the same tasks.

This book does not discuss the rated energy efficiency optimization of single-unit equipment, but rather focuses on the energy efficiency optimization of multi-unit systems. The so-called multi-unit system refers to a system composed of multiple energy-consuming equipment in order to complete the same task. It includes a pumping station composed of multiple pumps, a wind turbine station composed of multiple fans, a power plant composed of multiple generators, a hydropower station composed of several hydraulic generators, and a transformer and distribution station composed of multiple transformers. It also includes trains and automobiles driven by multiple electric motors, a refrigeration system composed of multiple water-cooling units, a heating system composed of multiple boilers, and a wind power hydrogen production station composed of multiple hydrogen generators, etc.

Energy efficiency optimization of multi-unit systems is a multi-variable, nonlinear, time-varying, integer, real-number optimization problem with constraints. Its analysis and application involve multiple disciplines and majors such as mathematics, computers, electrical engineering, and automation. Since it is difficult to establish an accurate mathematical model of a multi-unit system, conventional optimization methods are difficult to function. This is a hot and difficult issue in the field of "dual carbon".

The quantum optimization method and energy efficiency prediction theory proposed in this book solve the energy efficiency optimization problem of multi-unit systems. They have been successfully applied in nearly a thousand pumping stations in the fields of municipal administration, steel, chemical industry, pharmaceuticals, fertilizers, petroleum, construction, and other fields. This method is simple and practical, with clear concepts. It is an assistant and guide for electrical engineers and automation engineers to achieve efficient operation of engineering projects. "Efficiency Peak" is the technical support for "Carbon Peak", and energy-saving automation is an inevitable choice for future electrical control engineering.

This book is mainly written by Fulai Yao and Yaming Yao. Parts of Chap. 14 are assisted by Dr. Robert Gao. Parts of Chaps. 15 and 16 are assisted by Dr. Qingbin Gao. Parts of Chaps. 18 and 22 are assisted by Mr. Shengyong Zhang. Parts of Chaps. 19 and 21 are assisted in writing by Dr. Chengyu Cao. Parts of Chap. 20 are assisted in writing by Dr. Hexu Sun. Parts of Chap. 23 are assisted in writing by senior engineer Bosheng Yao and senior engineer Yanfang Zhang. Parts of Chap. 24 was assisted in writing by Dr. Yuxi Gu. Ms. Amy Kong and Ms. Yolanda Wang completed the data collection and typing of some chapters. Dr. Qingbin Gao polished and proofread the entire content of this book. We would like to express our sincere gratitude! Some of the contents in Chap. 8 are not suitable for beginners to master. Readers can freely choose to study according to their needs.

Tianjin, China   Fulai Yao
Beijing, China   Yaming Yao

Contents

About the Authors

Fulai Yao is "Gold Medal Author" and "Author of Top 100 Works of the 70th Anniversary" of China Machine Press. As the first author, he published 13 books, translated one book, published more than 100 papers in Chinese and English, and was the first inventor of more than 100 patents. He has completed a large number of electrical control projects. His book *Electrical Automation Engineer Crash Course* has been reprinted more than 20 times as a bestseller, and the book *Automation Equipment and Engineering Design. Installation. Debug. Fault Diagnosis* has been reprinted more than 10 times. One of the companies he founded was ranked 29th among the top 100 unlisted companies in China by Forbes in 2014.

Yaming Yao is the director and engineer of the energy-saving engineering Department of Beijing IAO Company. He has completed many optimization and energy-saving renovation projects of water supply systems, and researched and completed scientific and technological projects such as air water extraction desert plant irrigation devices and multi-station laser die-cutting machines. He has participated in the writing of 10 books, published four papers, and is the inventor of 13 patented technologies.

Chapter 1
Energy-Saving Theory, Technology, and Double Carbon Target

Excessive carbon emissions lead to global warming and threaten human survival. On November 13, 2021, at the United Nations Climate Change Conference, all parties completed the implementation details of the Paris Agreement. The agreement is signed by 178 parties around the world. The long-term goal is to limit the increase in global average temperature to less than 2 °C compared with the pre-industrial period, and strive to limit the increase in temperature to less than 1.5 °C. So far, the "carbon peaking and carbon neutrality" ("double carbon") strategy with the goal of reducing carbon dioxide emissions has become the consensus of all mankind.

1.1 Energy Efficiency Is the Number One Fuel

The use of fossil energy sources such as coal, natural gas and oil is a major source of carbon dioxide. In order to reduce the use of fossil energy, we need green and efficient energy production for power generation systems; for energy consumption systems, we need energy conservation and high energy efficiency.

The International Energy Agency (IEA) believes that "energy efficiency is the first fuel" and will make the greatest contribution to the climate goals of the Paris Agreement.

Energy efficiency peak is the technical support for "carbon peak"!

© The Author(s) 2024
F. Yao and Y. Yao, *Efficient Energy-Saving Control and Optimization for Multi-Unit Systems*, https://doi.org/10.1007/978-981-97-4492-3_1

1.2 Energy Saving Is One of the Important Purposes of Electrical Control

The purpose of electrical control and automation control is to replace labor, improve production and precision, and create a safe and comfortable environment. In the case of completing the same task, reducing energy consumption and improving energy efficiency are also one of its important purposes.

In addition to saving production costs, energy saving can also improve the living environment of human beings and reduce carbon dioxide emissions.

During the COVID-19 pandemic, our random survey of 501 secondary water supply pumping stations in an international metropolis (with a population of more than 20 million) showed that the average power waste is 54%. That is to say, if the energy efficiency optimization design method and the energy efficiency optimization control method are adopted, the power consumption of 200,000–300,000 sets of secondary water supply pumping stations in the city can be reduced by 54%!

1.3 Energy Saving Needs Are Everywhere

During the development of human society, various devices have been invented to help people use and change nature, including electric motors, transformers, diesel engines, steam engines, gas engines, water pumps, fans, ships, generators, automobiles, trains, airplanes, motor vehicles, steam turbines, boilers, etc.

In various fields of the national economy, including industry, agriculture, transportation, municipal administration, construction, etc., as long as energy is needed to maintain operation, energy conservation must be faced.

1.3.1 Single Device Energy Saving

As an individual device, according to its efficiency function, its energy saving status can be determined. The point of maximum efficiency corresponds to the optimum load. Taking a car as an example, the economic speed corresponds to the maximum efficiency point, which is the most fuel-efficient working point. It is easy to judge the energy saving status. The key issue is to obtain the efficiency curve of the equipment.

For energy-consuming objects such as human beings or man-made machines, each has an efficiency function, and its shape is generally shown in Fig. 1.1.

In Fig. 1.1, β is the load and η is the efficiency function. η_e is the maximum efficiency value corresponding to the optimal load β_e, and β_m is the maximum load.

For generator sets, if the load β is water intake or coal intake, each has an efficiency function whose shape is generally shown in Fig. 1.2.

Fig. 1.1 The efficiency
function η

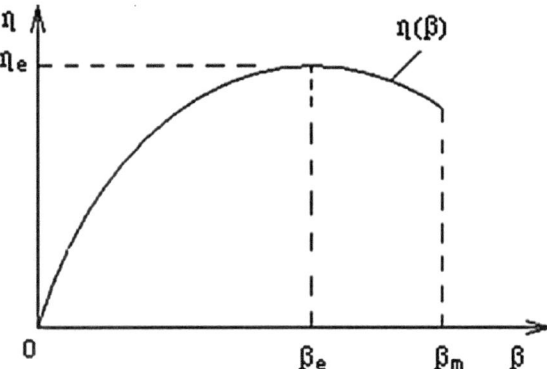

Fig. 1.2 The efficiency
function η

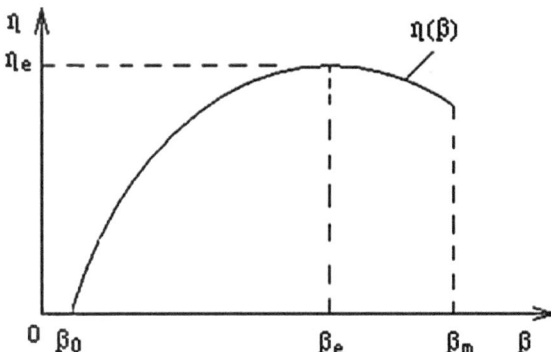

In Fig. 1.2, β is the power generation load, η is the efficiency function at rated speed, β_0 is the input load when electric energy starts to be generated, η_e is the maximum efficiency value corresponding to the optimal load β_e, and β_m is the maximum load.

According to the efficiency curve, we can judge whether the current load is optimal.

1.3.2 Multi-unit System Energy Saving

In reality, many systems are composed of multi-unit devices, called multi-unit systems. Judging the energy saving status of such a multi-unit system is not simple and intuitive (Figs. 1.3, 1.4 and 1.5).

(1) A hydropower station has multiple hydroelectric generating units. According to the flow and head of the river, how to schedule the number of operating units and how to control the load rate of each operating unit to obtain the maximum power generation?

Fig. 1.3 Power stations

Fig. 1.4 Power transmission

Fig. 1.5 Energy consumption

(2) There are multiple transmission lines supplying power in a region. Given the total load in the region, how to dispatch the transmission lines to maximize the overall efficiency of the power supply system?

(3) Given the total load in a power system with multiple transformers, how can these transformers be distributed to maximize the overall efficiency of the power system?

(4) A pumping station consists of multiple pumps. According to the required flow and head, how to arrange the number of operating units and how to adjust the

water volume of each operating pump to maximize the overall efficiency of the pumping station?

(5) A fan station consists of multiple fans. According to the required flow rate and pressure, how to arrange the number of running fans and how to adjust the flow rate of each running fan to maximize the overall efficiency of the fan station?

(6) A high-speed train is driven by multiple motors. Under given conditions, how to control the number of operating motors and adjust the output torque of each operating motor to minimize the overall power consumption?

(7) An electric car is driven by multiple motors, how to control the number of units running the motors and how to distribute the output torque of each running motor to minimize the power consumption and get the longest range?

(8) The propeller of a ship is driven by multiple motors. Facing a given situation, how to control the number of motors running and how to adjust the output torque of each motor to minimize power consumption and obtain the longest cruising distance?

(9) There are multiple boilers in the heating system, how to distribute the load of each boiler to maximize the overall efficiency of the system?

(10) Multiple motors drive a long conveyor, how to distribute the output torque of each motor so that the overall power consumption is the lowest?

As long as multiple devices or multiple people are required to cooperate to complete a job, there must be an energy saving problem, which is the overall efficiency optimization problem.

If you are given on-site operating data and equipment parameters, can you provide the optimal operating efficiency and the control method to achieve this efficiency? Obviously, this is not an easy job to do. Otherwise, the "carbon peaking" work will not be so tangled.

1.4   R&D Overview of Energy Efficiency Optimization

People always want to do the most work with the least cost, which is the origin of the efficiency optimization problem, and the way to achieve this goal is the efficiency optimization method.

If the optimization variables are independent parameters and do not change with time, for such a system, it is called a static optimization problem or a parameter optimization problem. The main task is to solve the extremum function. This optimization includes three basic elements, namely variables, objective function and constraints. The optimization process is to change and finally determine the optimization variables under the condition of satisfying the constraints, so that the objective function has a minimum or maximum value [1].

If the objective function is a linear function of the optimization variables, and the constraints are linear equations or inequalities of the optimization variables, then such optimization is classified as a linear programming problem [2]. If the objective function or constraints contain a nonlinear function of one or more optimization variables, it is classified as a nonlinear programming problem [3]. In nonlinear programming, if the objective function is a quadratic function of the optimization variables and the constraints are linear functions of the optimization variables, then the optimization problem is called a quadratic programming problem [4]. If the objective function is a convex function and the constrained feasible region is a convex set, it is called a convex optimization problem [5]. A notable feature of convex optimization is that the local optimal solution is the global optimal solution.

The most commonly used method to solve extreme value problems is the classic differential method in advanced mathematics [6, 7]. The optimal solution is obtained by taking the partial derivative of the objective function with respect to each variable so that it is zero.

If the objective function of a nonlinear optimization problem is too complex to be solved analytically, then we can resort to numerical methods [8]. In order to improve the computational efficiency on the basis of versatility, people have developed engineering numerical optimization methods [9, 10].

If the system changes from one working condition to another, the system parameters are a function of time, and the optimized objective function is a function of the function, that is, functional. Such a system, whose characteristics are expressed by differential equations, is a dynamic optimization problem, also known as an optimal control problem [11–13].

Solving dynamic optimization problems requires the use of variational methods [14–16]. Such a system has three essential elements: a mathematical model of the system, physical constraints, and performance metrics. The optimal control problem [17, 18] is summarized as: in a dynamic system, find the optimal control scheme from the allowable control methods, so that the system moves from one state to another, and the selected performance index is optimal. The essence of dynamic optimization is to find the extreme value of the function. According to different purposes, optimal control is divided into time optimal control, terminal optimal control, fuel optimal control and energy optimal control [19, 20]. Exact optimal control is used for deterministic systems, while stochastic optimal control is used for stochastic systems with random variables [21, 22]. When the control variables are unconstrained, the classical variational method is used to solve the problem. When the control variables are constrained, dynamic programming [23] or minimum principle [24] can be used to solve it. For linear systems represented by quadratic performance metrics, quadratic linear optimal methods can be used [25].

As the equipment ages or the environment changes, the mathematical model of the system will also change, and the original optimal system is no longer optimal. In order to solve this problem, many online optimization methods have been developed, such as rolling optimization algorithm of predictive control and steady-state hierarchical control method [26–29]. For complex multi-objective systems, it is difficult to establish an accurate optimization model, and genetic algorithms, neural network optimization methods, and fuzzy optimization methods are more superior than classical optimization methods [30–35]. Genetic algorithm is a search and optimization method. According to the biological evolution rules of the survival of the fittest, it gradually approaches the global optimal solution and suboptimal solution from the initial solution; the movement of the neural network is always in the direction of reducing energy, and finally reaches the system equilibrium point (that is, the energy minimum point); in the fuzzy optimization method, the control variable, the objective function and the constraint conditions may all be fuzzy, or only one aspect may be fuzzy. This method mainly uses cut sets or membership functions transform fuzzy problems into classical planning problems.

Many other optimization methods are widely studied as well, such as Lagrangian relaxation [36], quasi-Newton [37], recursive quadratic programming [38], equal increment principle [39], stochastic dynamic programming [40], ant colony optimization [41], particle swarm optimization [42], evolutionary strategy [43], simulated annealing [44], etc. [45–49].

With the proposal of dual carbon targets and the pursuit of energy-saving goals, the research on the efficiency optimization of various systems is becoming more and more extensive, such as the optimization and energy saving of urban water supply systems [50–54], heat power system optimization and energy saving [55–60], power system optimization and energy saving [61–65], motor system optimization and energy saving [66–69], transportation optimization scheduling system [70–72], etc. It can be said that as long as energy is used, there is a problem of optimizing energy saving.

A lot of overall optimal problems in human society, most of them are a combination of static optimization and dynamic optimization. For example, for a large-scale water supply system, the water supply of the water plant is randomly determined by a lot of users, and the water quantity and water pressure that meet the process requirements may change at any time, which requires dynamic optimization. If the water supply and water pressure are stable over a period, it is necessary to use a static optimization method to solve the problem. Large-scale gas transmission systems are similar. Under the influence of acceleration, deceleration, uphill, downhill, wind speed, passenger capacity and other factors, EMUs will also have static optimization and dynamic optimization problems. These systems have multiple motors to provide power, so the optimization of power switching also needs to be considered. For some fluid systems, it is still difficult to give an accurate mathematical model, and the characteristics of the equipment are mostly given in the form of discrete data or curves. For these complex systems, there are not only the problems of combining static optimization and dynamic optimization, but also the problems of how to extract the mathematical model of the system and how to establish the objective function.

At present, there is no unified conclusion on the optimization method for the universality of biological and artificial machines in nature. Another problem is that many typical control methods that are widely used cannot be said to be optimal, but people have not found out what the optimal control method is. For example, for pump systems that account for 20–25% of the world's electricity consumption, the conventional operation method is to use a single closed-loop control method to achieve control that meets process requirements, collect the actual values of process parameters, and compare them with the set values. Increase or decrease the speed of the pump, that is, adjust the load of the speed-regulating pump, and increase or decrease the number of running pumps according to the actual process parameters at full speed or zero speed. Although this method is adopted all over the world, it is not optimal. Because this method does not consider the energy consumption problem in the control process, it is impossible to naturally realize the operation with the lowest energy consumption under the process conditions. Even the South-to-North Water Diversion Project, the largest water conservancy project in the world, did not conduct acceptance checks on operational energy efficiency when the pumping stations were put into operation.

For the classical differential method of static optimization, the partial derivative of each variable is calculated, and it is equal to zero, and the optimal result is solved. However, due to the fact that some of the variables in the actual system are real numbers, such as the load borne by each device, and some are positive integers, such as how many devices or people are used or which device is used, etc., the optimal solution for integer variables may be real numbers with decimals, which doesn't match reality. Another problem is that for a total load required by the system and certain existing equipment, the optimal point of its total efficiency is not necessarily the point where the derivative is zero, but the best one among all the schemes that can be realized. Another problem is that many optimization methods have non-unique solutions, that is, the optimization results of this arrangement and that arrangement are the same. For many complex systems, because it is difficult to give an accurate mathematical description or mathematical model, it is sometimes difficult to give an accurate mathematical expression of the optimization problem whether it is a dynamic optimization method or a static optimization method.

1.5 Problems of Existing Energy Efficiency Optimization Methods

1. For a complex system, its accurate mathematical model is mostly a multi-variable, nonlinear, time-varying function. For such system, it is still difficult to obtain an accurate mathematical model or state equation, and its characteristics are usually provided by discrete data or curves.
2. Many practical optimization problems are a combination of static optimization and dynamic optimization. For example, for a large-scale urban water supply system, the water volume is randomly determined by urban residents, and the water pressure should maintain a demand value that changes with the change of water volume, which requires dynamic optimization. Static optimization is suitable if the water volume and pressure are fixed over time.
3. Many practical optimization problems are multi-objective. the above-mentioned water supply system must meet two objectives. the first objective is to maintain pressure, and the second objective is to save energy.

4. Many typical control methods, such as PID, are widely used in practical applications. However, they only meet the goals of process requirements and are not optimal because they do not consider the energy consumption of the system.
5. For the static optimization of the classical differential method, the partial derivative is calculated for each variable, and it is equal to 0 to obtain the optimal solution. However, some variables can only be positive integers, such as the number of operating equipment or the number of people, etc., if the optimal result of the variable is a real number with decimals, which is inconsistent with the reality.
6. Many optimization methods do not have a unique solution, which means that different schedules will get the same optimization results.
7. In the field of efficiency optimization, there is no unified general optimization method for human being and machines.

1.6 Quantum Optimization Method and Energy Efficiency Prediction Theory

The essence of energy saving is to keep a regulated system in the overall high efficiency area. When the overall efficiency is the highest value that can be achieved, we say that the system has been running in the most energy-saving state, so the essence of energy saving is to improve the overall operating efficiency of the system.

There are a lot of various systems in the industry, and the variables and solution methods involved in the efficiency optimization are different from what we have talked about in advanced mathematics in the past. One of the differences is that some of the variables are real numbers and some are positive integers. The real numbers are the load borne by each device, and the integers are how many devices or people are used, which device is used, etc. For the derivative of an integer variable, the optimal point may be a real number with decimals. It is obviously unrealistic. The second difference is that for a total load required by the system and certain existing equipment, the optimal point of its total efficiency is not necessarily the point where the derivative is zero, but the best one among all the schemes that can be realized. Another difference is that there are non-unique solutions at many optimization points, that is, the optimization results of this arrangement and that arrangement are the same.

Since the energy efficiency function is nonlinear and has a limit on the maximum input value, this is a nonlinear optimization problem with constraints. The number of optimized operating units is an integer, the optimal load distribution value is a real number, which is also an integer-real-number mixed optimization problem.

We call the method of solving such an efficiency optimization problem as the *quantum optimization method*.

Aiming at the existing problems, this book presents a quantum optimization method applicable to general-purpose devices, which is based on:

1. The efficiency function can be approximately regarded as a concave function passing through the origin, and the second derivative of the efficiency function is less than zero.
2. Assume the load rate of each working equipment is greater than zero, calculate partial derivatives for multiple variables, and deduce the optimal control method for each operating equipment.
3. Based on graphics, deduce the optimal switching method. When the total load increases, determine the optimal switching point of n and k_1 operating equipment, and when the total load decreases, determine the optimal switching point of n and k_2 operating equipment.

We call this theory to solve energy efficiency optimization as *energy efficiency predictive theory*.

This theory and related theorems proved in this book have been successfully applied to nearly a thousand pumping stations in the fields of building secondary water supply, urban water supply, steel, petrochemical, pharmaceutical and other fields.

It should be reminded that some control methods in this book are protected by patent laws in the United States, Japan, Germany and China.

Since participating in the research of "Simple Three-phase AC Motor Speed Regulating Device" in 1988 [73], the author has been paying great attention to energy efficiency optimization work, and has carried out some research and application work on actual energy saving optimization [74–129].

References

1. Li C, Ma G (2011) Optimal control. Science Press, Beijing
2. Zhang X (2009) Linear programming. Zhejiang University Press, Hangzhou
3. Avriel M (1976) Nonlinear programming analysis and methods. Prentice-Hall
4. Zhang Z (2006) Quadratic programming. Wuhan University Press, Wuhan
5. Boyd S, Vandenberghe L (2009) Convex optimization. Cambridge University Press, New York
6. Fan Y et al (1979) Lecture notes on advanced mathematics. People's Education Press, Shanghai
7. Compiled by the Mathematics Teaching and Research Office of China Institute of Mining and Technology (1980) Mathematics handbook. Science Press, Beijing
8. Li G (2005) Computational methods. Publishing House of Electronics Industry, Beijing
9. Schmit LA (1960) Structural design by systematic structural synthesis. In: Proceedings of the 2nd conference on electronic computation. American Society of Civil Engineers, New York, pp 105-22
10. Xia R (2000) Research progress in engineering numerical optimization methods. J Aeronaut Astronaut 21(6):488–491
11. He J (2007) Optimization method. Tsinghua University Press, Beijing
12. Wu C (2000) Theory and methods of optimal control. National Defense Industry Press, Beijing
13. Kirk DE (2004) Optimal control theory: an introduction. Dover Publications, New York
14. Qian W (1980) Variational method and finite element. Science Press, Beijing
15. Leitmann G (1981) The calculus of variations and optimal control. Springer, New York

16. Goltz E (1958) Method of variations. People's Education Press, Beijing
17. Huang P, Meng Y (2009) Optimization theory and methods. Tsinghua University Press, Beijing
18. Qin S, Zhang Z (1984) Optimal control. Publishing House of Electronics Industry, Beijing
19. Li G (2008) Optimal control theory and application. National Defense Industry Press, Beijing
20. Wang X, Lu J (2007) Optimization method and optimal control. Harbin Engineering University Press, Harbin
21. Beyer HG (2000) Evolutionary algorithms in noisy environments: theoretical issues and guidelines for practice. Comput Methods Appl Mech Eng 186(2–4):239–267
22. Wang L, Zheng D (2004) Simulated annealing approach based on hypothesis test for stochastic optimization problems. Control Decis 19(2):183–186
23. Bellman RE (1957) Dynamic programming. Princeton University Press, New Jersey
24. Pontryakin et al (1965) Mathematical theory of optimal processes. Shanghai Science and Technology Press, Shanghai
25. Kalman RE (1960) Contributions to the theory of optimal control. Bol Soc Mat Mex 5:102–119
26. Ruan X, Wan B (2002) The iterative learning control for dynamics in steady-state hierarchical optimization of nonlinear large-scale industrial processes. Syst Eng Theory Pract 22(6):16–20
27. Salazar LM, Rowe JE (2005) Particle swarm optimization and fitness sharing to solve multi-objective optimization problems. Congr Evol Comput 2:1204–1211
28. Gu J, Wan B (2001) Steady state hierarchical optimizing control for large-scale industrial processes with fuzzy parameters. IEEE Trans Syst Man Cybern Part C 31(3):1–9
29. Qian J, Zhao J, Xu Z (2007) Predictive control. Chemical Industry Press, Beijing
30. Holland JH (1992) Genetic algorithm. Sci Am 4:44
31. Chen G et al (2001) Genetic algorithm and its application. Posts and Telecom Press, Beijing
32. Basu M (2003) Hopfield neural networks for optimal scheduling of fixed head hydrothermal power station. Electr Power Syst Res 64(1):11–15
33. Orero S, Irving M (1998) A genetic algorithm modeling framework and solution technique for short term optimal hydrothermal scheduling. IEEE Trans Power Syst 5(2):1254–1265
34. Feng Y (1981) Fuzzy solution to multi-objective optimization problems. Sci Bull 17:1028–1031
35. Fang S (1997) Fuzzy mathematics and Fuzzy optimization. Science Press, Beijing
36. Balci H, Valenzuela J (2004) Scheduling electric power generators using particle swarm optimization combined with the Lagrangain relaxation method. Int J Appl Math Comput Sci 14:411–421
37. Giras T, Talukdar S (1981) Quasi-Newton method for optimal power flows. Int J Electr Power Energy Syst 3(2):59–64
38. Bartholomew-Biggs M (1987) Recursive quadratic programming method based on the augmented Lagrangian. Math Prog Study 31:21–41
39. Huang H, Peng D, Zhang Y, Liang Y (2013) Research on load optimal distribution based on equal incremental principle. J Comput Inf Syst 9(18):7477–7484
40. Feng L (1993) A parametric iteration method of stochastic dynamic programming for optimal dispatch of hydroelectric plants. In: IEEE 2nd international conference on advances in power system control, operation and management, Hong Kong, 1993, vol 12, pp 1304–1324
41. Lopezibanez M, Prasad T, Paechter B (2008) Ant colony optimization for optimal control of pumps in water distribution networks. J Water Resour Plan Manag 134(4):337–346
42. Kumar D, Reddy M (2007) Multipurpose reservoir operation using particle swarm optimization. J Water Resour Plan Manag 133(3):192–201
43. Yuan X, Zhang Y, Wang L et al (2008) An enhance differential evolution algorithm for daily optimal hydro generation scheduling. Comput Math Appl 55(11):2458–2468
44. Rajan A, Christober C (2011) Hydro-thermal unit commitment problem using simulated annealing embedded evolutionary programming approach. Int J Electr Power Energy Syst 33(4):939–946
45. Simopoulos N, Kavatza D, Vournas D (2006) Unit commitment by an enhanced simulated annealing algorithm. IEEE Trans Power System 21(1):68–76

46. Li J, Jiang X, Li G, Fang Y (2020) Two-stage optimal dispatching of active distribution network with two layer of source-network-load. Electr Energy Manag Technol 1:78–85
47. Zhou L, Lu Z (2021) Multi-microgrid optimal power flow dispatching and collaborative optimization control strategy. Mod Electr Power 38(5):473–482
48. Zhang J, Zhang Z, Wang Z et al (2022) Double layer robust optimal dispatching of micro-grid base on data-drive. Electr Drive 52(1):68–75
49. Ju Y, Chen X, Li J, Wang J (2023) Active and reactive power coordinated optimal dispatch of networked microgrid base on distributed deep reinforcement learning. Autom Electr Power Syst 47(1):115–125
50. Pezeshk S, Heiweg OJ, Oliver KE (1994) Optimal operation of ground-water supply distribution system. J Water Resour Plan Manag 120(5):573–586
51. Ormsbee L, Lansey K (1994) Optimal control of water supply pumping system. J Water Resour Plan Manag 120(2):237–253
52. Stadler H (2002) Energy saving by means of electrical drives. In: 3rd International conference on energy efficiency in motor driven systems
53. Fetyan KM, Younes MA, Hetal MA, Hallouda MM (2007) Energy saving of adjustable speed pump stations. In: Eleventh international water technology conference, 2007, Sharm El-Sheikh, Egypt
54. Zhang X, Liu X (2023) Research on energy-saving optimization design of water conservancy pumping stations, hydropower and water resources. (11):145–147
55. El-Sayed YM, Evans RB (1970) Thermoeconomics and the design of heat systems. J Eng Power 92(1):27–35
56. Frangopoulos CA (2003) Methods of energy systems optimization. In: Optimization of energy systems and processes, Gliwice, Poland, 2003
57. Valero A, Serra L, Lozano MA (1993) Structural theory of thermoeconomics. In: International symposium on thermodynamics and design, analysis and improvement of energy systems, New Orleans, LA, USA, pp 189–198
58. Silveira JL, Tuna CE (2003) Thermoeconomics analysis method for optimization of combined heat and power system—Part I. Prog Energy Combust Sci 29(6):479–485
59. Gogus YA (2005) Thermoeconomic optimization. Int J Energy Res 29(7):559–580
60. Sun P (2023) Heat load prediction and energy saving control method of secondary network in urban central heating system. District Heat (6):54–60
61. Mamandur KRC, Chenoweth RD (1981) Optimal control of reactive power flow for improvements in voltage profiles for real power loss minimization. IEEE Trans PAS 100(7):3185–3196
62. Lee KY (1998) Optimal reactive power planning using evolutionary programming, evolutionary strategy, genetic algorithm and linear programming. IEEE Trans Power Syst 13(1):101–108
63. Das DB, Patvardhan C (2002) Reactive power dispatch with a hybrid stochastic search technique. Int J Eletr Power Energy Syst 24(9):731–736
64. Abido MA (2002) Optimal power flow using particle swarm optimization. Electr Power Energy Syst 24:563–571
65. Han J, Cheng T, Xu J (2023) Optimization of solar-gas energy system considering grid renewable energy power penetration rates. Acta Energiae Solaris Sinica 44(7):80–87
66. Kim GS, Ha IJ, Ko MS (1992) Control of induction motors for both high dynamic performance and high power efficiency. IEEE Trans Ind Eletron 39(2):323–333
67. Abrahamsen F, Blaabjerg F (1998) On the energy optimized control of standard and high-efficiency induction motors in CT and HVAC application. IEEE Trans IA 34(4):822–831
68. Gribble JJ, Kjaer PC, Miller TJE (1999) Optimal commutation in average torque control of switched reluctance motors. IEE Proc Electron Power Appl 146(1):2–10
69. Spiegel RJ, Turnerb MW, Mccormick VE (2003) Fuzzy-logic-based controllers for efficiency optimization of inverter-fed induction motor drives. Fuzzy Sets Syst 137(3):387–401
70. Zhang F, Yan L, Fan Y (2002) Optimizing dispatching of public traffic vehicles in intelligent transport systems. J Beijing Univ Aeronaut Astronaut 28(6):707–710

71. Guly B, Demet O (1999) A tube search algorithm for vehicle routing problem. Comput Oper Res 26(3):255–270
72. Wu Y, He S, Zhou H (2023) Comprehensive optimization model and algorithm of operation plan for smart port station of heavy haul railway. J Transp Syst Eng Inf Technol 23(6):215–226
73. Yao F Simple three-phase AC motor speed regulating device, China, Patent Number: 90225362.X
74. Yao F Water pump fan operating efficiency control method for speed regulator, China, Patent Number: 02159869.X
75. Yao F Water pump and fan operation efficiency control method for industrial controllers and configuration software, China, Patent Number: 03103247.8
76. Yao F Speed regulation and switching method for controlling parallel energy-saving operation of water pumps and fans, China, Patent Number: 2008100994276
77. Yao F Control and scheduling method for overall high-efficiency and energy-saving operation of general equipment, China, Patent Number: 201010265930.1
78. Yao F Power-saving optimization operation method and switching point determining method for water pump unit, China, Patent Number: 201911064017.2
79. Yao F Power-saving optimization operation method and switching point determining method for water pump unit, United States, Patent Number: 17/339,381
80. Yao F Power-saving optimization operation method and switching point determining method for water pump unit, Germany, Patent Number: 11 2020 000 196.2
81. Yao F Power-saving optimization operation method and switching point determining method for water pump unit, Japan, Patent Number: 2021-526462
82. Yao F (2022) Finding electricity-saving potentiality of pump house using target-power technology. World Invert (11):56–59
83. Yao F (2003) Significance of determination of object power waste for large-scale water transfer projects. Adv Sci Technol Water Resour 23(5):56–57
84. Yao F (2004) Study on energy-saving criterion of pump station, drainage and irrigation machinery. 22(6):37–39
85. Yao F (2002) Calculation of energy-conservation ratio in pump station. Energy Eng (6):33–36
86. Yao F (2005) Sticking point and countermeasure of electricity waste in large-scale pumping station. Pump Station Technol (4):35–38
87. Yao F, Zhang Y (1998) Energy saving calculation and system design of water pump frequency conversion speed regulation. Science Press, Beijing
88. Yao F, Zhang Y (2009) Electrical energy saving control method and practice. China Electric Power Press, Beijing
89. Yao F, Sun H (2011) Frequency converter and energy-saving control practical technology crash course. Publishing House of Electronics Industry, Beijing
90. Yao F, Sun H (2012) Efficiency optimal control and dispatching method for general equipment. China Machine Press, Beijing
91. Yao F, Zhang Y (2015) Electrical energy-saving technology and application. China Electric Power Press, Beijing
92. Yao F, Sun H (2011) Optimal control in constant-speed pumping stations. In: The international conference on management science and intelligent control (ICMSIC2011), Aug 24–26, 2011 in Anhui, China
93. Yao F, Sun H (2011) Optimal control in constant-speed fan stations. In: The IEEE joint international information technology and artificial intelligence conference, Aug 20–22, 2011 in Chongqing, China
94. Yao F, Sun H (2011) Optimal switch for a common system. In: IEEE power engineering and automation conference (PEAM 2011), September 8–9, 2011 in Wuhan, China
95. Yao F, Sun H (2011) Optimal switch in constant-speed fan stations. In: 2011 The 2nd international conference on data storage and data engineering, Xi'an, China, May 13–15, 2011
96. Yao F, Sun H (2011) Optimal switch in variable-speed fan stations. In: 2011 International conference of renewable energy sources and environmental materials, May 20–22, 2011, Shanghai, China

97. Yao F, Sun H (2011) Optimal switch for a class of system. In: 2011 International conference on electronics and optoelectronics (ICEOE 2011), July 29–31, 2011, Dalian, China
98. Yao F, Sun H (2011) A class of optimizations and energy-saving application. In: The 3rd WASE international conference on information engineering (WASE ICIE 2011), Aug 13–14, 2011, Xi'an, China
99. Yao F, Sun H (2011) Optimal control for a common system. In: 2011 3rd IEEE international conference on information management and engineering (IEEE ICIME 2011), May 21–22, 2011, Zhengzhou, China
100. Yao F, Sun H (2011) Optimal switch in variable-speed pumping stations. In: 2011 IEEE 2nd international conference on computing, control and industrial engineering (CCIE 2011), August 20–21, 2011, Wuhan, China
101. Yao F, Sun H (2011) The stability criterion of pumps. In: 2011 International conference on electronic & mechanical engineering and information technology (EMEIT 2011), 12–14 August, Harbin, China
102. Yao F, Sun H (2011) Optimal control for a class of system. In: 2011 4th IEEE international conference on computer science and information technology (IEEE ICCSIT 2011), Chengdu, China, June 10–12, 2011
103. Yao F, Sun H (2011) Optimal control in variable-speed fan stations. In: IEEE the 18th international conference on industrial engineering and engineering management, September 3–5, 2011, ChangChun, P.R. China
104. Yao F, Sun H (2011) Optimal control in variable-speed pumping stations. In: The 2011 IEEE international conference on mechatronics and automation (ICMA 2011), August 7–10, 2011, Beijing, China
105. Yao F, Sun H (2011) Optimal switch in constant-speed pumping stations. In: The 2011 international conference on artificial intelligence and computational intelligence (AICI'11), Sep 24–25, 2011,Taiyuan, China
106. Yao F, Sun H (2011) The stability criterion of fans. In: The international conference on economic management and information engineering technology 2011 (ICEMIE2011), December 26–27, 2011, Xi'an, China
107. Yao F, Sun H (2011) Optimal control and switch in high speed train. In: The 3rd international conference on computational and information sciences (ICCIS2011) will be held in Chengdu, China, from October 21 to 23, 2011
108. Yao F, Sun H (2011) Optimization and energy-saving of a pumping station. In: The 2011 international conference on information security and intelligence control (ISIC2011), Jilin, China, during August. 13–15, 2011
109. Yao F, Sun H (2011) Optimal control and optimal switch in conveyors driven by multi-motor. J Nanjing Univ Sci Technol 35(Sup):127–131
110. Yao F, Sun H (2010) Optimization method and energy-saving project application of a function. J Hebei Acad Sci 27(3):50–53
111. Yao F (2012) Efficiency optimization method and optimal control of mechanical and electrical equipment. Hebei University of Technology
112. Yao F, Cao C (2014) Energy optimization of the heat supply system with different boilers. In: 2014 IEEE workshop on advanced research and technology industry application, Canada
113. Yao F, Cao C (2015) Energy optimization of the wire part driven by three different motors in paper machine. Appl Mech Mater 716–717 Switzerland
114. Yao F, Cao C (2014) Energy optimization of the scraper conveyor driven by two different motors. In: 2014 International conference on control engineering and automation, China
115. Yao F, Cao C (2014) Efficiency optimization in the efficiency similarity fan stations. In: 2014 Mechatronics control and electronic engineering Part 3, France
116. Yao F, Gao Q (2017) Optimal control and switching mechanism of a power station with identical generators. In: Proceedings of the 10th ASME dynamic systems and control conference, Tysons Corner, Virginia, USA
117. Yao F, Gao Q (2017) Efficiency optimization of a power station with different generators. In: REM 2017: renewable energy integration with mini/micro grid systems, applied energy symposium and forum 2017, Tianjin, China, 2017

118. Yao F, Gao Q (2017) Optimal control and switch in a hydraulic power station. In: IEEE IAEAC 2017, Proceedings of 2017 IEEE 2nd advanced information technology, electronic and automation control conference, Chongqing, China
119. Yao F (2021) Dr. Yao's talk on energy saving in pumping station, city and town water supply. 219(1):20–24
120. Yao F (2021) History and trend of energy-saving standards and control methods for pumping station, city and town water supply. 220(2):9–14
121. Yao F (2021) Mathematical proof of predictive theory of pump station energy efficiency, city and town water supply. 221(3):10–14
122. Yao F (2021) Stability of pumping stations and rationality of energy efficiency standards, city and town water supply. 222(4):10–17
123. Yao F (2021) Cases and analysis of power waste in pumping stations, city and town water supply. 223(5):12–16
124. Yao F (2021) Optimization of energy efficiency dispatching and selection rule of speed regulating pump in urban water plant, city and town water supply. 224(6):6–10
125. Yao F, Gao R (2023) Energy efficiency optimization of multi-unit system. In: 3rd International conference on energy, power and electrical engineering, 2023, Wuhan, China
126. Yao F, Zhang S (2023) Optimal distribution of power grid load. In: 2nd International seminar on energy, power and electrical technology, 2023, Malaysia
127. Yao F, Yao Y (2023) Optimal control and scheduling of distribution stations. In: 10th International forum on electrical engineering and automation, 2023, Nanjing, China
128. Yao F, Yao Y (2023) Energy efficiency optimization of pumping stations. In: 3rd International conference on statistics, applied mathematics and computing science, 2023, Nanjing, China
129. Yao F, Yao Y, Zhang S (2023) Optimal dispatching of wind power hydrogen production system. In: 3rd International conference on new energy and power engineering, 2023, Hangzhou, China

Chapter 2
Energy Conversion and Overall Energy Efficiency

We divide the energy system into three parts: generation, transmission and consumption. When generating electricity, improve energy efficiency and maximize power generation. In terms of power transmission and energy consumption, we need to improve energy efficiency and minimize energy consumption.

2.1 Energy Form of the Power Station

The power station converts raw energy into electrical energy, which can be expressed as

$$W_t = W_0 \eta_0 \tag{2.1}$$

where W_0 represents the original energy, that is, the ideal work, η_0 represents the overall energy efficiency of the power station, and W_t is the power generation of the power station. In the case of a certain ideal work W_0, the higher the η_0, the greater the W_t, and the more energy output by the power station.

2.1.1 Convert Potential Energy to Electrical Energy

A hydropower station or energy storage power station has n hydroelectric generating units. The working head of each hydro-generator set in the parallel system is the same. The potential energy of water in a river or reservoir is used to generate electricity, expressed as

$$W_t = W_0 \eta_o = \frac{Q_t H}{k_1} \eta_0 = \frac{H}{k_1} \sum_{i=1}^{n} Q_i \eta_i = \frac{Q_t H}{k_1} \sum_{i=1}^{n} \frac{Q_i}{Q_t} \eta_i = W_0 \sum_{i=1}^{n} \theta_i \eta_i \qquad (2.2)$$

$$W_0 = \frac{Q_t H}{k_1}$$

where k_1 is a constant, W_0 is the potential energy of water flowing through the hydro-generator unit, W_t is the electric energy output by the power station, n is the total number of hydro-generator units. Q_t is the total water flow rate through the hydro-generator unit. H is the water head, η_0 is the overall energy efficiency of the hydropower station, η_i is the operating energy efficiency of the i-th hydro-generator set, Q_i is the flow rate through the i-th hydro-generator set. θ_i is the load rate of the i-th hydro-generator set, as the load rate of the i-th hydroelectric generating unit, expressed as

$$Q_t = \sum_{i=1}^{n} Q_i \qquad (2.3)$$

$$\theta_i = \frac{Q_i}{Q_t}$$

$$\sum_{i=1}^{n} \theta_i = 1$$

$$\eta_0 = \sum_{i=1}^{n} \theta_i \eta_i$$

For a fixed Q_t and H, the higher the η_0, the larger the W_t.

2.1.2   Convert Heat Energy to Electricity

Thermal power plants consist of steam boilers and turbo generators that convert the thermal energy of coal, natural gas or oil into electricity. The electrical energy produced by a thermal power plant is expressed as

$$W_t = \sum_{i=1}^{n} (W_i \eta_i) = W_{t0} \sum_{i=1}^{n} \frac{W_i}{W_{t0}} \eta_i = W_{t0} \sum_{i=1}^{n} \theta_i \eta_i = W_{t0} \eta_o \qquad (2.4)$$

$$W_{t0} = \sum_{i=1}^{n} W_i$$

where n is the total number of generating units in thermal power plants, W_{t0} is the thermal energy contained in coal, natural gas or oil used in thermal power plants, W_t is the total electric energy output by thermal power plants. η_0 is the comprehensive energy efficiency of thermal power plants, and η_i is the i-th generator operating efficiency, θ_i is the load rate of the i-th generator, the expression is

$$\theta_i = \frac{W_i}{W_{t0}} \tag{2.5}$$

$$\sum_{i=1}^{n} \theta_i = 1$$

$$\eta_0 = \sum_{i=1}^{n} \theta_i \eta_i$$

2.1.3 Wind Power Hydrogen Production System

Due to the involvement of chemical reactions, fission reactions and other factors, some systems cannot directly convert energy. Such systems can only maximize the output as the optimal goal of the overall energy efficiency of the system. For example, electrolyzing water to produce hydrogen, you cannot say that all the hydrogen energy produced is converted from the electricity consumption of the electrolyzer.

Assuming that there are m hydrogen generators in the hydrogen production station, the total input power is W_t, the electric energy input by the i-th device is W_i, and the overall efficiency of the system is η_t. The overall energy efficiency expression of the system is

$$\eta_t = \sum_{i=1}^{m} \frac{W_i}{W_t} \eta_i(W_i) = \sum_{i=1}^{m} \theta_i \eta_i(\theta_i)$$
$$s.t. \sum_{i=1}^{m} W_i = W_t > 0 \tag{2.6}$$
$$W_{imax} \geq W_i > 0$$

where W_{imax} is the maximum load of the hydrogen generator.

If expressed in terms of hydrogen production, this optimization problem can also be transformed into

$$W_t = \sum_{i=1}^{n} W_i = k_0 \sum_{i=1}^{n} \frac{Q_i}{\eta_i} = k_0 Q_t \sum_{i=1}^{n} \frac{Q_i}{Q_t} \cdot \frac{1}{\eta_i} = k_0 Q_t \sum_{i=1}^{n} \theta_i \frac{1}{\eta_i} = k_0 Q_t \frac{1}{\eta_o} \quad (2.7)$$

$$\eta_0 = \frac{1}{\sum_{i=1}^{n} \theta_i \frac{1}{\eta_i}}$$

where k_0 is an energy conversion constant, W_t is the energy consumption of the system, W_i is the energy consumption of the i-th hydrogen generator, Q_t is the total hydrogen production of the system. Q_i is the hydrogen production provided by the i-th hydrogen generator, n represents the total number of hydrogen generators in the system. η_0 represents the overall energy efficiency of the system, η_i represents the operating efficiency of the i-th hydrogen generator, and θ_i represents the load rate of the i-th hydrogen generator, the expression is

$$Q_t = \sum_{i=1}^{n} Q_i \quad (2.8)$$

$$\theta_i = \frac{Q_i}{Q_t}$$

$$\sum_{i=1}^{n} \theta_i = 1$$

For a fixed Q_t, the higher the η_0, the smaller the W_t.

2.2 Power Dispatch and Distribution

The high-voltage power generated by power stations needs to be transmitted to customers through many transmission lines and a large number of transformers. So we need to consider the issue of power dispatch and distribution.

2.2.1 Power Distribution

There are n transformers in the power supply system, and all transformers are connected in parallel. For a parallel system, the line voltage of each transformer is U_0. The total apparent power S_t and the total output current I_0 of all transformers, the line current of the i-th transformer is I_i, and the total energy consumption W_t of n transformers is expressed as

$$W_t = \sqrt{3} \sum_{i=1}^{n} \frac{U_0 I_i}{\eta_i} = \sqrt{3} U_0 \sum_{i=1}^{n} \frac{U_0 I_i}{\eta_i} = \sqrt{3} U_0 I_0 \sum_{i=1}^{n} \frac{I_i}{I_0} \cdot \frac{1}{\eta_i} = \sqrt{3} U_0 I_0 \sum_{i=1}^{n} \theta_i \frac{1}{\eta_i}$$

$$= S_t \sum_{i=1}^{n} \theta_i \frac{1}{\eta_i} = \frac{S_t}{\eta_0} \tag{2.9}$$

$$I_0 = \sum_{i=1}^{n} I_i$$

$$S_t = \sqrt{3} U I_0$$

$$\eta_0 = \frac{1}{\sum_{i=1}^{n} \theta_i \frac{1}{\eta_i}}$$

Among them, η_0 represents the overall efficiency of the power supply system, θ_i represents the load rate of the i-th transformer, expressed as

$$\sum_{i=1}^{n} \theta_i = 1 \tag{2.10}$$

$$\theta_i = \frac{I_i}{I_0}$$

2.2.2 Power Dispatch

There are n transmission lines supplying power in an area, and the power grids are connected in parallel. The total apparent power required is S_t, and the line voltage U of each transmission line is the same. The total current I_0, the resistance and current of the i-th line are R_i and I_i respectively, then the total energy consumption W_t of the n transmission lines is expressed as

$$W_t = 3 \sum_{i=1}^{n} I_i^2 R_i = 3 I_0^2 \sum_{i=1}^{n} \left(\frac{I_i}{I_0} \right)^2 R_i = 3 I_0^2 \sum_{i=1}^{n} \theta_i^2 R_i \tag{2.11}$$

$$I_0 = \sum_{i=1}^{n} I_i$$

$$S_t = \sqrt{3} U I_0$$

where θ_i represents the load rate of the i-th transmission line, expressed as

$$\sum_{i=1}^{n} \theta_i = 1 \tag{2.12}$$

$$\theta_i = \frac{I_i}{I_0}$$

2.3 Energy Consumption System

Energy consumption system transforms electricity energy into target energy, ideal work W_0, W_t is the total energy consumption of the system, W_t is expressed as

$$W_t = W_0 \frac{1}{\eta_0} \tag{2.13}$$

where W_0 represents the target energy which is the ideal work, η_0 denotes the overall efficiency of the energy consumption system. It is seen that for a fixed ideal work W_0, the higher η_0 is, the smaller W_t will be, so that the system needs less energy.

2.3.1 Gaining Potential Energy

In pumping stations and energy storage power stations, pumps are used to obtain the potential energy of water and work in parallel or in series.

1. **Working in Parallel**

A pumping station or a storage power station has n pumps. Each pump in a parallel system gets the same head. Using electric energy to obtain the potential energy of water. Expressed as follows:

$$W_t = W_0 \frac{1}{\eta_0} = \frac{Q_t H}{k_1} \cdot \frac{1}{\eta_0} = \frac{H}{k_1} \sum_{i=1}^{n} Q_i \frac{1}{\eta_i} = \frac{Q_t H}{k_1} \sum_{i=1}^{n} \frac{Q_i}{Q_t} \cdot \frac{1}{\eta_i} = W_0 \sum_{i=1}^{n} \theta_i \frac{1}{\eta_i} \tag{2.14}$$

$$W_0 = \frac{Q_t H}{k_1}$$

$$\eta_0 = \frac{1}{\sum_{i=1}^{n} \theta_i \frac{1}{\eta_i}}$$

where k_1 is a constant, W_0 is the obtained water potential energy, W_t is the electric energy consumption of the pumping station or energy storage power station, n is the total number of pumps. Q_t is the total flow rate of water delivered, H is the obtained water head. η_0 is the overall energy efficiency of the pumping station or energy storage power station, η_i is the operating efficiency of the i-th pump. Q_i is the flow rate of the i-th pump, θ_i is the load rate of the i-th pump. Expressed as follows:

$$Q_t = \sum_{i=1}^{n} Q_i \qquad (2.15)$$

$$\theta_i = \frac{Q_i}{Q_t}$$

$$\sum_{i=1}^{n} \theta_i = 1$$

When Q_t and H are fixed, the higher η_0 is, the smaller W_t is.

2. Running in Series

There are n pumps in a pumping station, and the flow rate of each pump in the series system is the same, and the potential energy of water is obtained by using electric energy, which is expressed as

$$W_t = W_0 \frac{1}{\eta_0} = \frac{QH_t}{k_1} \cdot \frac{1}{\eta_0} = \frac{Q}{k_1} \sum_{i=1}^{n} H_i \frac{1}{\eta_i} = \frac{QH_t}{k_1} \sum_{i=1}^{n} \frac{H_i}{H_t} \cdot \frac{1}{\eta_i} = W_0 \sum_{i=1}^{n} \theta_i \frac{1}{\eta_i}$$

$$(2.16)$$

$$W_0 = \frac{QH_t}{k_1}$$

$$\eta_0 = \frac{1}{\sum_{i=1}^{n} \theta_i \frac{1}{\eta_i}}$$

where k_1 is a constant, W_0 is the potential energy of water obtained, W_t is the power consumption of the pumping station, n is the total number of water pumps, Q is the flow rate of water delivered by the pumps. H_t is the total water head obtained. η_0 is the overall efficiency of the pumping station, η_i is the operating efficiency of the i-th pump, H_i is the lift of the i-th pump, θ_i is the load rate of the i-th pump, the expression is

$$H_t = \sum_{i=1}^{n} H_i \qquad (2.17)$$

$$\theta_i = \frac{H_i}{H_t}$$

$$\sum_{i=1}^{n} \theta_i = 1$$

For a fixed Q and H_t, the higher the η_0, the smaller the W_t.

2.3.2 Provide Pressure Energy

Fans, compressors and blowers are used to obtain pressure energy of air or other gases. Take a fan station as an example, there are n fans in the station. Each fan in a parallel system provides the same pressure. Using electric energy to obtain the pressure energy of air, the expression is

$$W_t = W_0 \frac{1}{\eta_o} = \frac{Q_t P}{k_2} \cdot \frac{1}{\eta_0} = \frac{P}{k_2} \sum_{i=1}^{n} Q_i \frac{1}{\eta_i} = \frac{Q_t P}{k_2} \sum_{i=1}^{n} \frac{Q_i}{Q_t} \cdot \frac{1}{\eta_i} = W_0 \sum_{i=1}^{n} \theta_i \frac{1}{\eta_i}$$

(2.18)

$$W_0 = \frac{Q_t P}{k_2}$$

$$\eta_0 = \frac{1}{\sum_{i=1}^{n} \theta_i \frac{1}{\eta_i}}$$

where k_2 is a constant, and W_0 represents the pressure energy of the output air of the fan station. W_t is the power consumption of the fan station, Q_t is the air flow rate provided by the fan station, P is the air pressure, n is the total number of fans, and η_0 is the overall energy efficiency of the fan station. η_i represents the operating energy efficiency of the i-th fan, Q_i represents the flow rate of the i-th fan, θ_i represents the load rate of the i-th fan, the expression is

$$Q_t = \sum_{i=1}^{n} Q_i$$

(2.19)

$$\theta_i = \frac{Q_i}{Q_t}$$

$$\sum_{i=1}^{n} \theta_i = 1$$

For a fixed Q_t and P, the higher the η_0, the smaller the W_t.

2.3.3 Provide Cold and Heat Energy

1. Central Air Conditioning System

There are many central air-conditioning systems in public buildings, which consume a lot of energy. There are n chillers in the central air-conditioning system to provide cold energy. The total power consumption is expressed as

$$W_t = \sum_{i=1}^{n} W_i = k_3 \Delta t \sum_{i=1}^{n} \frac{Q_i}{\eta_i} = k_3 \Delta t Q_t \sum_{i=1}^{n} \frac{Q_i}{Q_t} \cdot \frac{1}{\eta_i} = k_3 \Delta t Q_t \sum_{i=1}^{n} \theta_i \frac{1}{\eta_i}$$

$$= k_3 \Delta t Q_t \frac{1}{\eta_o} \tag{2.20}$$

$$\eta_0 = \frac{1}{\sum_{i=1}^{n} \theta_i \frac{1}{\eta_i}}$$

where W_t is the power consumption of the system, k_3 is a constant, Δt is the temperature difference between the inlet and outlet fluids of the system, W_i is the power consumption of the i-th chiller. Q_t is the total cooling or heat required by the building, Q_i is the cooling or heat provided by the i-th chiller, n represents the total number of chillers in the central air-conditioning system, η_0 represents the overall energy efficiency of the system, η_i represents the operating efficiency of the i-th chiller, and θ_i represents the load rate of the i-th chiller, the expression is

$$Q_t = \sum_{i=1}^{n} Q_i \tag{2.21}$$

$$\theta_i = \frac{Q_i}{Q_t}$$

$$\sum_{i=1}^{n} \theta_i = 1$$

For a fixed Q_t, the higher the η_0, the smaller the W_t.

2. Central Heating System

There are many boilers in the city district heating system to provide heat energy. There are n boilers in the system, which are used to provide heat energy, and the total energy consumption is expressed as

$$W_t = \sum_{i=1}^{n} W_i = k_4 \Delta t \sum_{i=1}^{n} \frac{Q_i}{\eta_i} = k_4 \Delta t Q_t \sum_{i=1}^{n} \frac{Q_i}{Q_t} \cdot \frac{1}{\eta_i} = k_4 \Delta t Q_t \sum_{i=1}^{n} \theta_i \frac{1}{\eta_i} = k_4 \Delta t Q_t \frac{1}{\eta_o}$$

$$\tag{2.22}$$

$$\eta_0 = \frac{1}{\sum_{i=1}^{n} \theta_i \frac{1}{\eta_i}}$$

where W_t is the energy consumption of the system, k_4 is a constant, Δt is the temperature difference between the inlet and outlet fluids of the system, W_i is the energy consumption of the i-th boiler. Q_t is the total flow rate of the system, Q_i is the flow rate provided by the i-th boiler, n is the total number of boilers in the system. η_0 is the overall energy efficiency of the system, η_i indicates the operating energy efficiency of the i-th boiler, θ_i indicates the load rate of the i-th boiler, the expression is

$$Q_t = \sum_{i=1}^{n} Q_i \tag{2.23}$$

$$\theta_i = \frac{Q_i}{Q_t}$$

$$\sum_{i=1}^{n} \theta_i = 1$$

For a fixed Q_t, the higher the η_0, the smaller the W_t.

2.3.4 Motion System

Many motion systems, such as high-speed trains, subway trains, electric vehicles, ships, long conveyors, and paper machines, may be driven by multiple motors.

1. **High-Speed Trains and Subway Trains**

It is driven by n motors to overcome friction, air resistance and weight potential energy. The speed of each carriage is the same. For a given vehicle speed V_0, the power consumption of a high-speed train can be expressed as

$$W_t = W_0 \frac{1}{\eta_0} = k_5 \sum_{i=1}^{n} \frac{F_i V_0}{\eta_i} = k_5 V_0 \sum_{i=1}^{n} \frac{F_i}{\eta_i} = k_5 V_0 F_t \sum_{i=1}^{n} \frac{F_i}{F_t} \cdot \frac{1}{\eta_i} = W_0 \sum_{i=1}^{n} \theta_i \cdot \frac{1}{\eta_i} \tag{2.24}$$

$$W_0 = k_5 F_t V_0$$

$$\eta_0 = \frac{1}{\sum_{i=1}^{n} \theta_i \frac{1}{\eta_i}}$$

where k_5 is a constant, V_0 is the speed of the high-speed train, F_t is the total traction force of the high-speed train, W_0 is the ideal work, W_t is the total power consumption of the high-speed train. n is the total number of motors, and η_0 is the overall energy efficiency of the high-speed train. η_i represents the operating energy efficiency of the i-th motor, F_i represents the traction force of the i-th motor, θ_i represents the load rate of the i-th motor, the expression is

$$F = \sum_{i=1}^{n} F_i \qquad (2.25)$$

$$\theta_i = \frac{F_i}{F}$$

$$\sum_{i=1}^{n} \theta_i = 1$$

For a fixed F_t and V_0, the higher the η_0, the smaller the W_t.

V_0 corresponds to the rotational speed n_0 of the motor, and F_t corresponds to the total torque M_t of all the motors. The power consumption of high-speed trains can also be expressed as

$$W_t = W_0 \frac{1}{\eta_0} = k_6 n_0 \sum_{i=1}^{n} \frac{M_i}{\eta_i} = k_6 n_0 M_t \sum_{i=1}^{n} \frac{M_i}{M_t} \cdot \frac{1}{\eta_i} = W_0 \sum_{i=1}^{n} \theta_i \cdot \frac{1}{\eta_i} \qquad (2.26)$$

$$W_0 = k_6 n_0 M_t$$

$$\eta_0 = \frac{1}{\sum_{i=1}^{n} \theta_i \frac{1}{\eta_i}}$$

where k_6 is a constant, n_0 is the speed of each motor, the same, M_t is the total torque of all motors, M_i is the output torque of the i-th motor, θ_i is the load rate of the i-th motor, the expression is

$$M_t = \sum_{i=1}^{n} M_i \qquad (2.27)$$

$$\theta_i = \frac{M_i}{M_t}$$

$$\sum_{i=1}^{n} \theta_i = 1$$

2. Electric Vehicles

It is driven by n motors to overcome friction, air resistance and weight potential energy. Each motor has the same speed. If it is driven by a motor, we can replace this motor with n small motors.

Similarly, for a given vehicle speed V_0 (corresponding to the rotational speed n_0 of the electric motors), the total tractive force F_t corresponds to the total torque M_t of all electric motors. The electric energy consumption of electric vehicles can also be expressed as

$$W_t = W_0 \frac{1}{\eta_0} = k_6 n_0 \sum_{i=1}^{n} \frac{M_i}{\eta_i} = k_6 n_0 M_t \sum_{i=1}^{n} \frac{M_i}{M_t} \cdot \frac{1}{\eta_i} = W_0 \sum_{i=1}^{n} \theta_i \cdot \frac{1}{\eta_i} \qquad (2.28)$$

$$W_0 = k_6 n_0 M_t$$

$$\eta_0 = \frac{1}{\sum_{i=1}^{n} \theta_i \frac{1}{\eta_i}}$$

where n_0 is the speed of each motor, the same, W_0 is the ideal work, W_t is the total electric energy consumption of the electric vehicle, M_t is the total torque of all motors, M_i is the output torque of the i-th motor, and η_0 represents the electric vehicle's overall energy efficiency, η_i represents the operating energy efficiency of the i-th motor, θ_i represents the load rate of the i-th motor, the expression is

$$M_t = \sum_{i=1}^{n} M_i \qquad (2.29)$$

$$\theta_i = \frac{M_i}{M_t}$$

$$\sum_{i=1}^{n} \theta_i = 1$$

For a fixed M_t and n_0, the higher the η_0, the smaller the W_t.

3. Electric Boat

It is driven by n motors to overcome friction, water resistance and wind resistance. Each motor has the same speed. If it is driven by a motor, we can replace this motor with n small motors.

Similarly, the power consumption of an electrical boat can be expressed as

$$W_t = W_0 \frac{1}{\eta_0} = k_6 n_0 \sum_{i=1}^{n} \frac{M_i}{\eta_i} = k_6 n_0 M_t \sum_{i=1}^{n} \frac{M_i}{M_t} \cdot \frac{1}{\eta_i} = W_0 \sum_{i=1}^{n} \theta_i \cdot \frac{1}{\eta_i} \qquad (2.30)$$

$$W_0 = k_6 n_0 M_t$$

$$\eta_0 = \frac{1}{\sum_{i=1}^{n} \theta_i \frac{1}{\eta_i}}$$

where n_0 is the speed of each motor, W_0 is the ideal work, W_t is the total power consumption of the ship, M_t is the total torque of all motors, M_i is the output torque of the i-th motor. η_0 is the overall energy efficiency of the ship, η_i represents the operating efficiency of the i-th motor, θ_i represents the load rate of the i-th motor, the expression is

$$M_t = \sum_{i=1}^{n} M_i \tag{2.31}$$

$$\theta_i = \frac{M_i}{M_t}$$

$$\sum_{i=1}^{n} \theta_i = 1$$

For a fixed M_t and n_0, the higher the η_0, the smaller the W_t.

4. Conveyor Belt and Scraper Conveyor

In order to make the conveyor belt evenly stressed, it is generally necessary to use multiple motors to drive a long conveyor belt, this type of optimization does not present the problem of choosing the best number of motors to run.

There is a forming wire in the wire section of the paper machine, which is jointly driven by n motors. The speed of each motor must be the same (regardless of the reducer), only the torque of each motor is adjusted. This situation is similar with the high-speed train above.

There are many long scraper conveyors in coal mines, usually driven by two motors at both ends. The speed of each motor must be the same, just adjust the torque of each motor so that the force is even. This situation is also similar with high-speed rail.

Ports have many long conveyor belts for transporting raw materials. For uniform force, one belt is driven by two motors.

Similarly, the power consumption can be expressed as

$$W_t = W_0 \frac{1}{\eta_0} = k_6 n_0 \sum_{i=1}^{n} \frac{M_i}{\eta_i} = k_6 n_0 M_t \sum_{i=1}^{n} \frac{M_i}{M_t} \cdot \frac{1}{\eta_i} = W_0 \sum_{i=1}^{n} \theta_i \cdot \frac{1}{\eta_i} \tag{2.32}$$

$$W_0 = k_6 n_0 M_t$$

$$\eta_0 = \frac{1}{\sum_{i=1}^{n} \theta_i \frac{1}{\eta_i}}$$

where η_0 is the speed of each motor, which are the same as each other, W_0 is the ideal work, W_t is the total power consumption of the belt. M_t is the total torque of all motors, M_i is the output torque of the i-th motor. η_0 is the overall energy efficiency of the system. η_i is the operating energy efficiency of the i-th motor, θ_i is the load rate of the i-th motor, the expression is

$$M_t = \sum_{i=1}^{n} M_i \tag{2.33}$$

$$\theta_i = \frac{M_i}{M_t}$$

$$\sum_{i=1}^{n} \theta_i = 1$$

2.3.5 Manpower Scheduling

The manpower scheduling mentioned here is aimed at completing tasks while keeping everyone in the team relaxed and not left behind.

1. A Group of n People

There are n individuals in a military operation. Since one must eat even when not working, there is no question of choosing the optimal number of people. When the machine is turned off, it consumes no energy. In this respect, humans are different from machines. The total weight of the items to be carried is Q_t, and Q_i is the weight of the items carried by the i-th person. The total energy consumption is expressed as

$$W_t = W_0 \frac{1}{\eta_0} = k_7 Q_t \sum_{i=1}^{n} \frac{Q_i}{Q_t} \cdot \frac{1}{\eta_i} = W_0 \sum_{i=1}^{n} \theta_i \frac{1}{\eta_i} \tag{2.34}$$

$$W_0 = k_7 Q_t$$

$$\eta_0 = \frac{1}{\sum_{i=1}^{n} \theta_i \frac{1}{\eta_i}}$$

where k_7 is a constant, W_0 represents the ideal work, W_t represents the total body energy consumption of the group, n represents the total number of people. η_0 represents the overall energy efficiency of the group, η_i represents the operation energy efficiency of the i-th person, θ_i represents the i-th person's load rate, expressed as

$$Q_t = \sum_{i=1}^{n} Q_i \tag{2.35}$$

$$\theta_i = \frac{Q_i}{Q_t}$$

$$\sum_{i=1}^{n} \theta_i = 1$$

For a fixed Q_t, the higher the η_0, the smaller the W_t.

2. **Teams of Two**

Two people use a pole to carry an object from point A to point B. The total weight of things to be moved is Q_t, and Q_i is the weight of things carried by the i-th person. The total energy consumption is expressed as

$$W_t = W_0 \frac{1}{\eta_0} = k_7 \sum_{i=1}^{2} Q_i \frac{1}{\eta_i} = k_7 Q_t \sum_{i=1}^{2} \frac{Q_i}{Q_t} \cdot \frac{1}{\eta_i} = W_0 \sum_{i=1}^{2} \theta_i \frac{1}{\eta_i} \tag{2.36}$$

$$W_0 = k_7 Q_t$$

$$\eta_0 = \frac{1}{\sum_{i=1}^{2} \theta_i \frac{1}{\eta_i}}$$

where k_7 is a constant, W_0 represents the ideal work, W_t is the total energy consumption of the group, n represents the total number of people. η_0 represents the overall efficiency of the group, η_i represents the work efficiency of the i-th individual, and θ_i represents the load rate of the i-th individual, expressed as

$$Q_t = \sum_{i=1}^{2} Q_i \tag{2.37}$$

$$\theta_i = \frac{Q_i}{Q_t}$$

$$\sum_{i=1}^{2} \theta_i = 1$$

For fixed Q_t and the road conditions of two points A and B, the higher η_0 is, the smaller W_t is.

2.4   Overall Energy Efficiency and Weighted Energy Efficiency

The η_0 represents the overall operating energy efficiency of the system.
In power stations, it is expressed as

$$\eta_0 = \sum_{i=1}^{n} \theta_i \eta_i \tag{2.38}$$

In the energy consumption system, it can be expressed as

$$\eta_0 = \frac{1}{\sum_{i=1}^{n} \theta_i \frac{1}{\eta_i}} \tag{2.39}$$

The η_0 is also called the weighted energy efficiency of the system.

2.5   Efficiency Function

In nature, both humans and artificial machines have efficiency functions. In general, the shape of the efficiency function η is shown in Fig. 2.1.

In Fig. 2.1, β is the load, and η is the efficiency function. η_e is the efficiency at β_e, which is the maximum efficiency, β_m is the maximum load, and

Fig. 2.1 Efficiency function η

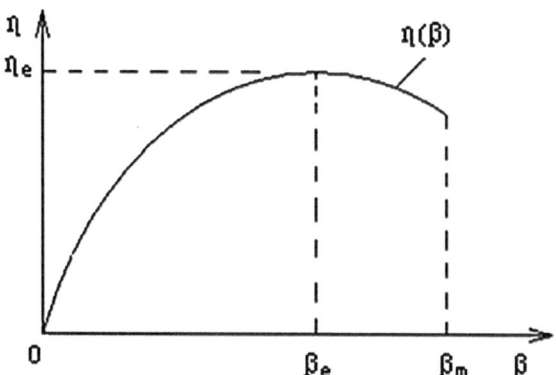

$$\beta_m \geq \beta \geq 0 \qquad (2.40)$$

$$\eta(\beta) \geq 0$$

$$\eta(0) = 0$$

$$\eta''(\beta) < 0$$

Efficiency can be expressed as

$$\eta(\beta) = \beta \sum_{i=1}^{\infty} \left(a_i \beta^{i-1} \right) \qquad (2.41)$$

In the energy consumption system, when $\beta = 0$, the energy consumption W_t is not necessarily zero. $W_t = W_0/\eta = 0/0$, which can be derived by the L'Hopital's Law. The efficiency function is a concave function.

2.6 Unification of Optimization of Power Generation and Energy Consumption

The total power generation W_t of the generalized power station can also be expressed as

$$W_t = \sum_{i=1}^{n} W_i \qquad (2.42)$$

where n is the total number of generators, W_t is the total power generation of the power station, and W_i is the power generation of the i-th generator.

The total input energy W_{t0} consumed by the generalized power station is expressed as

$$W_{t0} = \sum_{i=1}^{n} W_{i0} = \sum_{i=1}^{n} \frac{W_i}{\eta_i} = W_t \sum_{i=1}^{n} \frac{W_i}{W_t} \frac{1}{\eta_i} = W_t \sum_{i=1}^{n} \theta_i \frac{1}{\eta_i} = W_t \frac{1}{\eta_0} \qquad (2.43)$$

$$\theta_i = \frac{W_i}{W_t}$$

$$\eta_0 = \frac{1}{\sum_{i=1}^{n} \theta_i \frac{1}{\eta_i}}$$

$$\sum_{i=1}^{n} \theta_i = 1$$

where W_{i0} represents the energy consumed by the i-th generator, η_0 represents the overall energy efficiency of generalized power station. θ_i represents the load rate of the i-th generator, and η_i represents the operating energy efficiency of the i-th generator.

For the maximization problem of the total power generation W_t

$$\max W_t \qquad\qquad (2.44)$$

equivalent to the minimization problem of the total consumed energy W_{t0}

$$\min W_{t0} \qquad\qquad (2.45)$$

The optimization of these systems is to solve two problems, one is to determine the optimal number of operating units, and the other is how to distribute the load of each unit.

2.7 Not Working Is Different from Shutting Down

Regarding equipment optimization and manpower optimization, there are several interesting phenomena that need to be noticed. When everyone travels together, it is the most labor-saving arrangement to carry items evenly, and there is no problem of changing the number of people. If you arrange a group of people to complete a job, you need to consider the number of people, because even if the people sent out do not work, they still need to eat. This is the difference between people and machines. When the machine is turned off, it will no longer consume energy; in addition, the machine does not work (such as the pressure-holding operation when the water supply system has zero flow) is not the same as the machine does not work when it is turned off. Energy is still consumed when the machine is on and not working.

Chapter 3
Overall Structure and Fieldbus of Energy Saving Control System

The so-called energy efficiency optimization of the power generation systems, power transmission systems, and energy consumption systems we mentioned in the previous chapters are actually to solve two problems. One is to determine how many devices to use, and the other is to allocate the load rate of each device.

3.1 The Four Components of the Energy-Saving Control System

In order to solve these two problems, it is necessary to collect on-site process data and equipment operation data, which requires sensors to complete this task. Commonly used sensors include: pressure, flow, temperature, speed, torque and other sensors, and then the process data, energy parameters and equipment operation data measured by the sensors are sent to the controller. Common controllers include: PLC programmable controllers, DCS distributed controllers or other controllers. Through the analysis and calculation of the software program in the controller, the switch control signal is sent to the actuator to realize the start and stop switching of the equipment, send the analog control signal to the actuator to control the load distribution of each operating equipment. Commonly used actuators include: governors, frequency converters, clutches, valves, etc. In order for the operator to have an intuitive understanding of the process status and equipment operation, it is necessary to use a touch screen or a computer. Use the touch screen software to program the touch screen, or use the universal configuration software to program the PC. After completing the software programming, these operating data can be displayed on the screen for monitoring. At the same time, data input windows and buttons should be left on the screen to realize the operator's modification and control of process parameters and control parameters, as well as start and stop of equipment.

Simply put, the energy-saving control system includes four parts:

© The Author(s) 2024
F. Yao and Y. Yao, *Efficient Energy-Saving Control and Optimization for Multi-Unit Systems*, https://doi.org/10.1007/978-981-97-4492-3_3

(1) Sensors,
(2) Actuators,
(3) Controller and software,
(4) Touch screen and programming software (or PC and configuration software).

3.2 Several Structures of Energy-Saving Control System

3.2.1 Single Controller Structure

For the situation where the energy-saving control system is relatively small and the equipment is relatively concentrated, one controller (such as PLC) can generally be used for energy-saving control.

For example, the pumping station for secondary water supply, 2–5 sets of 3–15 kW water pumps, are small in size and are generally placed in the basement of a building, occupying a very small space. For such a pumping station energy-saving control system, it can be installed in 1–2 control cabinets on site. As shown in Fig. 3.1, the pressure sensor at the inlet and outlet of the pipeline, the flow rate sensor at the outlet of the pipeline are installed on the pipeline. The air circuit-breaker, intermediate relay, AC contactor, power sensor, frequency converter, PLC programmable controller are installed inside the cabinet, and the touch screen is installed on the cabinet door.

The control structure of this small energy-saving system is shown in Fig. 3.2.

Figure 3.2 is a block diagram of an energy-saving control system for a small pumping station.

In Fig. 3.2, the HMI is a touch screen, and it is connected to the PLC through the programming network port RJ45. PLC has 3 analog input signals, pumping station input pressure P1, pumping station output pressure P2 and pumping station flow F1.

Fig. 3.1 Secondary water supply pump station

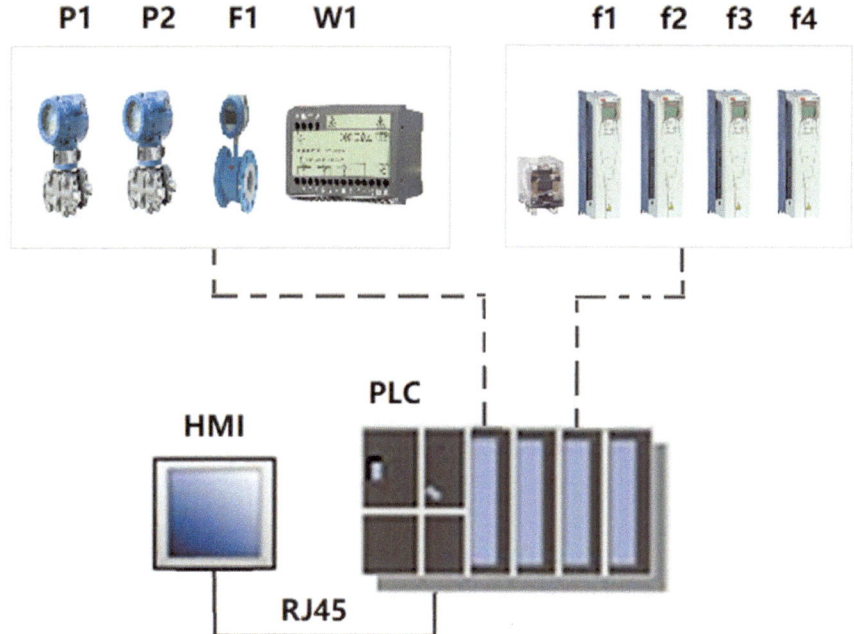

Fig. 3.2 A small energy-saving control system

PLC reads the data of power synthesis module W1, voltage, current and power data through RS485 communication port, and the four of PLC Digital output DO and 4 analog output AO control the start-stop and frequency of 4 frequency converters. 4 frequency converters drive the motors of 4 water pumps.

3.2.2 Multi-Controller Structure

The same pumping station, for a medium or large pumping station energy-saving control system, for example, a water plant pumping station with a daily water supply of 650,000 tons and a circulating water pumping station in the steel industry. As shown in Fig. 3.3, it is much more complicated. Each pump may require the installation of flow rate sensors and pressure sensors, as well as vibration sensors and temperature sensors, so that the working status, fault diagnosis and energy efficiency of each pump can be monitored online at any time. The volume of the pump is large, and the footprint of the pipeline is also large. The sensors are scattered and installed in different motors, pumps and relevant parts of the pipeline.

Figure 3.3 water plant pumping station and steel mill circulating water pumping station.

Fig. 3.3 Pumping stations

For a large-scale energy-saving system, such as a thermal power plant, a hydropower station, or a large pumping station, it may involve multiple workshops, multiple workstations, or multiple large-scale equipment. If there is a certain distance between workshops, workstations, or large-scale equipment, the network monitoring of multi-PLC and multi-PC can be realized by using Profibus-DP bus or industrial Ethernet.

3.2.2.1   Profibus-DP Fieldbus

The Profibus-DP bus uses optical fiber transmission to realize long-distance communication at a high communication speed. Since the optical signal is not subject to external electromagnetic interference, the anti-interference ability of the bus can be improved.

Install an optical fiber module (OLM: Optical Link Module) on a DP network segment to connect each PLC and PC with a DP bus interface to the OLM, as shown in Fig. 3.4. The OLM module converts the DP electrical signal of the device into an optical signal and transmits the signal using an optical cable. The electrical signals between the OLM fiber optic modules are electrically isolated. The OLM module automatically recognizes the communication speed of the DP bus, from 1.6Kbps to 12Mbps. Through the fiber optic module, the total length of the communication distance can be greatly improved. Using glass fiber, OLM fiber optic modules, the communication distance between them can reach 10–15 km.

In Fig. 3.3.4, 3 PLCs and 1 PC are connected together with Profibus-DP field bus. 1#PLC is in an independent workshop and uses an OLM module alone. 2#PLC and 3#PLC are in the same workshop, relatively close to each other. They are arranged in a DP network segment, and share an OLM module. The PC in the control center is far away from the two workshops and uses an OLM module alone. Various sensor signals enter three PLCs respectively, and the control signals of the three PLCs are sent to the actuators in the two workshops respectively.

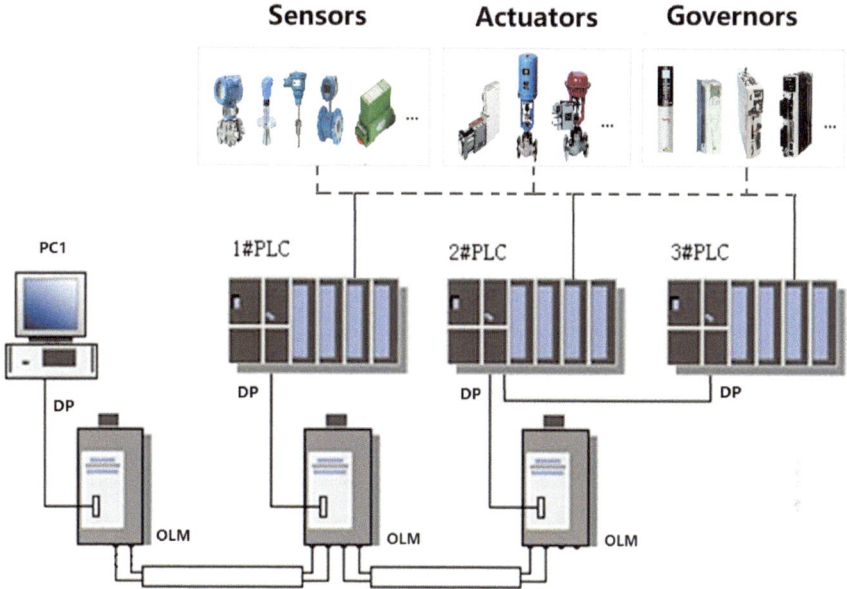

Fig. 3.4 Profibus-DP bus-optical fiber transmission

3.2.2.2 Industrial Ethernet

Selecting PLC and PC with industrial Ethernet interface can make use of the existing Ethernet equipment and structure to realize the networking and information transmission of the energy-saving control system, which is very convenient, as shown in Fig. 3.5.

In Fig. 3.3.5, 3 PLCs and 1 PC form a star network structure. The PN port on the PLC in the figure is the Ethernet interface, and the DP port is the Profibus-DP field bus interface.

3.2.2.3 Notes on DP Fieldbus

Address Occupied by Repeaters

Beginners should pay attention to the use of RS485 repeaters. The two ends of the RS485 repeater are electrically isolated. In the DP bus, although the DP address is not arranged for the RS485 repeater in the hardware configurationof the PLC device, but it needs to occupy the DP address.

In the DP bus, when there are more than 32 connected devices, it should be divided into several sections with no more than 32 devices, and the sections are connected with RS485 repeaters; as shown in Fig. 3.6.

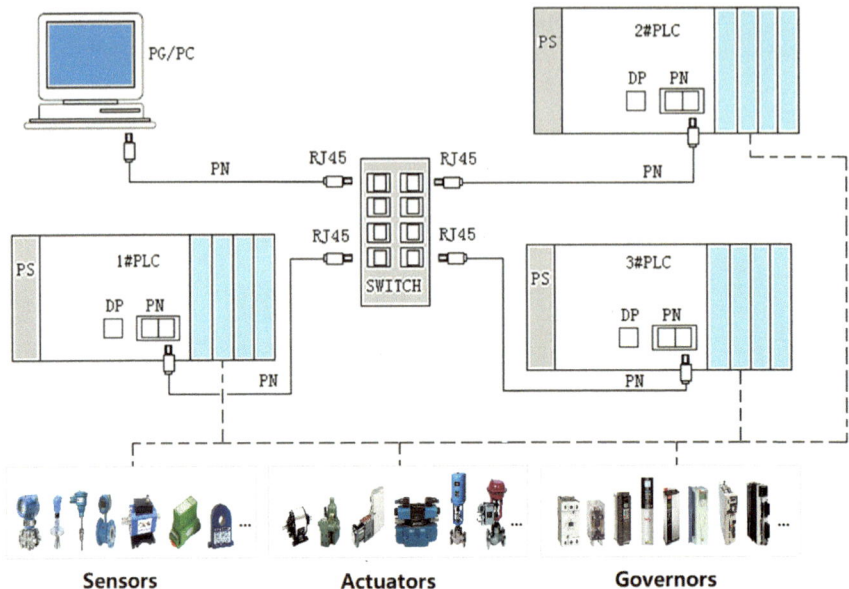

Fig. 3.5 Industrial Ethernet - star network structure

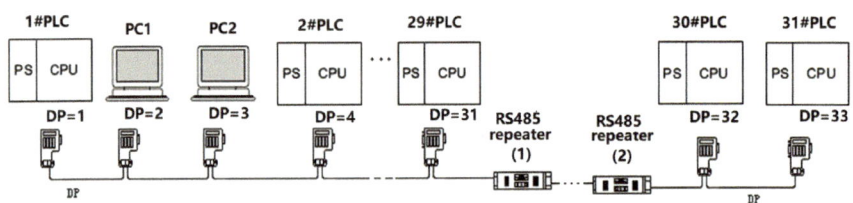

Fig. 3.6 DP bus

In Fig. 3.6, since there are 33 devices occupying DP addresses, which exceeds the maximum allowable number of DP addresses of 32 in a network segment, it is necessary to add an RS485 repeater to divide the DP bus into two network segments, so that the total number of devices in each network segment does not exceed 32. The added RS485 repeater also occupies the DP address, there are 35 DP devices in Fig. 3..3.6.

For each section of the bus, the terminal resistors of the bus connectors at both ends should be set to "ON", and the rest should be set to "OFF". The maximum distance allowed between two repeaters is 1 km.

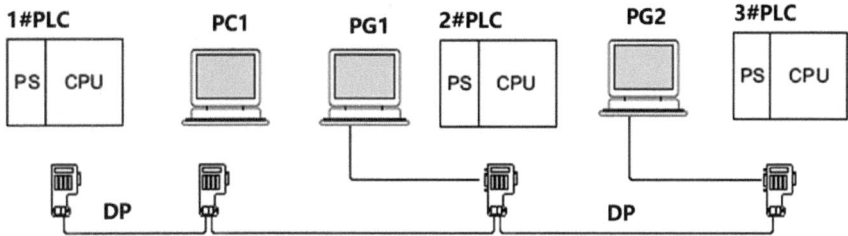

Fig. 3.7   Computer PG or PC connected to DP bus

The Total Length of the PG Cable

Connect several programming computers PG to the DP bus through "connecting cables", as shown in Fig. 3.7. In the figure, PG1 and PG2 are connected to the bus using a single cable connection, which is the "connecting cable".

If you use PROFIBUS bus cables to make your own "connecting cables", you need to pay attention to the length requirements of these "connecting cables". When the communication speed is equal to or greater than 3Mbps, please use standard connecting cables. In a network segment of the DP bus, there are certain requirements for the maximum total length of these "connecting cables", the length of each segment and the total number. Please refer to the instructions of the relevant PLC manufacturer. Cables that are too long can cause communication failures.

Fork Problem of DP Bus

The workshops of an enterprise are scattered, and some workshops are distributed in the opposite direction of the main workshop. At this time, attention should be paid to the layout of the DP bus. Assume that the distribution of the workshops is shown in Fig. 3.8.

In Fig. 3.8, one water source workshop is in the south of the main factory area, and the other water source is in the east of the main factory area. With the design method in Fig. 3.8, there will be no problems in the "main factory area" and "2# water source" workshops. But there will be a problem in the "1# water source" workshop, the communication is often disconnected, and there is no data in the "1# water source" workshop. The main reason is that the RS485 repeater (1) is in the middle of the DP bus network, and the fork is too long. It's like a "connection cable" that is too long causing communication problems.

There are two solutions, depending on the distance between 1#PLC and 2#PLC in the main factory area, method 1, as shown in Fig. 3.9.

Putting the bifurcated RS485 repeater (1) on one end of the DP bus can avoid communication problems. Method 2 is shown in Fig. 3.10.

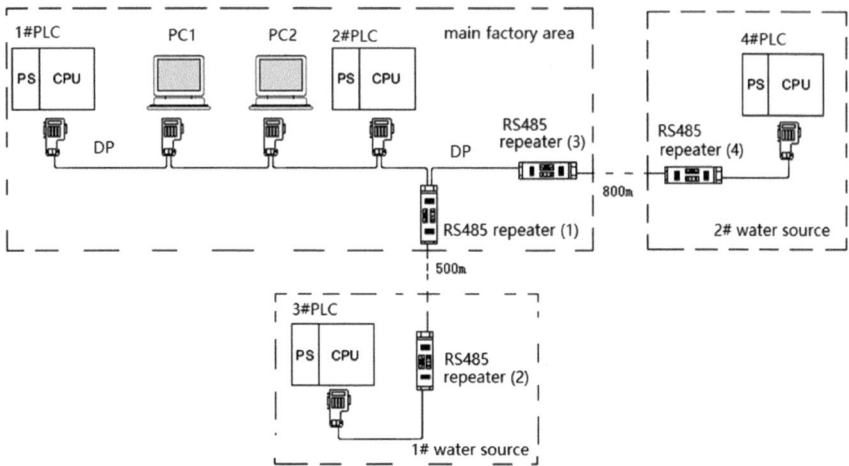

Fig. 3.8 Communication problem

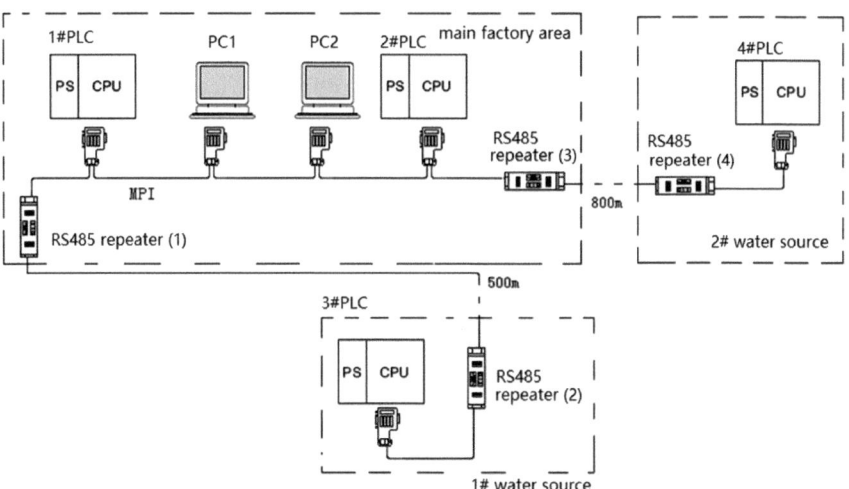

Fig. 3.9 Solution 1

The communication line between the "main factory area" and "1# water source" workshop can use "2 pairs of shielded twisted pair" cables, and add 2 RS485 repeaters, and the cost will increase.

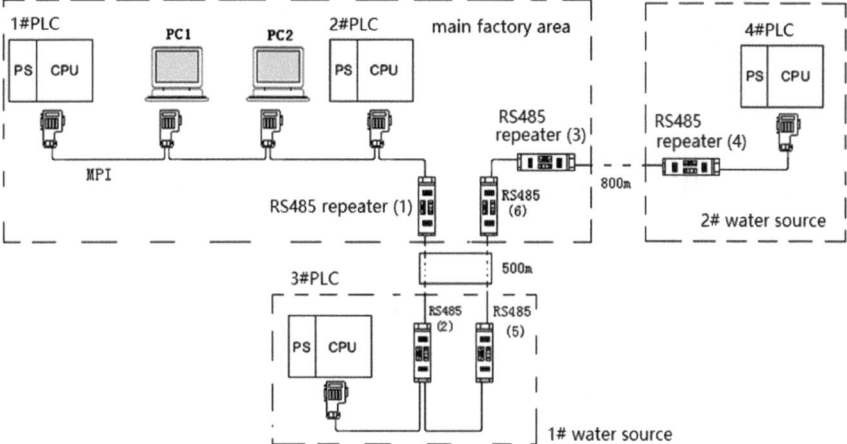

Fig. 3.10 Solution 2

Fig. 3.11 One PC and two PLCs with DP ports

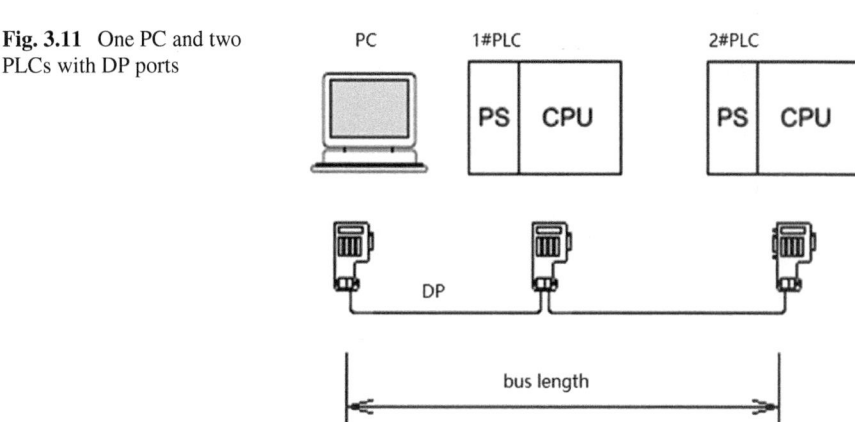

DP Bus Communication Speed and Distance

When using DP bus communication, the longest cable length allowed by a network segment is related to the communication speed. When the longest distance is exceeded, an RS485 repeater is required to extend the communication distance. Beginners need to pay attention to these details, otherwise the entire DP bus communication problems may arise.

One PC and two PLCs with DP ports are shown in Fig. 3.11. The terminating resistors on the DP plugs at both ends of the DP bus need to be set to "ON", and the DP plug in the middle is set to "OFF".

The maximum length of a bus cable allowed by a DP bus is related to the communication speed of the DP bus. If the communication speed is 187.5Kbps, the length is 1 km; if the communication speed is 3Mbps, the length is 100 m.

If the optical fiber module (OLM) is used to convert the electrical signal of the DP bus into an optical signal, the communication distance between two optical fiber modules can reach 10-15 km.

3.3 The Four Key Points of Industrial Bus and Industrial Ethernet Applications

Taking Siemens' industrial automation PLC product, S7-300 as an example, the description is as follows:

1. The hardware structure of industrial bus and industrial ethernet communication Industrial bus: one communication line connected in series.

(1) MPI bus: This bus is the bus that comes with all Siemens S7-300PLC CPU modules, and it is not open to third parties. Equal but opposite electromotive forces approximately cancel each other, so the twisted pair has a certain ability to resist electromagnetic interference.

(2) It is necessary to pay attention to the number of PCs in an MPI bus. According to the author's experience in a project many years ago, when more than 3 PCs are used in an MPI bus network, sometimes the PCs that are added after appearing cannot be connected. It is not normal, and it will be normal again after turning off one unit. The current situation has not been checked, and this is only for readers' reference.

(3) Profibus-DP bus: some Siemens S7-300PLC CPU modules have their own bus. CPU modules without DP interface need to insert a dedicated DP module on the guide rail. This bus is open to third-party profibus devices, and the medium is 1 pair twisted pair wires (2) or optical fiber. When optical fiber is connected, it needs to use optical fiber module to connect, and the transmission distance can be very long. Because the optical fiber is not affected by the electromagnetic field, it has strong anti-electromagnetic interference ability.

(4) Siemens industrial Ethernet (one communication line connected in series or star radiation structure):

(5) Profinet Industrial Ethernet: 4 pairs of twisted pairs (8, namely network cables) or optical fibers, need to use switches for connection, open to third parties, when using optical fibers, it has strong anti-electromagnetic interference ability and long transmission distance.

2. After the hardware is set up, use Siemens STEP7 software to configure the PLC hardware:

(1) For MPI and DP buses, assign an MPI or DP address to each PLC connected to the bus, which cannot be repeated.

(2) For industrial Ethernet, assign a URL to each PLC connected to the Ethernet, which cannot be repeated.

(3) Download it to all PLCs in the bus or network, so that the PLCs can know each other which address PLCs exist.

3. In the hardware configuration stage, define the automatic transmission of data between PLCs, without programming in the PLC program:

(1) Define the address and quantity of data to be transmitted to each other, define the sender PLC, define the receiver PLC, define the sending address, define the receiving address, define the number of sent data, define the number of received data, and realize data transmission without programming.
(2) Download the hardware configuration information to all PLCs on the bus or network.

4. Program with Siemens STEP7 software to realize data transmission between PLCs.

In the program, write the data transmission module, define the address of the PLC to send data, the amount of data to be transmitted, and the address of the receiving PLC, to realize the mutual data transmission between PLCs. This data transmission method is clear and clear. Online inspection and monitoring are more convenient.

Chapter 4
Commonly Used Energy Parameter Sensors

From the discussion in Chapter Two, we know that energy-related parameters include: liquid pressure, liquid level, liquid flow, temperature, torque, speed, rotational speed, voltage, current, power, etc. These parameters are measured by sensors, and turned into standard analog signals or digital signals, then output to the energy-saving control system. We will discuss these sensors in this chapter.

4.1 Liquid, Gas Pressure Sensor and Liquid Level Sensor

The pressure of liquids and gases represents the pressure energy they have.

Liquid and gas pressure sensors are among the most widely used sensors. Liquid and gas pressure sensors are used to measure the pressure in pipes and containers. Pressure sensors use resistance strain gauges, semiconductor strain gauges, piezoresistive, inductive, capacitive, magnetically controlled, ceramic piezoelectric, etc. The U-shaped manometer uses the height of the liquid to directly indicate the level of air pressure.

For the strain gauge type pressure sensor, the strain gauge is bonded to the measuring body. The pressure of the measuring body changes, and the strain gauge also deforms together, so that the resistance value of the strain gauge changes, the voltage on the strain changes. The bridge circuit extracts the differential signal, amplifies it, and gives the corresponding measurement signal. The principle of the strain gauge sensor is shown in Fig. 4.1.

For the pressure sensor using the pressure measuring tube, the measured liquid will produce mechanical deformation after passing through the pressure measuring tube. The measuring elements such as the potentiometer are driven by the mechanical deformation to give an electrical signal proportional to the pressure of the measured liquid.

© The Author(s) 2024
F. Yao and Y. Yao, *Efficient Energy-Saving Control and Optimization for Multi-Unit Systems*, https://doi.org/10.1007/978-981-97-4492-3_4

Fig. 4.1 Strain gauge sensor

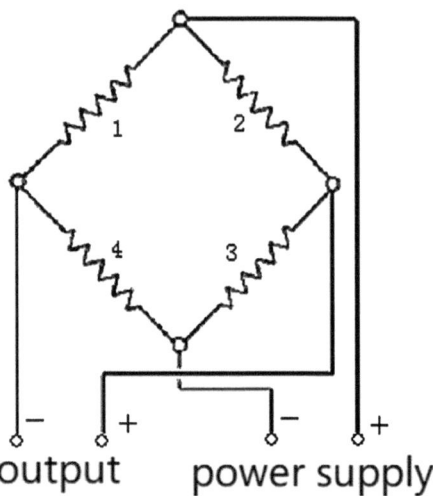

For corrosion resistance, impact resistance and high temperature resistance, there are diaphragm type, oil filled, shockproof and high temperature resistant pressure gauges. In order to meet different pressure measurement ranges, there are pressure gauges with positive pressure, negative pressure, differential pressure, positive and negative pressure and other ranges. In order to measure different media, there are pressure gauges for different purposes. A transmitter capable of measuring the pressure difference between two pipes is called a differential pressure transmitter. If one end of the sensor is connected to atmosphere, the other end can measure the gauge pressure of the pipe.

For continuous measurement pressure transducers, the simplest is a remote pressure gauge that outputs a signal of change in resistance, similar to a potentiometer. When the pressure of the pipe or container changes, the resistance value between the middle tap of the potentiometer and the fixed end also changes accordingly.

It should be noted that for the pressure gauge installed at the outlet of the well. If the check valve is not tight when the pump stops running, it will cause negative pressure. When the pump supplies water to a low place, negative pressure also occurs when the pump stops. For pressure sensors that only have a positive pressure display function, they are often damaged. You should choose a pressure gauge or pressure sensor with positive and negative ranges.

When the pressure sensor outputs standard 0–10 mA, 4–20 mA, 0–5 V, 1–5 V signals, it is called a pressure transmitter. There are two wiring methods for pressure transmitters: 2-wire and 4-wire. The main parameters of the pressure sensor include measurement range, output signal type, explosion-proof grade, protection grade, etc.

The shape of a common pressure transmitter is shown in Fig. 4.2. The shape of a common pressure gauge is shown in Fig. 4.3.

The height of the liquid surface represents the potential energy of the liquid.

Liquid level sensors are divided into switch type and continuous measurement type according to different signals, and are divided into contact type and non-contact liquid level sensors according to whether the sensor is in contact with the liquid to be measured.

Contact liquid level sensors include plug-in liquid level transmitters, static pressure liquid level transmitters, and differential pressure liquid level transmitters. Differential pressure transmitters measure the difference in pressure between the top of the liquid level and the bottom of the liquid. Float driven rotary encoder changes with liquid level. Capacitive liquid level gauges use the dielectric constant

Fig. 4.2 Common pressure sensors

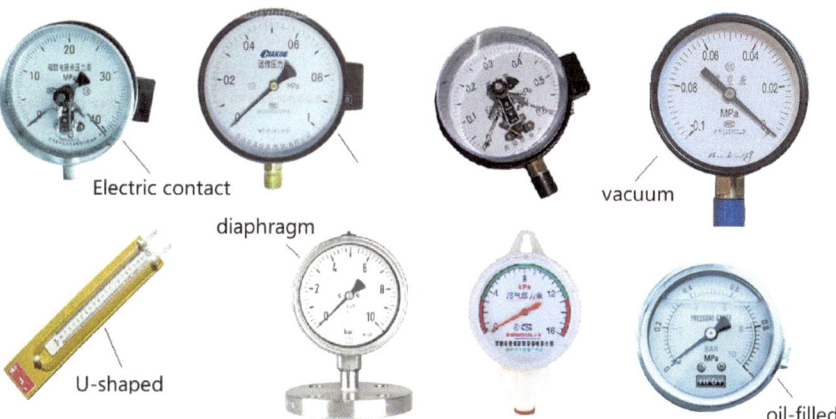

Fig. 4.3 Common pressure gauges

of a capacitor to measure liquid level as the liquid level changes. A liquid level sensor that utilizes a magnet on a float that changes as the liquid level rises and falls. Sensors that use the principle of magnetostriction and use electrode height to measure liquid level, etc.

Non-contact liquid level sensors include ultrasonic liquid level gauges that use ultrasonic echoes to measure liquid level heights, radar level gauges that use radar reflected waves to measure liquid levels, and ray level gauges that use the principle of nuclear radiation, etc.

The liquid level sensor for continuous measurement is also called liquid level gauge, which can be composed of strain gauge, capacitor plate, magnetic element, ultrasonic wave, radar, infrared ray, etc.

The measuring principle of the strain gauge type is similar to that of a pressure sensor. Four strain gauges connected to the measurement chamber form an electrical bridge. Due to the pressure generated by the liquid level, the sensor probe will deform the measuring chamber. The depth of the liquid is measured based on the deformation of the measuring chamber.

The principle of ultrasonic liquid level sensor and radar liquid level sensor is similar. In both measurement methods, the sensor and the liquid cannot come into contact, or they can be placed outside the pressure vessel.

For submersible liquid level sensors, in order to eliminate the influence of atmospheric pressure on the liquid level measurement, there is a gas guide tube on the sensor probe, which is placed in the liquid leading from the cable to keep the reference pressure chamber connected to the ambient pressure. Be careful not to block or disrupt this airway during installation.

For open containers, install a static pressure liquid level transmitter directly at the bottom of the container to measure the liquid level directly.

For a closed container with pressure, a differential pressure liquid level sensor is required to obtain the liquid level value by measuring the pressure difference between the liquid surface and the liquid bottom.

The main parameters of the liquid level sensor include measuring range, output signal type, etc., when the output signal of the liquid level sensor is a standard 0 ~ 10 mA, 4 ~ 20 mA, 0 ~ 5 V, 1 ~ 5 V signal, it is also called a liquid level transmitter.

If the transmitter uses only two wires to complete the power supply and measurement signal return at the same time, the transmitter is called a two-wire transmitter, and if the power line and signal line are separated, it is called a four-wire transmitter. A 4-wire sensor can also be 3-wire if the OV and signal negative terminals are shared.

Figure 4.4 shows the appearance of the liquid level sensor.

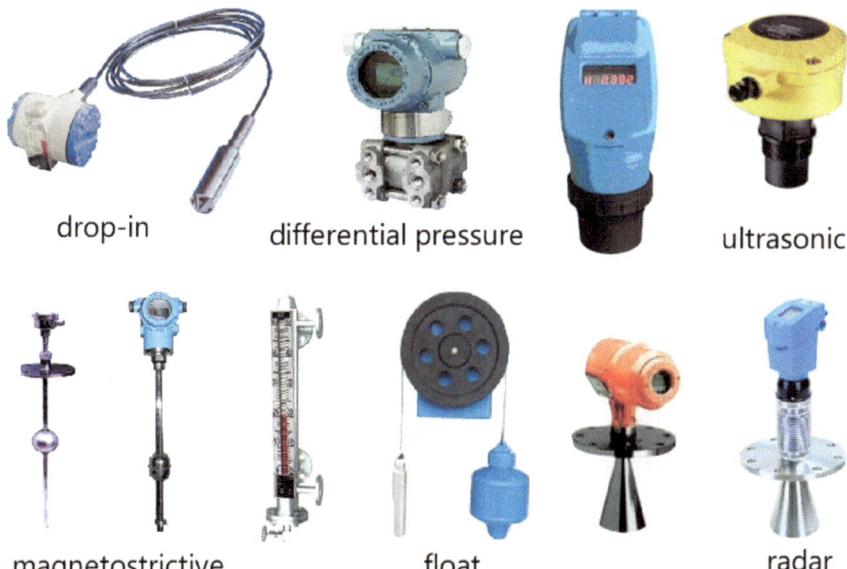

drop-in   differential pressure   ultrasonic

magnetostrictive   float   radar

Fig. 4.4 Liquid level sensor

4.2 Temperature Sensor

The temperature sensor is also one of the most widely used sensors. The temperature sensor is used to measure the temperature of liquid, gas, solid or thermal radiation. Typical temperature sensor types include thermal resistance, thermocouple, semiconductor, bimetal, pressure, glass liquid (such as thermometers), optical (infrared), radiation, colorimetric thermometers, etc., of which temperature sensors (thermometers) of thermal resistance, thermocouple, semiconductor, bimetal, pressure, and glass liquid type are the measurement methods that contact the measured substance, optical, radiation, colorimetric, etc. are non-contact temperature sensors.

Thermal resistance and thermocouple temperature sensors are the most used temperature sensors. In thousands of households, glass liquid thermometers may be widely used. In stations and airports with dense floating populations, non-contact infrared thermometers are the best choice. Appears more convenient.

The main material of the thermal resistance is metal. The resistance value of metals with different components will change with the change of temperature. Using this principle, the temperature of the measured object can be calculated by measuring the change of metal resistance value. Its main measurement range −200–500 °C, the linearity and accuracy of platinum (Pt) resistance is better than that of copper resistance (Cu), thermal resistance temperature sensors include Pt10, Pt100, Pt1000, Cu50, Cu100, NTC, etc.

Two different conductors A and B, or two different semiconductors A and B are combined to form a thermocouple. If the temperature t1 at the junction is different, the

Fig. 4.5 Principle and appearance of semiconductor coolers

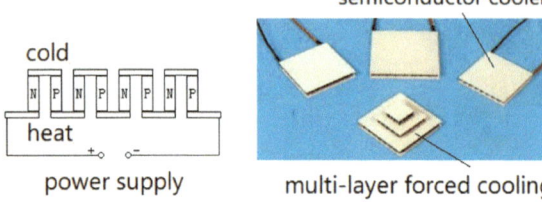

cold

heat

power supply

semiconductor cooler

multi-layer forced cooling

electromotive force at both ends of the conductor or semiconductor will be different. This phenomenon is called Seebeck effect. Thermocouple temperature sensors have different divisions such as K, S, E, B, J, N, T, R, WRE, etc., and are used to measure different temperature ranges.

On the contrary, if different conductors A and B, or different semiconductors A and B, pass current, cooling and heating will occur, this phenomenon is called the Peltier effect. P-type and N-type semiconductors are connected to form a galvanic pair, and direct current is applied to both ends of the electric refrigeration sheet, so that one side of the electric refrigeration sheet becomes cold and the other side becomes hot. The cold surface can be used to make an electronic refrigerator or water chiller, can also be used for heat dissipation of other electronic devices.

The principle of the semiconductor cooler and the appearance of the actual product are shown in Fig. 4.5. The author has used this principle to invent a body temperature battery for watches and a physical battery for heat meters. Using the way of producing integrated circuits, a large number of PN junctions are connected in series. The temperature of the lower sides is high, and the temperature of the upper sides is low. This is a physical battery.

Temperature sensors vary greatly depending on the application occasions. The sensor for measuring room temperature and air temperature may be a small metal bead, which is thinner than a wire. When measuring the temperature in a pipeline, a protective shell is required, as well as accessories such as an oil cup that is easy to disassemble and maintain, so sometimes it is difficult to judge whether it is a temperature sensor based on its appearance.

Temperature sensors that can output standard signals of 0–10 mA, 4–20 mA, 0–5 V, and 1–5 V are called temperature transmitters. The main parameters of the temperature transmitter are measurement range, output signal type, signal accuracy, linearity, two-wire system or four-wire system, protection method and explosion-proof or not.

The appearance of the temperature sensor is shown in Fig. 4.6.

The temperature transmission module and display transmission unit matched with the temperature sensor can be installed externally on site to match the temperature measuring element, or can be built into the temperature sensor, or can be centrally installed in the instrument cabinet with guide rails. The temperature transmission modules are shown in Fig. 4.7.

The appearance of temperature sensors such as electric contact thermometers, colorimetric thermometers, and glass thermometers is shown in Fig. 4.8.

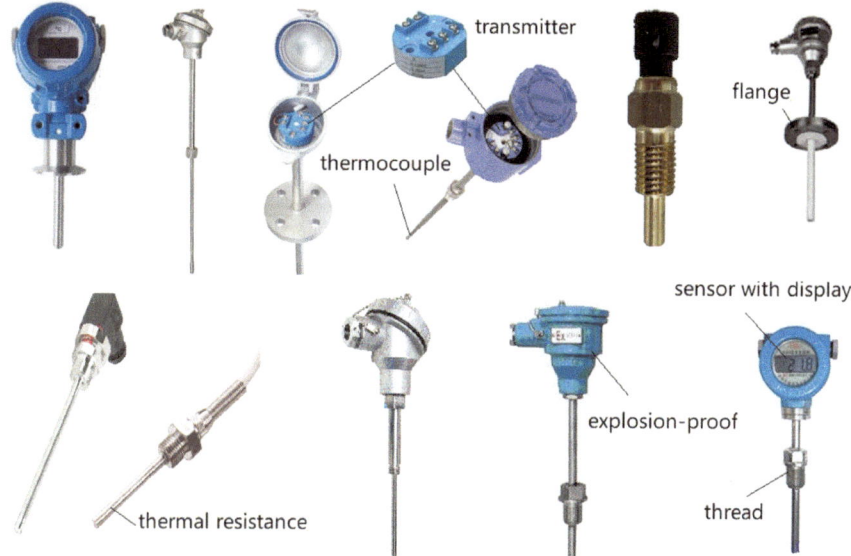

Fig. 4.6 Temperature sensor

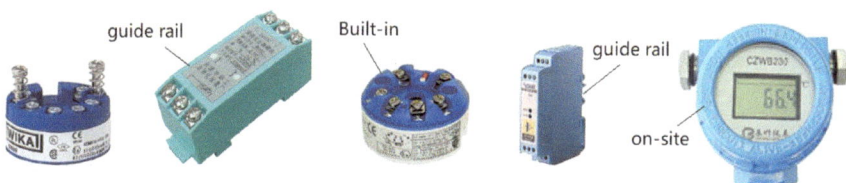

Fig. 4.7 Temperature Transmitting Module

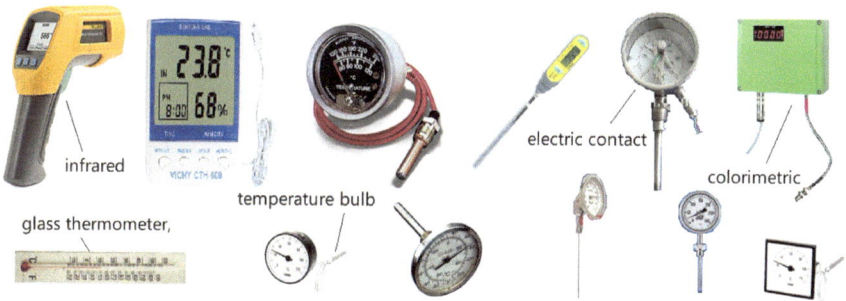

Fig. 4.8 Electric contact thermometer, colorimetric thermometer, glass thermometer

4.3 Flow Sensor

Flow sensors are mainly used to measure the flow of liquid or gas in pipelines, or the flow of liquid in open channels. Common flow sensors include orifice plate type, turbine type, electromagnetic type, ultrasonic type, vortex type, rotor type, open channel type, etc.

In addition, there is the target type, in which the target for measuring pressure is placed in the center of the pipe. Annubar type inserts a tube with a hole in the middle of the pipeline, and measure the pressure difference before and after the tube. Verabar types measure the differential pressure before and after the plate with holes. Venturi type has a tube that shrinks and then diffuses in the center of the pipeline to measure the pressure difference of different sections. Pitot tube type inserts a pitot tube with a hole in the tube head to measure differential pressure. Cone type places a cone inside the middle of the pipeline to measure differential pressure. Oval gear uses fluid to drive the gear to rotate, measure flow rate. Volumetric type, rotor type, and other different forms of flow sensors.

The simplest flow sensor is a flow switch. Its principle is to put a small baffle in the pipeline. When the fluid passes through, the baffle is pushed to make the baffle move. The baffle moves a micro switch to send a switch signal. For the flow rate switch, the action value of the flow rate can be set, when the flow rate is greater than the set value, the contact switch will act.

Turbine flow sensors exploit the effect of liquid in a pipe on a turbine or propeller placed in the pipe. The speed of the turbine or propeller is directly proportional to the flow rate or flow rate. The greater the flow, the faster the speed. The flow rate is calculated by measuring the speed of the turbine and the known pipe diameter. The common water meters found in thousands of households are basically turbine type.

The orifice plate flowmeter is to place a plate with a hole in the middle of the pipeline. When the fluid (gas or liquid) flows through the orifice plate, due to the throttling loss, a pressure difference is generated on both sides of the orifice plate. The square root value is proportional to the flow rate. The differential pressure transmitter is used to measure the pressure difference, and then the flow rate of the fluid in the pipeline can be obtained through calculation. The orifice flowmeter has a high degree of reliability because it has no moving parts. The disadvantage is Waste of energy.

The electromagnetic flow sensor is used to measure the flow of conductive liquid. The conductive liquid flows through the pipeline with the induction coil. According to the law of electromagnetic induction, the different flow speed of the conductive liquid induces a voltage perpendicular to the magnetic field and proportional to the flow speed. Using this parameter change, the flow rate of the conductive liquid is measured.

The ultrasonic flow sensor is composed of a transmitting probe and a receiving probe. The flow of the liquid in the pipeline affects the sound velocity of the ultrasonic wave in the pipeline. The liquid flow rate is measured by measuring the sound velocity change measured by the ultrasonic receiving probe. At the same time, the flow rate of the liquid in the pipe is based on the known pipe diameter. This is the time-of-flight ultrasonic flowmeter, used to measure liquids such as clear water; another ultrasonic flowmeter uses the particles in the fluid to reflect ultrasonic waves to form the Doppler effect, so that the ultrasonic frequency reflected by the particles change, uses the change in frequency to calculate the flow velocity of the fluid. This ultrasonic flowmeter is used to measure fluids such as sewage; the ultrasonic flowmeter is installed outside the pipeline without resistance loss, and the cost has little to do with the diameter of the pipeline to be measured.

The vortex street flow sensor is used to form a vortex behind the measuring rod when the fluid flows through the measuring rod. The flow rate and the formed vortex are in a certain proportional relationship.

The rotor type flow sensor is similar to putting a float (rotor) in a glass tube. The diameter of the tube is large and the bottom is small. When the flow rate is large, the float will rise. It is used for flow measurement in vertical pipelines or for direct observation. If magnetism or other objects that can be sensed by the outside are added inside the float, the flow signal can also be transmitted.

The main parameters of the flow sensor are the measurement range, the minimum flow velocity of the fluid, the output signal type, the protection level, the explosion-proof requirements, etc.

For engineers and technicians, special attention should be paid to the minimum flow velocity requirements, otherwise the flow results measured at low flow rates may appear larger error. Flow rate sensors that can output standard signals of 0–10 mA, 4–20 mA, 0–5 V, 1–5 V are called flow rate transmitters.

The shapes of flow sensors such as turbines, orifice plates, vortex streets, ultrasonic waves, and rotors are shown in Fig. 4.9. The appearance of target type, Annubar, gear and other flowmeters are shown in Fig. 4.10.

4.4 Force Sensor

Sensors used to measure tension, pressure, and gravity are all force sensors.

Force sensors include resistance strain gauge type, transformer type and silicon semiconductor type.

For strain gauge force sensors, the strain gauge resistors are pasted on the parts of the measuring body that are subject to tension and compression. The measuring body is deformed by the external force, resulting in changes in the resistance of the 4 sets of strain gauges on it, which makes the output of the bridge circuit unbalanced, generate a signal voltage output, and measure the change of the voltage to indirectly measure the magnitude of the force.

For a transformer-type force sensor, the iron core is connected to the measuring body, and the excitation coil and the measuring coil are coupled through the iron core. When an external force is applied to the measuring body, the deformation of the measuring body will cause the iron core to move, thereby changing the excitation

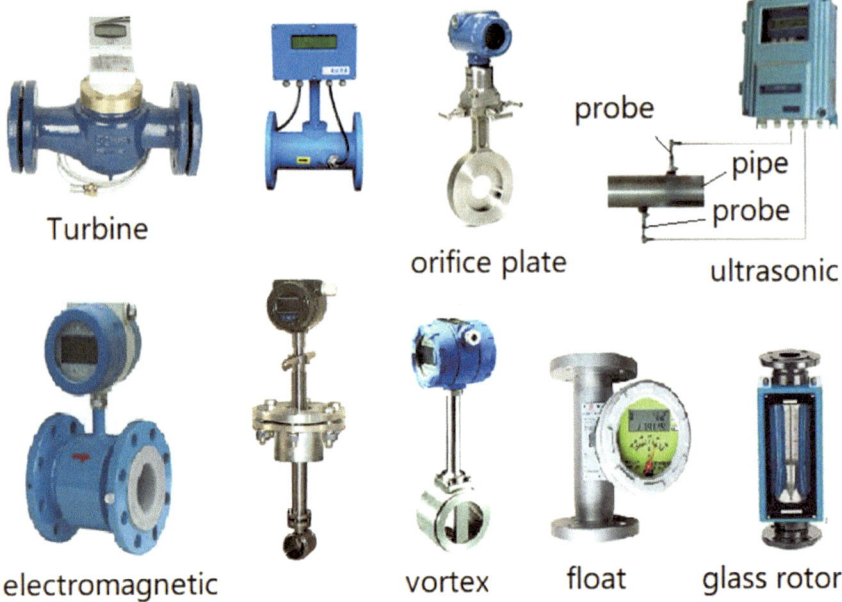

Fig. 4.9   Flow sensor

Fig. 4.10   Target, Annubar, and gear flowmeters

coil to the measuring coil. The change of the electrical signal of the measuring coil reflects the magnitude of the action force.

A semiconductor-type force sensor measures force by using changes in the electrical properties of a semiconductor when it is deformed under pressure.

Load cells are often used in the fields of batching, material conveying, and machinery manufacturing. In the conveying and processing of strip materials, it is often necessary to measure the tension of the tape, which requires a tension sensor.

Fig. 4.11 Force and load cell

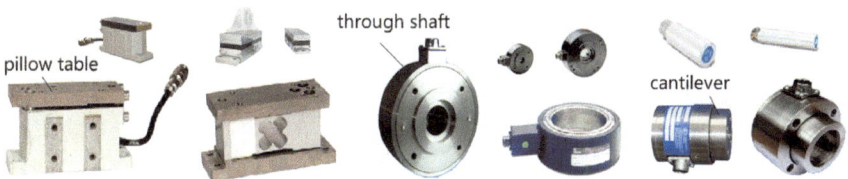

Fig. 4.12 Appearance of common tension sensors

The appearance of common sensors for measuring force and weight is shown in Fig. 4.11. There are several types of common tension sensors, such as pedestal type, cantilever type and shaft-through type, and their appearance is shown in Fig. 4.12.

4.5 Speed Sensor

Rotational speed is also a common parameter of energy. The sensors for measuring the signal of rotational speed are mainly rotary encoders, resolvers and tachogenerators.

Rotary encoders can be divided into absolute type and incremental type. The rotation angle position of the absolute type encoder corresponds to the output rotation angle position one by one. The incremental type encoder sends out a pulse every time a certain angle is increased, and is measured by measuring the number of pulses. Corresponding to the angle value, the output of the incremental encoder is mostly two-phase A and B plus a zero-phase Z, or two-phase output of A and B, as shown in Fig. 4.13. The absolute encoder needs to be subdivided at different positions, there are many grating gaps on the inner disk grating from the center of the circle to the outermost circle that are staggered according to certain rules. The number of output lines varies with the accuracy. Generally speaking, the number of lines is more than that of incremental encoders. The output lines of the absolute value encoder output by serial communication can be only 3. The encoder measuring 16 absolute positions is shown in Fig. 4.13. The input shaft of the rotary encoder can be a hollow shaft or a solid shaft. The encoder of the hollow input shaft has a spring leaf for fixed installation. When the output shaft of the load is not concentric with the hole of

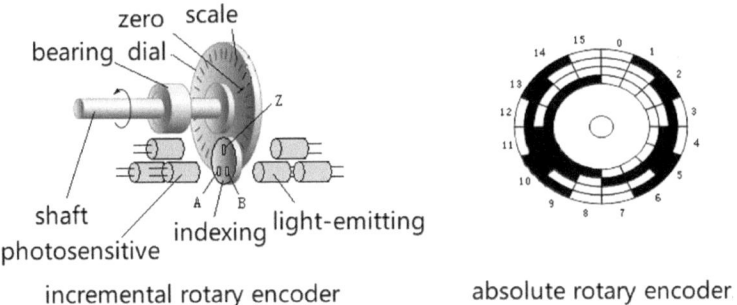

Fig. 4.13 Principle of rotary encoder

the encoder, the spring leaf provides cushioning. There are fixed holes on the shaft encoder for installation, and the rotary encoder is a common sensor for measuring angles, especially in servo motors.

There are main scale and zero scale on the dial. There is only one zero scale for one circle. The main scale is evenly distributed along the circumference according to the required resolution, such as 2500. There are three gaps A, B, Z on the dial, where the distance between A and B is half the distance between the two scales on the main scale. The input shaft drives the dial to rotate. The light emitted by the light-emitting element passes through the scale and index plate on the dial and is received by the photosensitive element. It receives a Z signal for each revolution. The A and B signals each have 2500 pulses in one revolution. Since the signals between A and B have a 90° difference, the actual resolution of the encoder is 4 times the number of 2500 scales in one circle, 10,000 pulses/axis.

Resolver (also known as synchronous resolver) is based on the principle of transformer, by applying an excitation signal to the stator winding (primary side), generally an AC voltage signal above 400 Hz, and detecting the phase angle of the secondary winding side on the rotor with the rotation angle periodic changes occur to determine the angle and speed.

The self-aligning machine (also called self-synchronizing machine) is similar to the rotary transformer. It is equivalent to a wound AC motor, and the angle can be measured by measuring the phase change of the winding voltage signal.

The tachogenerator is used to measure the speed of the rotating object, also known as the speed sensor. There are AC tachogenerator and DC tachogenerator. The output voltage is proportional to the speed. The structure of the DC tachogenerator is similar to that of the DC motor. The rotor of the machine is in the shape of a hollow cup, one phase winding on the stator is excited, and the other phase winding outputs an AC voltage with a constant frequency.

The main parameters of the speed sensor are resolution, range, accuracy and linearity, etc., and the main parameters of the encoder are the number of pulses per revolution, the number of output phases, and the signal type.

The appearance of the speed sensor is shown in Fig. 4.14.

Fig. 4.14 Speed sensor

4.6 Torque and Speed Torque Compound Sensor

Torque is a commonly used energy parameter. In the case of driving equipment such as electric motors or internal combustion engines, in order to calculate the input and output power, in addition to measuring the rotational speed, the output torque of these equipment must also be measured. The product of torque and rotational speed is proportional to the output power of these equipment.

Common torque sensors and torque speed composite sensors, as shown in Fig. 4.15. There are an input shaft and an output shaft on both sides of the torque sensor, or an input flange and an output flange. The input shaft transmits the torque to the output shaft through the torque sensor in the middle, which is the same as the force sensor mentioned above. Similarly, the torque sensor is deformed by the torque, and the torque signal is measured according to the deformation. If the rotational speed of the shaft is also measured, the power is also measured. Based on such convenience, in addition to sensors measuring static torque, most dynamic torque sensors are torque-speed composite.

The main parameters of torque, speed and power sensors are measurement range, output signal size and type, explosion-proof grade, protection grade, etc.

4.7 Voltage Transmitter

The calculation of the power supply is inseparable from the measurement of the power supply voltage.

In electrical systems and automation systems such as power transmission, power distribution, and power equipment protection, it is necessary to measure the voltage, current, power factor, and power of the power supply line in order to perform power distribution or take other protective measures. The measurement of these parameters is used for single-phase, three-phase three-wire, three-phase four-wire, etc.

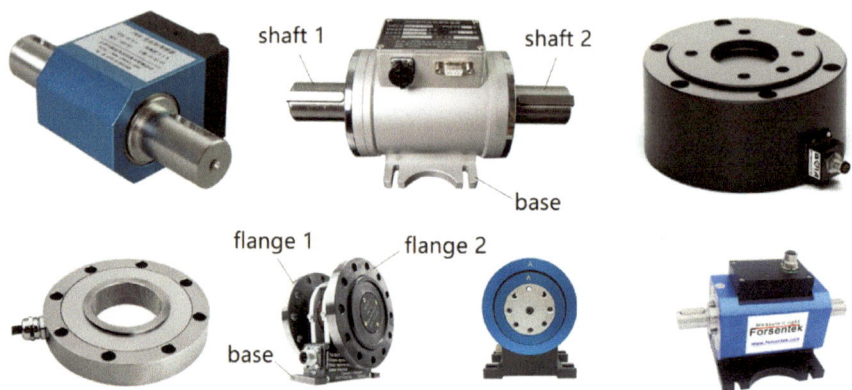

Fig. 4.15 Torque, speed and power integrated sensor

Most of the currently used voltage, current, power factor, and power transmitters are integrated.

Voltage transmitter: convert the voltage signal of the line into a linear and isolated 4–20 mA (or 0–10 mA, 0–20 mA, 1–5 V, 0–5 V, 0–10 V, etc.) or digital signal. If the voltage signal of the AC line is measured, and the voltage is very high, it is necessary to use a voltage transformer to convert the high voltage into a low voltage signal, and then input it into the voltage transmitter. The voltage input has 0–220 V, 0–380 V direct input mode and PT secondary input mode. Fuses are added to the primary side of the voltage transformer to avoid the influence of the primary circuit on the secondary side when the secondary short circuit is too large. So, the secondary side is not allowed to be short-circuited and one end is grounded. The shape of the voltage transmitter is shown in Fig. 4.16.

4.8   Current Transducer

The calculation of the supply power is inseparable from the measurement of the supply current.

Current transmitter: convert the current signal of the line into a linear and isolated 4–20 mA (or 0–10 mA, 0–20 mA, 1–5 V, 0–5 V, 0–10 V, etc.) or digital communication signal. If the current of the AC line is large, it is necessary to use a current transformer to convert the large current into a small current signal, and then input it into the current transmitter. The current input has 0–5A direct input mode and CT secondary input. The secondary side of the current transformer is not allowed to be open to avoid high voltage on the secondary side, so the secondary side is not allowed to install a fuse and must be grounded at one end. The shape of the current transmitter is shown in Figs. 4.3, 4.4, 4.5, 4.6, 4.7, 4.8, 4.9, 4.10, 4.11, 4.12, 4.13, 4.14, 4.15, 4.16 and 4.17.

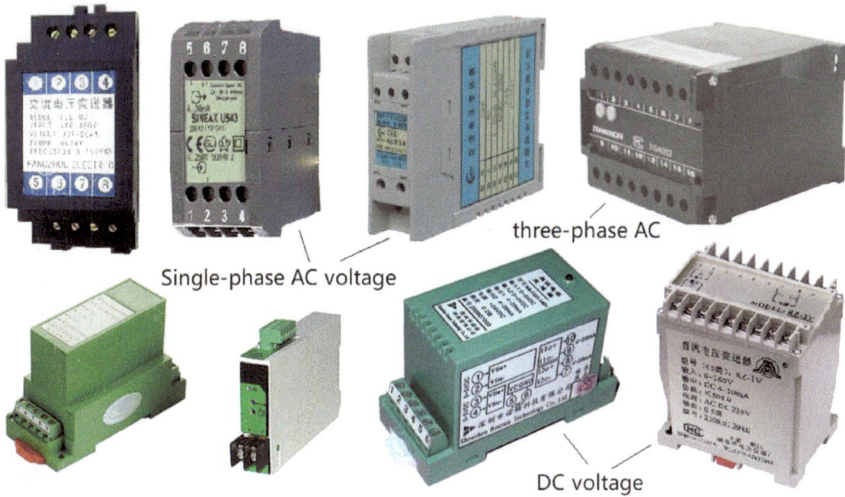

Fig. 4.16 Voltage Transmitter

Fig. 4.17 Current transmitter

4.9 Power Factor Transmitter and Supply Power Transmitter

Power factor transmitter: change the power factor of the AC line (that is, the cosine function of the phase angle between the voltage and the current, with a value of 0–1) into analog standard signal or digital communication signal output, such as 4–20 mA, 0–10 mA, 0–20 mA, 1–5 V, 0–5 V, 0–10 V, etc.

If the current and voltage of the circuit are large, it is necessary to use a current transformer or voltage transformer to reduce the large current and large voltage. The current and voltage signals are input to the power factor transmitter. The power factor represents the functional relationship between the electric energy used in the line and the electric energy occupied, and its meaning can refer to the reactive power compensation part of Chap. 24.

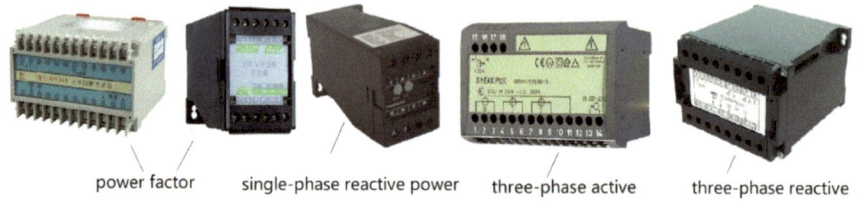

power factor   single-phase reactive power   three-phase active   three-phase reactive

Fig. 4.18 Power Factor Transmitter and Power Transmitter

Power supply power transmitter: Input the voltage and current of the line, calculate the power and turn it into analog standard signal or Digital communication signal output, such as 4–20 mA, 0–10 mA, 0–20 mA, 1–5 V, 0–5 V, 0–10 V, etc.

If the current and voltage of the AC circuit are large, it is necessary to use a current transformer or a voltage transformer to convert the large current and large voltage into a smaller current and voltage signal. Then input the power transmitter, AC line active power transmitter and reactive power transmitter, the concept of active power and reactive power can refer to Chap. 24 of reactive power compensation.

The appearance of power factor transmitter and power transmitter is shown in Fig. 4.18.

Chapter 5
Valves and Clutches Commonly Used in Energy-Saving Systems

In energy-saving systems, clutches and valves are used for equipment switching and load distribution.

5.1 Magnetic Powder Clutch and Magnetic Powder Brake

The magnetic powder clutch places magnetic powder between the active part and the passive part, and adjusts the attraction and distribution of the magnetic powder inside the magnetic powder clutch by changing the excitation voltage of the electromagnetic coil. Use magnetic powder to transmit torque between the driving shaft and the driven shaft. When the speed of the driving shaft is constant, the speed of the driven shaft can be controlled. The principle of the magnetic powder brake is the same as that of the magnetic powder clutch, except that it fixes the speed of the driving shaft to zero. The rotational speed and output torque of the driven shaft are adjustable.

Magnetic powder clutches and magnetic powder brakes can be used for breaking control and speed control.

Magnetic powder clutches and magnetic powder brakes are widely used in automatic production lines such as printing machines, die-cutting machines, paper machines, compound machines, wire drawing machines, coating machines, winding machines, metal sheets, strips, and films. They are mainly used for automatic tension control. For unwinding or rewinding control, generally the magnetic powder brake is used for unwinding control, and the magnetic powder clutch is used for rewinding control.

Magnetic powder clutches can also be used in power transmission mechanisms such as motors, engines, electric mechanisms, and reducers. One shaft of the magnetic powder clutch is connected to the motor side, and the other shaft is connected to the load side. Changing the excitation voltage of the magnetic powder clutch can adjust

© The Author(s) 2024
F. Yao and Y. Yao, *Efficient Energy-Saving Control and Optimization for Multi-Unit Systems*, https://doi.org/10.1007/978-981-97-4492-3_5

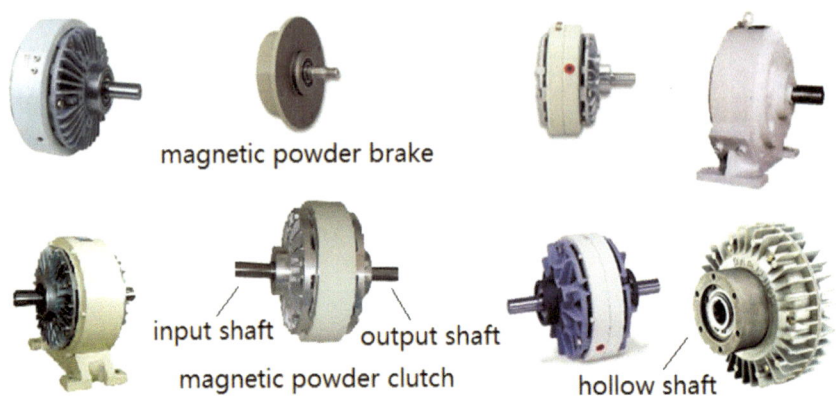

Fig. 5.1   Magnetic powder clutch and magnetic powder brake

the torque and speed of the output shaft, or realize separate control of driving shaft and driven shaft.

A magnetic powder brake generally has only one output shaft, and a magnetic powder clutch has an input shaft and an output shaft. The appearance of the magnetic powder clutch and magnetic powder brake is shown in Fig. 5.1.

5.2   Electromagnetic Clutch and Electromagnetic Brake

There are coils, passive friction plates and active friction plates inside the electromagnetic clutch and electromagnetic brake. There is an armature on the passive friction plate. The passive friction plate is an elastic disc structure. The active friction plate is connected to the rotating power shaft. When the coil is powered off, the passive friction plate separates from the active friction plate under the action of elastic force, and the clutch is in a separated state. The coil is energized to generate magnetic force, overcomes the elastic force of the shrapnel, the active friction plate and the active friction plate are attracted together, and they are in the engaged state. Some electromagnetic clutches and electromagnetic brakes are integrated. The appearance of the clutch and electromagnetic brake is shown in Fig. 5.2.

5.3   Electro-Hydraulic Proportional Valve

By changing the duty ratio of the voltage applied to the solenoid valve coil, that is, changing the average voltage on the solenoid valve coil, the flow area of the valve or the elastic force of the constant pressure spring is changed. Such a device is an electro-hydraulic proportional valve. In the hydraulic system, the manually adjusted

Fig. 5.2 Electromagnetic clutch and electromagnetic brake

throttle valve uses people to change the throttle area of the valve to adjust the flow rate. The manually controlled pressure regulating valve uses people to change the spring force of the constant pressure spring to adjust the output pressure. Changing the average voltage on the solenoid valve coil can continuously change the flow area of the valve, thereby continuously controlling the flow of hydraulic oil. Such an electro-hydraulic proportional valve is an electro-hydraulic proportional speed control valve. The principle is shown in Fig. 5.3. The power supply voltage of the electromagnetic coil can use PWM (pulse Wide modulation) method to continuously adjust, changing the duty cycle of the electromagnetic coil voltage, that is, changing the average voltage of the electromagnetic coil. Changing the thrust of the valve stem on the throttle valve core means changing the pressure from the P1 oil inlet to the P2 oil outlet, changing the area of the oil port throttle surface, thereby changing the flow rate of the hydraulic oil outlet.

Change the duty ratio of the voltage applied to the solenoid valve coil, change the elastic force of the constant pressure spring, and continuously control the output pressure of the hydraulic oil. Such an electro-hydraulic proportional valve is an electro-hydraulic proportional pressure valve. The principle is shown in Fig. 5.4. Change The duty cycle of the electromagnetic coil voltage, that is, the average voltage

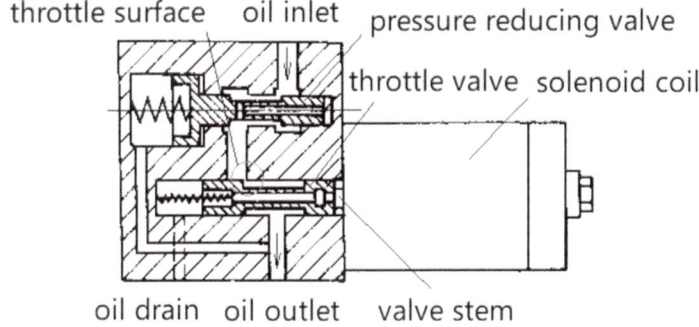

Fig. 5.3 Principle of electro-hydraulic proportional speed control valve

Fig. 5.4 Electro-hydraulic
proportional pressure valve

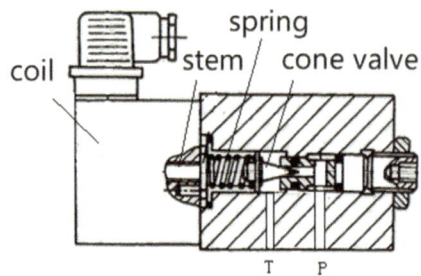

of the electromagnetic coil is changed, and the thrust of the spring on the valve stem
to the poppet valve is changed. The hydraulic oil is connected to the P port. When
the pressure of the P port is greater than the thrust of the spring to the poppet valve,
the poppet valve opens, the hydraulic oil is unloaded from the T port. The voltage
of the electromagnetic coil is changed, that is, the pressure of the hydraulic oil is
changed.

The shape of the electro-hydraulic proportional valve is shown in Fig. 5.5.

The electro-hydraulic proportional valve has lower requirements on the cleanli-
ness of the oil circuit than the electro-hydraulic servo valve. The control precision of
the electro-hydraulic proportional valve is higher, reaching the servo control level,
and it can also be used as an electro-hydraulic servo valve.

Fig. 5.5 Electro-hydraulic proportional valve

5.4 Electro-Hydraulic Servo Valve

Most of the electro-hydraulic servo valves use the amplification of the nozzle baffle to control the slide valve to perform power amplification, and use the weak electric signal sent by the automatic control system to control the precise movement of large parts of the hydraulic system. The weak electric signal controls the current of the electromagnetic coil. The change of the magnetic force of the electromagnetic coil brings about the change of the distance between the baffle and the nozzle, and at the same time changes the pressure in the space behind the nozzle. If this pressure is directly output, it is called a first-stage electro-hydraulic servo valve. The body is generally made of aluminum alloy, and its principle is shown in Fig. 5.6.

The hydraulic oil is connected to the input port P on both sides. One end enters the space connected to port A through the throttle hole, this space communicates with the nozzle on the left side. One end enters the space connected to port B through the throttle hole, and this space communicates with the nozzle on the right side. The hydraulic oil through the nozzle flows out through the O port. After the electromagnetic coil is energized, assuming that the left end of the armature is an N pole, this N pole acts on the NS pole of the magnet to make the armature move downward. Similarly, the right end of the armature is shown as S pole, which interacts with the N S pole of the magnet, so that the armature moves upward. The baffle connected to the armature rotates counterclockwise, the baffle is close to the right nozzle, the pressure of port B increases. The baffle and the left side nozzle are far away, the pressure of A port decreases, which changes the pressure and flow of B port and A port. A port and B port are respectively connected to the oil holes on both sides of the hydraulic cylinder to control the push rod of the hydraulic cylinder for precise movement.

The 2-stage electro-hydraulic servo valve uses the 1-stage electro-hydraulic servo valve as the pilot valve to control a 4-way hydraulic cylinder for power amplification, so that the power of the output port A and B is greater. The principle is shown in Fig. 5.7. There is a small ball under the baffle connected to the center of the slide valve to form a joint. The left and right movement of the baffle drives the slide valve to move left and right. The left and right movement of the baffle changes the

Fig. 5.6 First stage electro-hydraulic servo valve

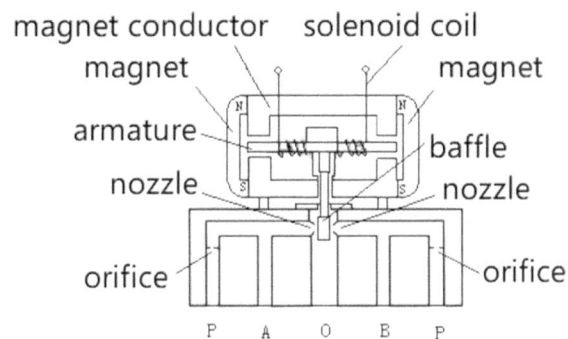

magnet conductor   solenoid coil
magnet
armature
nozzle
orifice

baffle
nozzle
orifice
magnet

P   A   O   B   P

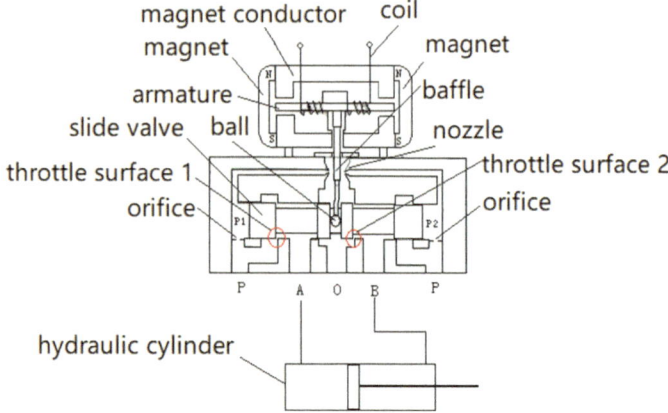

Fig. 5.7. 2-stage electro-hydraulic servo valve

pressure of P1 cavity and P2 cavity, while the pressure of P1 cavity and P2 cavity pushes the spool valve to move in reverse, and finally the spool valve balances at one position. Changing the throttling area 1 of A port (or B port) and P port, and changing the throttling area 2 of B port (or A port) and O port. Port A and port B are respectively connected to the oil holes on both sides of the hydraulic cylinder to control the cylinder rod of the hydraulic cylinder for precise movement.

The shape of the electro-hydraulic servo valve is shown in Fig. 5.8.

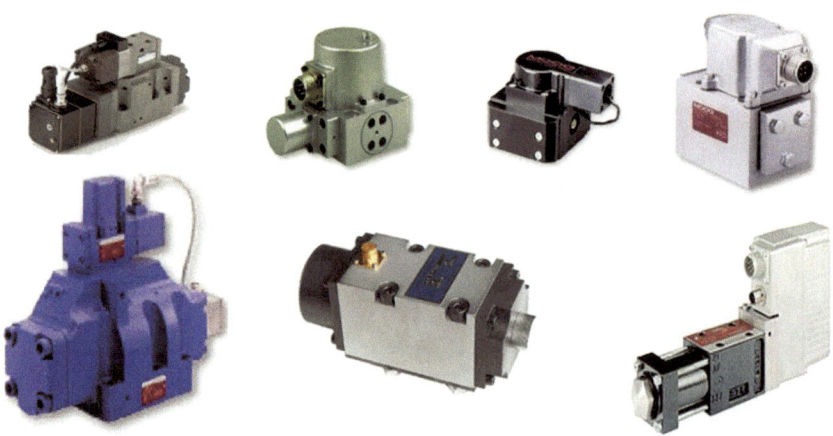

Fig. 5.8 Electro-hydraulic servo valve

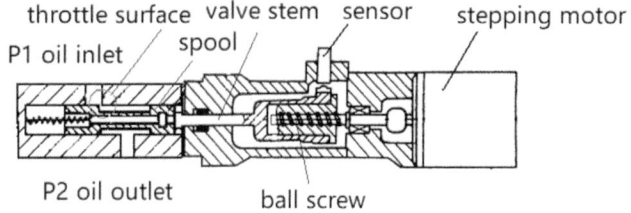

Fig. 5.9 Electro-hydraulic digital valve

5.5 Electro-Hydraulic Digital Valve

The electro-hydraulic digital valve uses the number and direction of digital pulses to control the hydraulic valve. Uses the stepping motor and ball screw to adjust the flow area of the valve or the spring force of the constant pressure spring to obtain more accurate flow or pressure. Using the motor and ball screw, changing the rotary motion of the stepping motor into a linear motion to continuously change the flow area of the valve or the elastic force of the constant pressure spring, thereby continuously controlling the flow of hydraulic oil, or continuously controlling the pressure of hydraulic oil. Its principle is shown in Fig. 5.9. The stepping motor drives the ball screw to rotate, the nut on the screw drives the valve stem to produce linear motion, and the valve stem pushes the valve core, thereby changing the pressure from P1 oil inlet to P2 oil outlet, the area of the throttle surface of the port changes the flow rate of the hydraulic oil outlet. The sensor is used to determine the zero position at the beginning, or to determine the zero position after the valve stem reaches one end.

5.6 Pneumatic and Hydraulic Directional Solenoid Valves

When it is required to control the on–off and reversing of multi-channel gases, it is necessary to use a multi-position multi-pass pneumatic reversing solenoid valve. The valve uses a slider as the valve core to switch channels. The pneumatic reversing solenoid valve is controlled by an electromagnetic coil and has two working positions. The electromagnetic coil is energized to one valve position, and the internal spring is used to reset to the other valve position. When the pneumatic reversing solenoid valve is controlled by two electromagnetic coils, there can be two working positions. The coil 1 is energized, working position 1, coil 2 is energized to change to another working position. There can also be three working positions, coil 1 is energized to working position 1, coil 2 is energized to change to another working position, both coil 1 and coil 2 are powered off in the middle bit.

Commonly used reversing solenoid valves: 2-position 2-way (2/2), 2-position 3-way (3/2), 3-position 3-way (3/3), 2-position 4-way (4/2), 3-position 4-way (4/3), 2-position 5-way (5/2), 3-position 5-way (5/3).

Fig. 5.10. 3-position 4-way intermediate closed reversing solenoid valve

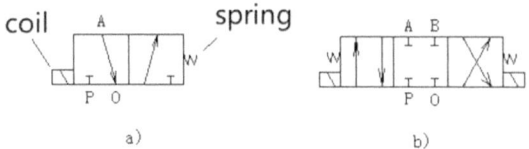

a)   b)

Generally, on the valve body, use P to represent the input port of compressed air, A to represent the output port, O to represent the exhaust port, B to represent another output port, O1 and O2 to represent two exhaust ports. The valve with 2 states is a 2-position reversing valve, and the reversing valve with 3 states is a 3-position reversing valve. With 2-position 3-way normally close (P and A disconnected) reversing solenoid valve and 3-position 4-way normally closed in the middle (P, A, B, O are all blocked) reversing solenoid valve is given as an example, as shown in Fig. 5.10. In the figure, the positions marked with letters are the normal working positions after power off.

Figure 5.10a is a 2-position 3-way (normally open) solenoid valve, single electromagnetic coil (mark on the left), spring return (mark on the left), in the normal (power-off) position, the compressed air input port P and output Port A is disconnected, A is connected to exhaust port O, after reversing, the spool moves to the right position, P is connected to A, and exhaust port O is disconnected from A.

Figure 5.10b is a 3-position 4-way (normally closed in the middle) solenoid valve, double solenoid coils (one on the left and one on the right), in the middle position of the normal state (power-off), compressed air input port P and output port A, output port Both B and exhaust port O are disconnected.

After the left solenoid valve is energized, the input port P is connected to the output port A, and the output port B is connected to the exhaust port O.

After the right solenoid valve is energized, the input port P is connected to the output port B, the output port A is connected to the exhaust port O.

The switching frequency of some direct-acting high-speed 2-position 2-way solenoid valves with pneumatic reset can reach thousands of hertz. This kind of valve is often used for high-speed sorting. The switching frequency of the direct-acting high-speed 2-position 3-way switching valve with spring return can also reach several hundred Hz.

When the pressure of the compressed gas is high, the force required for the spool reversing is relatively large. In order not to use a more powerful electromagnetic coil, the pilot valve can be used to control the reversing valve. The working process is: first use the electromagnetic coil to control a smaller reversing valve, this valve is called the pilot valve. The electromagnetic power required by the pilot valve is small, and then the output port of the pilot valve is used to control the larger spool of the large reversing valve, and the switching control is performed after the power is amplified. The principle is shown in Fig. 5.11.

Figure 5.11a is a direct-acting type, directly using the electromagnetic coil to push the spool of the reversing valve, the electromagnetic coil is powered off, and under the action of the spring, the A port and the O port are connected, and the P port is

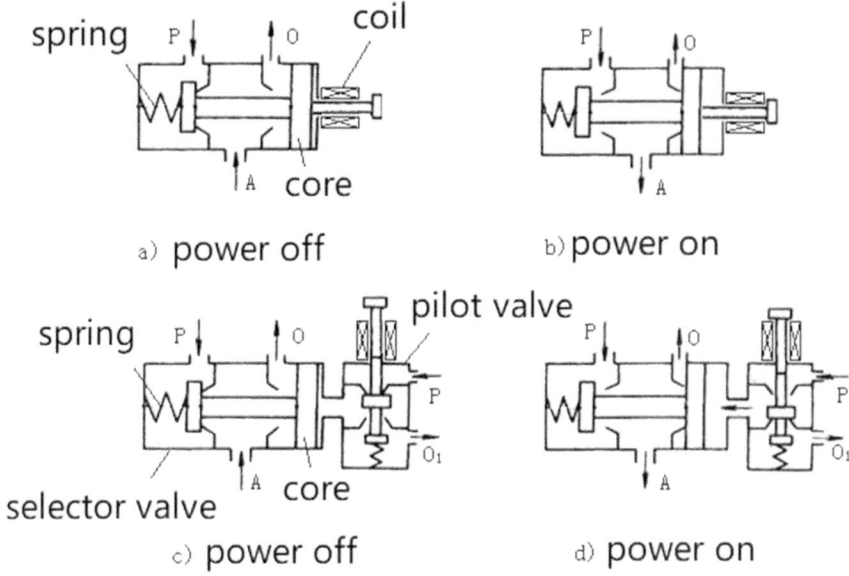

Fig. 5.11 Pilot valve controls reversing valve

disconnected, as shown in Fig. 5.11b, the electromagnetic coil is energized, the P port and the A port are connected, and the O port is disconnected.

Figure 5.11c is the pilot type. The solenoid coil is used to control the spool of the pilot valve first. When the pilot valve is powered off, the spool is connected to O1. Under the action of the spring, the A port and the O port are connected, and the P port is disconnected. In Fig. 5.11d, the pilot valve is energized, the P port of the pilot valve communicates with the spool, the spool moves to the left, the P port communicates with the A port, and the O port disconnects. Since the pilot valve bore is small, the power can be less.

The shape of the common pneumatic reversing solenoid valve is shown in Fig. 5.12.

When the on–off control of multiple liquids is required, a multi-position multi-pass hydraulic reversing solenoid valve is used. The hydraulic output power is greater than that of the air pressure. The shape of the common hydraulic reversing solenoid valve is shown in Fig. 5.13.

5.7 Solenoid Valve and Pneumatic Valve

Solenoid valve is a device that uses coil power on and off to control liquid, gas and steam on and off. The solenoid valve acts after power is turned on, and resets by spring or hydraulic pressure after power is turned off. Generally, the valve is

Fig. 5.12 Pneumatic reversing solenoid valve

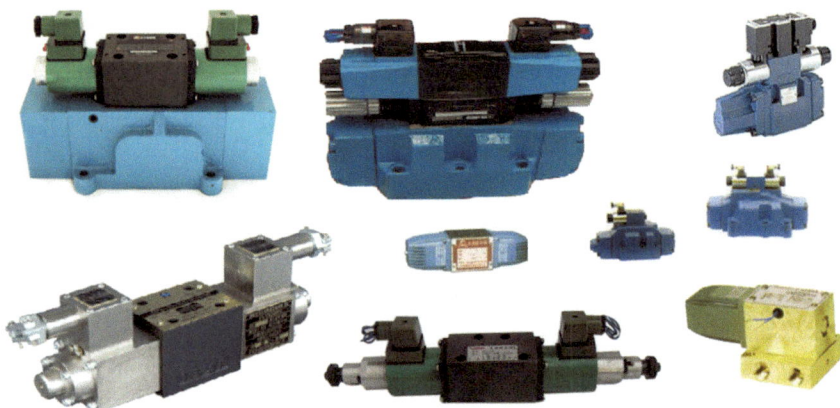

Fig. 5.13 Hydraulic reversing solenoid valve

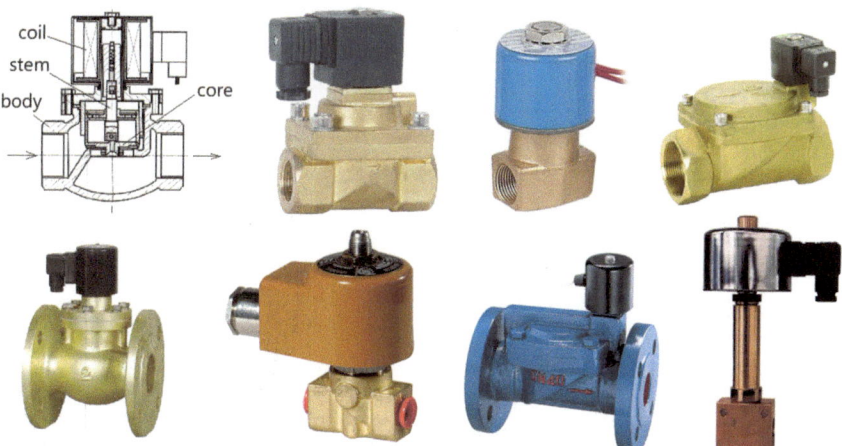

Fig. 5.14 Solenoid valve

marked with the direction of the fluid. The closing and opening of the solenoid valve are relatively fast, and are generally used in small-diameter pipelines. The solenoid valve is divided into two types: energized to close and energized to open. When the power is turned on, the medium must be shut off (such as gas), while other control processes may require that it is safe to open after a sudden power failure. There are two types of power supply voltages: AC and DC., air conditioners, water heaters, and IC water meters have many applications, and the shapes of common solenoid valves are shown in Fig. 5.14.

The pneumatic valve uses compressed air to control the diaphragm, bellows or cylinder, and controls the diaphragm, bellows or cylinder to drive the valve stem to move, push the valve core to close and open. Drive the valve stem to move, the valve stem is connected to the valve core to open or close. The bellows type is single-port air intake, the cylinder type can be single-port air intake or double-port air intake. The push rod of the cylinder is connected to the valve stem, and the valve stem is connected the valve core makes it open or close. Pneumatic valves are divided into two types: air-to-close and air-to-open. In flammable and explosive occasions, pneumatic valves are safer than solenoid valves because they do not have the problem of accidental ignition, and are generally used to control pneumatic valves with higher fluid pressure or larger pipe diameters. They have pilot valves inside, which amplify the force of switching valves. The shape of common pneumatic valves is shown in Fig. 5.15.

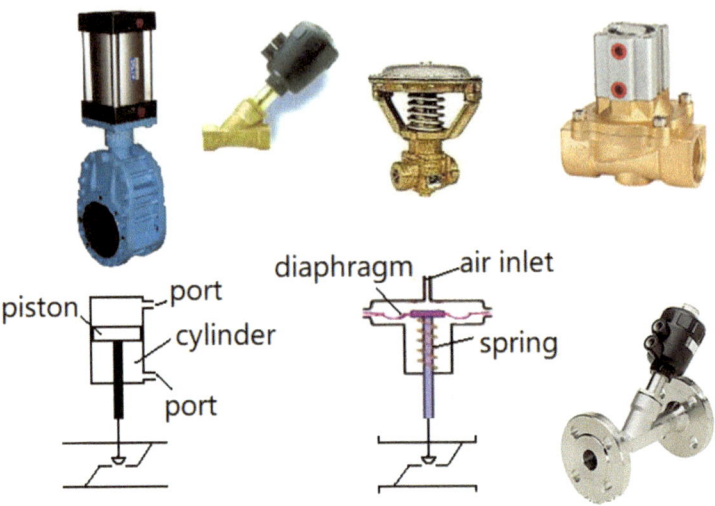

Fig. 5.15 Pneumatic valve

5.8 Electric Regulating Valve and Pneumatic Regulating Valve

When the position of the valve is only fully open and fully closed, which cannot meet the process requirements. We need to control the opening of the valve. At this time, an electric control valve or a pneumatic control valve is required.

The difference between the electric regulating valve (or electric valve, driven by an electric motor) and the solenoid valve is that the valve opening of the electric regulating valve can be controlled, rather than simply realizing on and off. The power of the electric control valve is the electric motor, and there is a process of closing and opening. It can be used in large-diameter pipelines or occasions where flow or pressure needs to be adjusted. The electric control valve with a valve positioner can use standard signals (0–5 V, 4–20 mA, etc.) to control the opening of the valve in the control system. The electric valve without valve regulator uses the forward and reverse of the valve motor and the feedback signal of the valve opening to control the opening of the valve. The appearance of the electric regulating valve is shown in Fig. 5.16. The electric damper (or louver) that regulates the low-pressure air duct (such as the air supply and induction of the boiler) has a larger volume and a different shape.

Motors, reducers or linkages form more commonly used linear strokes, angular strokes (0–90°) and multi-rotation (>360°) electric actuators, and the reducers are mostly worm wheel and gear structures. The electric actuator is mainly used for the control of various valves (gate valve, disc valve, ball valve, globe valve, etc.), and it is the actuator of the electric control valve, and its appearance is shown in Fig. 4.3.

Fig. 5.16   Electric regulating valve

In flammable and explosive occasions, the safety of the pneumatic control valve is higher than that of the electric control valve. The pneumatic control valve uses compressed air to control the diaphragm, bellows or cylinder, and the diaphragm, bellows or cylinder then drives the valve stem to push the valve core. The opening of the valve is controlled through the valve position feedback. The pneumatic control valve has two types: air opening and air closing. The choice of air opening and air closing will affect the positive and negative selection of the controller's control function. The shape of the pneumatic control valve is shown in Fig. 5.17.

5.9   Electric/Pneumatic Converter

Since the calculation and processing of the automatic control system are electrical signals, and some pneumatic systems require continuously adjustable air pressure, a device that converts electrical control signals into pneumatic control signals is required.

The electric/pneumatic converter converts the 4–20 mA control signal into a standard air pressure control signal of 0.2–1 kg/cm^2 by using nozzles, baffles, pneumatic amplifiers, signal control coils, iron cores, magnetic steel and levers. The principle of the electric/pneumatic converter is shown in Fig. 5.18. The signal coil is fed with

Fig. 5.17 Pneumatic regulating valve

Fig. 5.18 Principle of
electric/gas converter

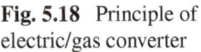

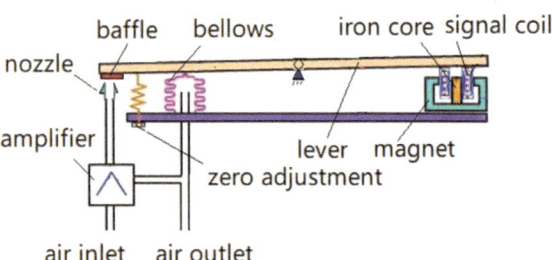

a 4–20 mA control signal. The greater the signal current, the greater the repulsion
between the coil and the magnet, the closer the distance between the baffle and the
nozzle, and the greater the outlet pressure. The smaller the signal current, the smaller
the repulsive force between the coil and the magnet, the farther the distance between
the baffle and the nozzle, the smaller the outlet pressure, so the current in the signal
coil is proportional to the outlet pressure. The outlet air pressure is fed back into the
Bellows, if the outlet pressure is too high, the bellows will push the lever to make the
baffle close to the nozzle, so that the outlet pressure will decrease. Adjust the zero
adjustment screw, so that 4 mA corresponds to the standard air pressure of 0.2 kg/
cm^2.

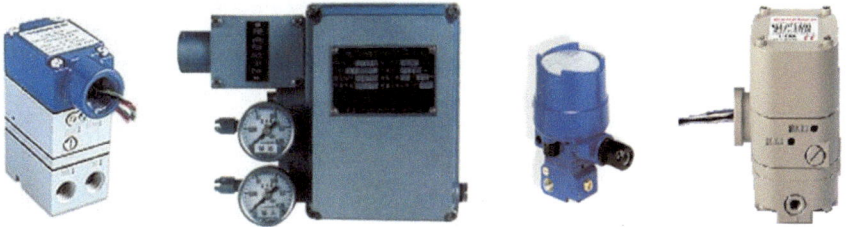

Fig. 5.19 Electrical converter

In fact, the electric/pneumatic converter can also be obtained by combining a small motor, a pressure regulating valve and a pressure sensor. The appearance of the electric/pneumatic converter is shown in Fig. 5.19.

5.10 Self-operated Regulating Valve

In some automatic control occasions, precise control is not required, and a self-operated regulating valve can be used to complete simple automatic constant pressure, constant temperature or constant current control without using PLC, PID and other controllers. There are many applications in gas, liquid and steam control in metallurgy, metallurgy and other occasions.

When using a self-operated regulating valve to control the temperature, a temperature ball is inserted into the pipe and the temperature ball is filled with a certain amount of liquid. The volume of the liquid expands and contracts at different temperatures, and the expanded and contracted liquid pushes the diaphragm or piston. After the movement of the diaphragm or piston is balanced with the internal spring force, the valve stem of the regulating valve is determined to rise or fall, and finally the opening of the valve is controlled to realize automatic temperature control.

The pressure control self-operated regulating valve uses the pressure of the liquid at the valve output port (or inlet) to push the diaphragm or piston to drive the valve stem up and down, and finally realizes the control of the opening of the valve, and automatically controls the pressure.

The self-operated regulating valve that controls the flow rate uses the pressure of the liquid at the input port and the output port of the valve, and the pressure difference between the two sides of the valve. The movement of the disc or piston drives the valve stem to rise and fall, and finally realizes the control of the opening of the valve and the automatic control of the flow rate.

The shape of the self-operated regulating valve is shown in Fig. 5.20.

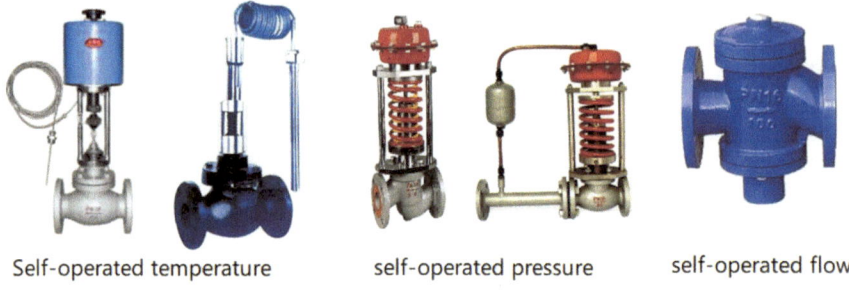

Self-operated temperature   self-operated pressure   self-operated flow

Fig. 5.20 Self-operated regulating valve

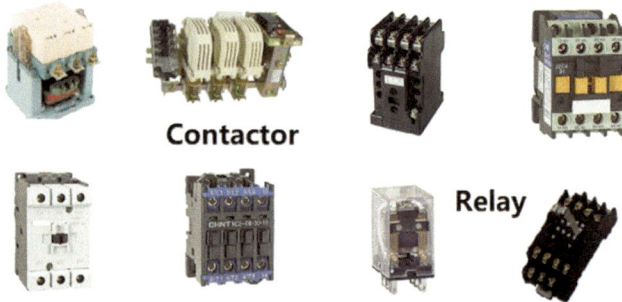

Fig. 5.21 Relay and AC contactor

5.11 Relays and Contactors

The on–off of the power supply and the device needs to be controlled by the contacts of the contactor. The start/stop, forward/reverse, function switching and other actions in the control circuit need to use the contacts of the relay for on–off control. These two devices are commonly used in energy-saving systems, as shown in Fig. 5.21.

5.12 Other Electric Devices

Motors, reducers and linkages can form complex movements, which can be applied to various automatic control occasions, such as fans driven by motors, water pumps, belt conveyors, scraper conveyors, impeller feeders, screw conveyors, etc.

General machinery includes feeders, oil pumps, quantitative pumps, peristaltic pumps, etc., and these mechanisms are widely used in automation systems as a whole or as non-standard products, and will not be listed here.

Chapter 6
Most Commonly Used Actuator–Motor

In energy-saving control, a large number of electric motors are used for load distribution and devices switching. A lot of rotary motions, linear motion or other forms of mechanical motion are mostly driven by motors. It can be said that motors are the most commonly used actuators in the field of electrical engineering and automation. After being connected to a suitable power supply, the motor generates rotary motion, and the linear motor produce linear motion.

6.1 Three-Phase AC Motor

Three-phase AC motor is connected to three-phase AC to generate rotation. It is one of the most widely used motors at present. There are many types of three-phase AC motors, including squirrel-cage three-phase AC asynchronous motors and wound rotor three-phase AC asynchronous motors, three-phase AC synchronous motor, three-phase AC permanent magnet synchronous motor, variable frequency motor, etc. The appearance of the three-phase AC motor is shown in Fig. 6.1.

6.1.1 Basic Principle of Three-Phase AC Asynchronous Motor

Knowing the basic knowledge of three-phase AC motors can better understand the control methods of three-phase AC motors.

At the beginning of the nineteenth century, the British scientist Faraday used a small magnetic rod to move through a closed-circuit coil, and found that there was a current in the coil. Since then, humans have discovered the phenomenon of electromagnetic induction. This phenomenon shows that mechanical energy and

© The Author(s) 2024
F. Yao and Y. Yao, *Efficient Energy-Saving Control and Optimization for Multi-Unit Systems*, https://doi.org/10.1007/978-981-97-4492-3_6

Fig. 6.1 Three-phase AC motor

electrical energy can be transformed into each other. It also shows that Inspired by the principle of mutual conversion between electricity and magnetism, generators and motors finally stepped onto the stage of mankind, revealing the arrival of the electrical age of mankind. Based on this principle, the American inventor Tesla invented the AC motor at the end of the nineteenth century.

There are two such experiments in middle school physics, one is as shown in Fig. 6.2. The U-shaped magnet is rotated clockwise by hand, and the two magnetic poles, N pole and S pole, form a magnetic field that also rotates at the same time. At this time, the aluminum in the middle of the magnet, the frame also turns in the same direction, and the faster the hand turns, the faster the aluminum frame turns.

The second experiment is shown in Fig. 6.3. UVW is three identical coils. The three coils are placed at 120° to each other. There is a rotatable aluminum frame in the middle of them. When the three coils are connected to three-phase AC, it can be seen that the aluminum frame rotates, which means that the three coils with three-phase alternating current also generate a rotating magnetic field, so the aluminum frame rotates.

So, why do three coils with three-phase alternating current generate a rotating magnetic field? Let us start with the characteristics of the three-phase alternating

Fig. 6.2 Converting magnetic field rotation into mechanical rotation

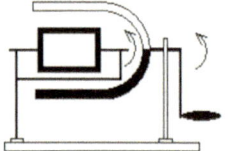

Fig. 6.3 Rotation principle
of three-phase AC motor

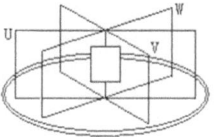

current, the expressions of the three-phase alternating current are as

$$i_U = I_m \cos\omega t \tag{6.1}$$

$$i_V = I_m \cos(\omega t - 120^0) \tag{6.2}$$

$$i_w = I_m \cos(\omega t - 240^0) \tag{6.3}$$

where I_m represents the maximum peak value of the current, ωt represents the electrical angle changing with time. i_U represents the current flowing through the U-phase coil, and i_V represents the current flowing through the V-phase coil, i_w represents the current flowing through the W-phase coil. Equations (6.1), (6.2) and (6.3) show that the UVW three-phase current has a time difference of 120° Angle, the waveform diagram of three-phase current is shown in 6–4.

Assume that the head end of the U-phase coil is U_1, the tail end of the U-phase coil is U_2. The first end of the V-phase coil is V_1, the end of the V-phase coil is V_2. The first end of the W-phase coil is W_1, and the end of the W-phase coil is W_2. When the U-phase current i_U is positive, it means that it flows in from the head end U_1 and flows out from the end U_2; when the U-phase current i_U is negative, it means it flows in from U_2 and flows out from U_1. Similarly, when the V-phase current i_V is positive, it means that it flows in from V_1 and flows out from V_2; when the V-phase current i_V is negative, it means that it flows in from V_2 and flows out from V_1. When the W-phase current i_w is positive, it means that it flows in from W_1 and flows out from W_2; when the W-phase current i_w is negative, it means it flows in from W_2 and flows out from W_1. When the current flows in, it is represented by ⊕ (similar to when we see the tail of an arrow, and the arrow is moving away from us). When the current flows out, it is represented by ⊙ (similar to the arrow we see when the arrow is coming towards us).

Taking the moment of $\omega t = 0°$ and $\omega t = 60°$ as an example, analyze the current flow direction in the three UVW coils and the change of the magnetic field caused by the current flow direction.

Draw the flow direction of the current in the three UVW coils when $\omega t = 0°$, as shown in Fig. 6.5.

According to Fig. 6.4, when $\omega t = 0°$, the U-phase current is positive (and the maximum value), the V-phase current is negative, and the W-phase current is negative, so in Fig. 6.5, the U coil current flows from U_1, flows out from U_2; the V coil current flows in from V_2 and flows out from V_1; the W coil current flows in from W_2 and flows

Fig. 6.4 Waveform diagram
of three-phase current

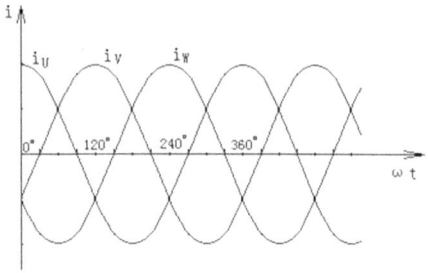

out from W_1. According to the right-hand spiral rule, the composite magnetic field
formed by the three-phase winding is a two-pole magnetic field, and the direction of
the magnetic field is from top to bottom, with the N pole at the top and the S pole at
the bottom.

When $\omega t = 60°$, draw the flow direction of the current in the three UVW coils, as
shown in Fig. 6.6.

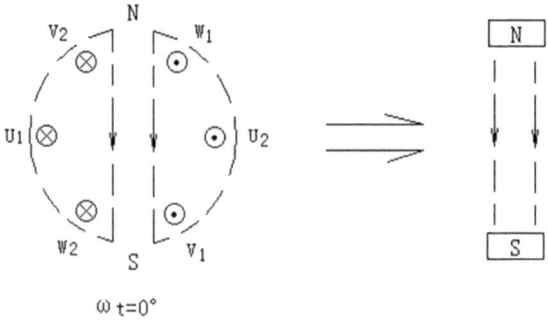

Fig. 6.5 Magnetic field position where $\omega t = 0°$

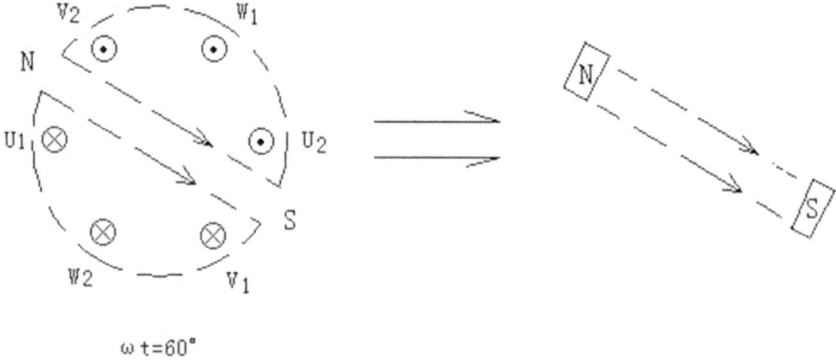

Fig. 6.6 Magnetic field position where $\omega t = 60°$

It can be seen from Figs. 6.5 and 6.6 that although the three UVW coils do not move, the magnetic field formed after they are supplied with alternating current is rotating counterclockwise, which is equivalent to artificially turning the U-shaped magnet in Fig. 6.2, the magnetic field formed by the rotation of N pole and one S pole (1 pair of magnetic poles). This is the principle of the rotating magnetic field of a three-phase AC motor with the number of pole pairs equal to 1.

The rotation speed n0 (also called synchronous speed) of the stator magnetic field of the three-phase AC motor is

$$n_0 = \frac{60 \times f}{p} \tag{6.4}$$

where f is the frequency of the power supply of the three-phase AC motor, p is the number of pole pairs of the three-phase AC motor, and the number of pole pairs generally ranges from 1 to 5.

6.1.2 Several External and Internal Wiring Methods of Three-Phase AC Motors

1. The simplest and most common external wiring method of a three-phase AC motor is to leave 2 taps for each phase winding, the head end and the tail end, U_1, U_2, V_1, V_2, W_1, W_2, total 6 taps. Connect U_2, V_2, and W_2, 3 taps together to form a neutral point N, lead out U_1, V_1, W_1 to connect to power supply U, V, W to form a "wye (Y) connection". As shown in Fig. 6.7.

Connect U_1 and W_2, 2 taps to together, U_1 is connected to the U-phase power supply. V_1 and U_2, 2 taps are connected together, V_1 is connected to the V-phase power supply. W_1 and V_2, 2 taps are connected together, and W_1 is connected to

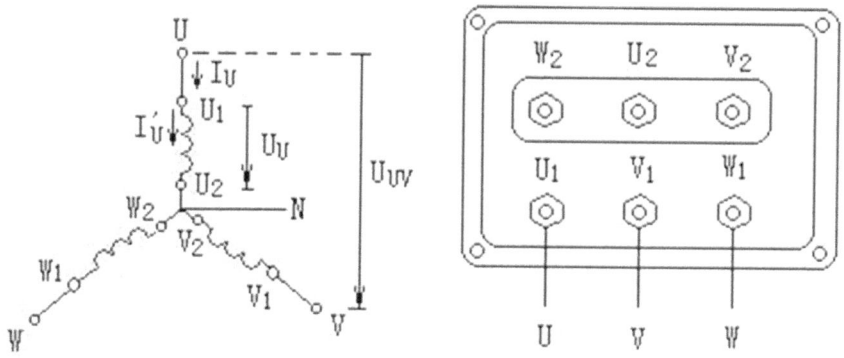

Fig. 6.7 Wye (Y) connection

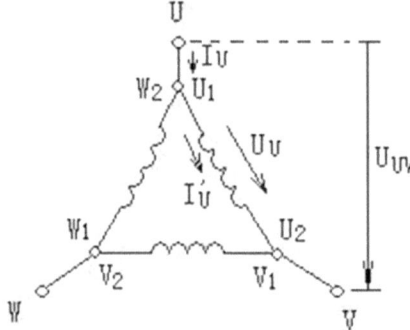 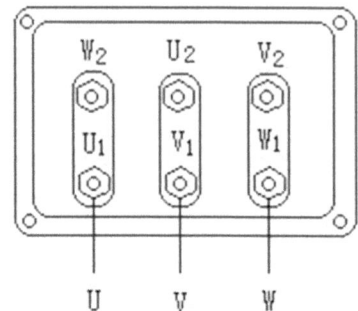

Fig. 6.8 Delta connection

the W-phase power supply, thus forming a "delta connection". As shown in Fig. 6.8, "wye (Y) connection" is represented by the symbol Y, and "delta connection" is represented by the symbol △.

Y-connection and △-connection correspond to different rated working voltages of motors, the rated working voltage of Y-connection is high, and the rated working voltage of △-connection is low. When working at the same voltage, the output power of the Y connection is low. Using the AC contactor to switch the two connection methods can also realize the Y-△ step-down start of the high-power motor.

The line voltage of the three-phase AC motor: it refers to the voltage between the two terminals of the three-phase power supply. Generally, the three-phase power supply voltage we refer to refers to the line voltage.

The phase voltage of the three-phase AC motor: it refers to the voltage on each phase winding. If the Y-shaped connection is adopted, the phase voltage U_U of the U-phase winding is equal to the voltage between the U terminal and the neutral point N, which is the line voltage $1/\sqrt{3}$; If the △-shaped connection is adopted, the phase voltage U_U of the U-phase winding is equal to the line voltage.

The line current of the three-phase AC motor: it refers to the current flowing through each power line, such as the U line current I_U.

The phase current of the three-phase AC motor: it refers to the current flowing through each phase winding, such as the phase current I'_U of the U-phase winding, $I_U = I'_U$ in the Y-shaped connection method, and $I_U = I'_U \times \sqrt{3}$ in the △-shaped connection method.

2. Multi-speed three-phase AC motors that can change the number of pole pairs have different numbers of shifting gears, and the number of terminals and wiring methods are also different.

Take the 2-speed variable speed motor as an example to illustrate the corresponding △/2Y speed change method. Each phase winding of this motor is composed of two sets of coils in series, and a terminal 2U, 2 V and 2W is drawn out at the middle connection point of the two sets of coils. In order to keep the speed direction of the motor unchanged after the pole change, the 2U and 2W leads are reversed. The

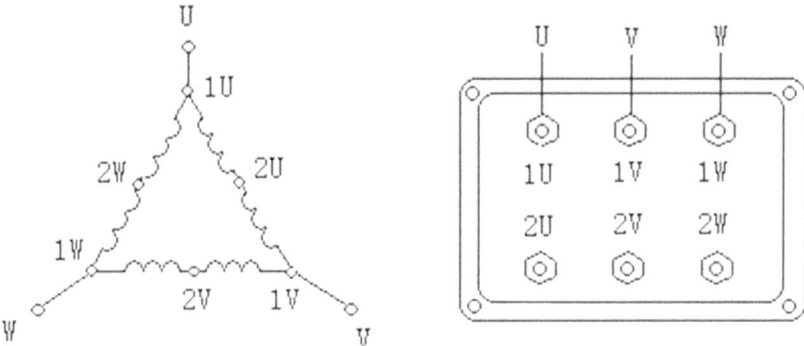

Fig. 6.9 Delta connection with large number of pole pairs

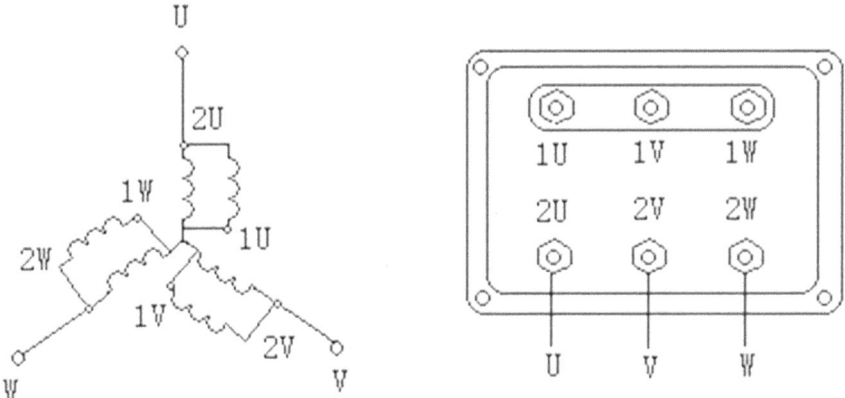

Fig. 6.10 2Y connection with small number of pole pairs

ends of the three-phase windings are connected in pairs and 3 terminals 1U, 1 V and 1W are drawn out, and there are 6 terminals in total. When the motor is connected in △ shape, 1U, 1 V, and 1W are connected to the power supply U, V, W. At this time, the number of pole pairs of the motor is large and the speed is low, as shown in Fig. 6.9. When the motor is converted to 2Y connection, the poles of the motor as the logarithm becomes smaller, the motor speed increases, as shown in Fig. 6.10.

6.1.3 Calculation of Rated Torque of Three-Phase AC Motor

The rated torque T_e of the three-phase AC motor must be greater than or equal to the torque required by the load to ensure the normal operation of the equipment. The starting torque T_q of the three-phase AC motor must be greater than the starting torque required by the load to ensure the equipment. It can start normally. When the power supply frequency f is constant, the rated torque T_e, starting torque T_q and maximum torque T_M of the motor are proportional to the power supply voltage U, that is:

$$T_e \propto U^2 \tag{6.5}$$

$$T_q \propto U^2 \tag{6.6}$$

$$T_M \propto U^2 \tag{6.7}$$

For this reason, in the application of frequency converter, if the driven motor cannot start or run by dragging the load, the way of increasing the output voltage is often used to increase the starting torque or increase the running torque.

The method of calculating the rated torque T_e according to the rated power P_e and the rated speed n_e is as follows:

$$T_e = 9550 \times \frac{P_e}{n_e} \tag{6.8}$$

where P is kilowatts (Kw), n_e is rotation speed per minute (rpm), and T_e is N.m.

6.1.4 Three-Phase Permanent Magnet Synchronous AC Motor

If the aluminum frame is replaced with a permanent magnet with high magnetic field strength, as shown in Fig. 6.11, it becomes a three-phase permanent magnet synchronous AC motor. Because the rotor itself has electromagnetic force with the stator magnetic field, which is no longer like an asynchronous motor that requires the relative motion of the rotor and the stator's rotating magnetic field to generate electromagnetic force. So as long as the torque of the AC motor dragging the external load does not exceed the rated value, the rotational speed of the permanent magnet rotor and the speed of the stator's rotating magnetic field will maintain a more precise synchronization. In many occasions, using this kind of motor, the speed of the motor can be precisely controlled only by changing the power supply frequency, and speed measurement and feedback can often be omitted.

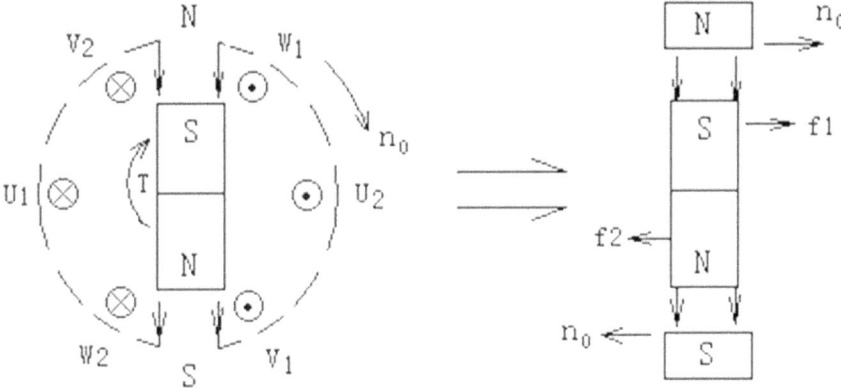

Fig. 6.11 Three-phase permanent magnet synchronous AC motor

6.1.5 Three-Phase AC Synchronous Motor

If the above permanent magnet is replaced with an electromagnet formed by introducing direct current with a slip ring and a brush, as shown in Fig. 6.12. This becomes a three-phase AC synchronous motor, which is the same as a permanent magnet synchronous AC motor. Because the rotor itself is an electromagnet with fixed magnetic poles do not require relative motion between the rotor and the stator's rotating magnetic field to generate electromagnetic force. Therefore, as long as the external load torque of this synchronous AC motor does not exceed the rated value, the rotational speed of the rotor is the same as the speed of the stator's rotating magnetic field. Just stay in sync.

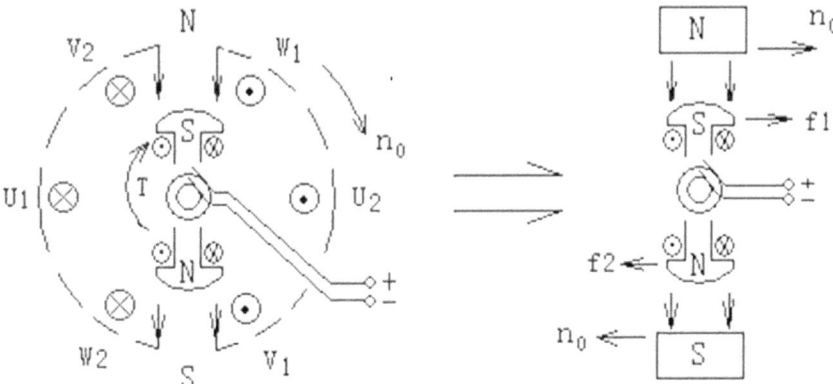

Fig. 6.12 Three-phase AC synchronous motor

Fig. 6.13 Wound rotor
three-phase AC
asynchronous motor

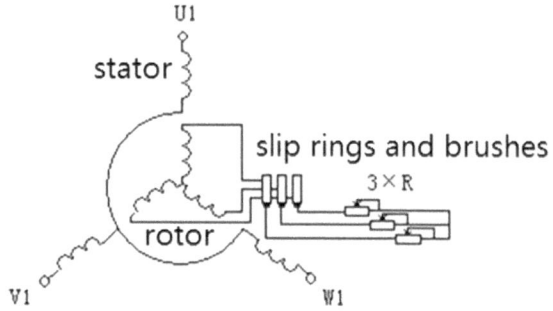

6.1.6 Three-Phase AC Asynchronous Motor with Wound Rotor

If the aluminum frame is replaced with a winding that can be used to lead out the end of the slip ring, as shown in Fig. 6.13, this becomes a three-phase AC asynchronous motor with a wound rotor. Since the starting current of a high-power three-phase AC motor is very large, it is affected due to the limitation of the development of power electronic devices, it was difficult to start and adjust the speed of the early high-power cage type three-phase AC motor. But the three-phase wound AC asynchronous motor can be adjusted by adjusting the size of the rotor series resistance R or the series connection electromotive potential can better solve the problem of starting and speed regulation.

6.1.7 Three-Phase Frequency Conversion Speed Regulation Motor

With the society's increasingly urgent requirements for energy saving and environmental protection, frequency converters for speed regulation of three-phase AC motors are widely used. If the motor is required to run at low speed for a long time, the current in the motor has to be reduced, that is, the torque output by the motor is reduced, which is also called reducing the load capacity of the motor. For this problem, the fan driven by the rotor of the motor can be replaced by a fan powered by an independent external power supply, so that the cooling fan can always run at high speed; the cooling area and cooling capacity of the three-phase AC motor shell can be increased. Another problem is that the starting torque of ordinary three-phase AC motors (when the slip rate is large) is small. In order to increase the starting torque, the rotor structure of the motor is redesigned to increase the starting torque. After these improvements, the ordinary three-phase AC The motor becomes a three-phase variable-frequency speed-regulating motor. The operating frequency of the specially designed variable-frequency motor is higher than that of the ordinary three-phase AC

Fig. 6.14 Three-phase variable frequency speed regulating motor

motor. The fan part of the motor is large and has an independent terminal box, and some variable frequency motors are equipped with a rotary encoder or resolver to measure the rotor speed and position, so some variable frequency motors have three terminal boxes. Three-phase variable frequency speed regulating motors are shown in Fig. 6.14.

6.2 Single-Phase AC Motors

Single-phase AC motors are the most widely used in the civilian field. Almost all motors in household appliances are single-phase AC motors. In the industrial field, they are only used in some occasions that require small output power of the motor, such as cooling fans of control cabinets and control equipment.

A single-phase AC motor adopts the principle of phase separation. There are main winding AX and auxiliary winding AY on the stator. The main winding and auxiliary winding are 90° apart in space. The main and auxiliary windings are connected in parallel, and a suitable starting capacitor C is connected in series on the auxiliary winding. Make the current i_b phase of the auxiliary winding lead the current i_a phase of the main winding, as shown in Fig. 6.15, so that a rotating magnetic field is generated on the stator, and the rotor of the motor can rotate. This kind of motor is called a capacitor single-phase motor.

At t = 0, i_a = 0, according to the left-hand rule, the magnetic field is up N down S. At t = 0.25, i_b = 0, according to the left-hand rule, the magnetic field is right N left S. The magnetic field rotates clockwise 90°. The analysis of other points is the same.

Another single-phase AC asynchronous motor adopts the principle of shaded poles, which divides the stator poles into two parts. The large part is the main magnetic field, and the small part (1/5–1/3) of the magnetic poles is surrounded by short-circuit

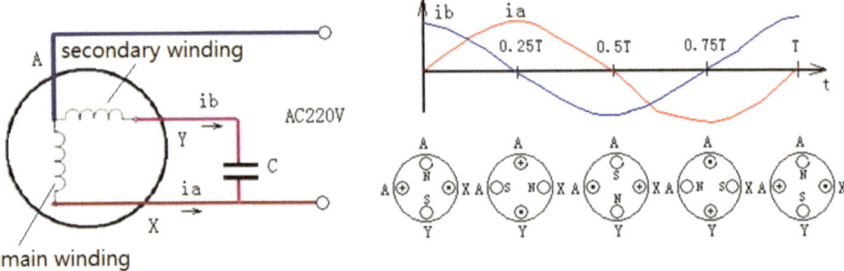

Fig. 6.15 Split-phase single-phase AC motor

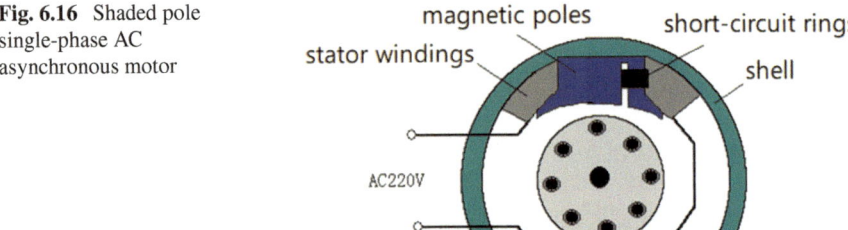

Fig. 6.16 Shaded pole
single-phase AC
asynchronous motor

copper rings, as shown in the Fig. 6.16. When the stator coil is fed with single-phase alternating current, the magnetic field in the magnetic pole part covered by the short-circuit copper ring will produce hysteresis, which is equivalent to covering the magnetic pole of this part. The magnetic field lags behind the magnetic pole of the part without the short-circuit ring Part of the magnetic field causes the magnetic field in the stator poles to produce a rotating effect. This type of motor is called a single-phase shaded pole motor.

6.3 DC Motors

DC motors are widely used in electric vehicles, automated production lines, printing, steel rolling, papermaking, vending machines, machine tools and other fields. DC motors have magnetic poles, armatures, mechanical commutators and brushes. The magnetic poles can be composed of permanent magnets. Such a DC motor is called a permanent magnet DC motor. It can also use the field winding to form the magnetic poles. The brushes and the mechanical commutator are in sliding contact. The rotation principle of the DC motor is shown in Fig. 6.17.

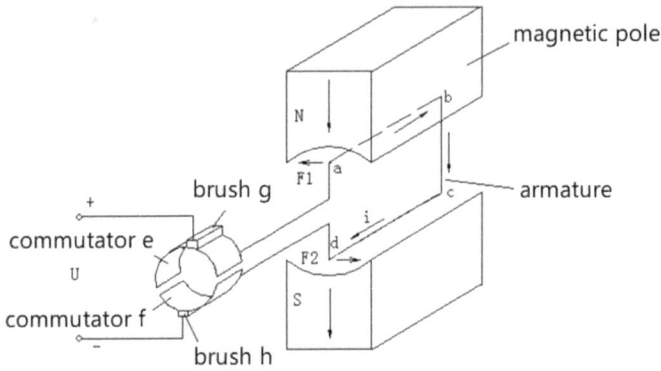

Fig. 6.17 Rotation principle of DC motor

In Fig. 6.17, brush g and brush h are connected to DC power supply U, brush g and brush h are slidingly connected with commutator segment e and commutator segment f, and conductor ab and conductor cd on the armature are energized, to generate current i, according to the left-hand rule, the conductor ab under the N pole will produce a leftward force F1, the conductor cd under the S pole will produce a rightward force F2, and the armature will produce counterclockwise rotation. When the conductor ab and When the corresponding commutator piece e turns to contact with the brush h, the conductor cd and the corresponding commutator piece f turn to contact with the brush g, the current direction of the conductor ab and the conductor cd is reversed, and the conductor in contact with the brush g The current direction of cd is inward, and the current direction of the conductor ab in contact with the brush h is outward. According to the left-hand rule, the conductor cd under the N pole produces a force to the left, and the conductor cd under the S pole produces a force to the right. The armature still rotates counterclockwise, and the motor rotates continuously.

The function of the mechanical commutator in the DC motor is to keep the direction of the current in the armature coil under the magnetic pole unchanged, that is, to keep the direction of the force on the armature coil unchanged, thereby producing continuous rotational motion.

According to the relationship between the field winding and the armature winding, there are other excitation, parallel (shunt) excitation, series excitation and compound excitation DC motors, as shown in Fig. 6.18.

The field winding of separate excited DC motor is powered by an independent power supply, the field winding of a shunt excited DC motor is connected in parallel with the armature winding and uses the same DC power supply. The field winding of a series excited DC motor is connected in series with the armature winding. One excitation winding of compound (combined) excitation DC motor is connected in series with armature winding, one excitation winding is connected in parallel with armature winding, powered by a DC power supply.

The shape of the DC motor is shown in Fig. 6.19.

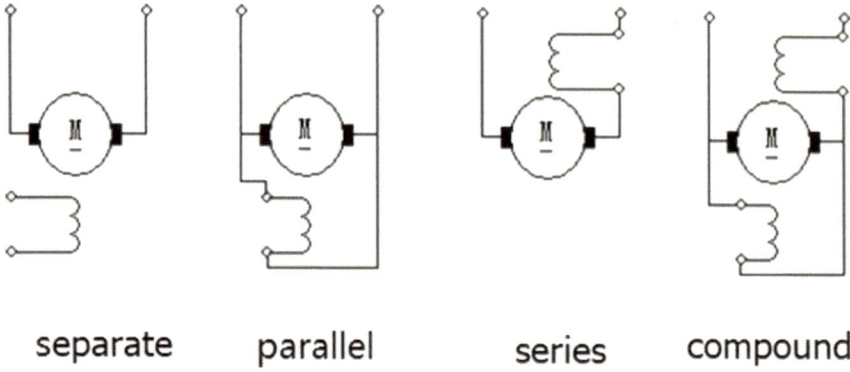

separate   parallel   series   compound

Fig. 6.18 Separate excitation, parallel excitation, series excitation and compound excitation

Fig. 6.19 DC motor

6.4 Brushless DC Motor

DC brushless motors are widely used in the fields of electric bicycles, instruments, household appliances, computer peripherals, small rotating machinery, cooling fans, small water pumps, etc., its principle is basically the same as that of the DC motor mentioned above, except that there are no brushes and mechanical commutators, and position sensors and power electronic devices are used for commutation, so they are also called commutator-free motors and commutator-free DC motors. Since there are no vulnerable parts such as sliding brushes, it is easy to maintain and there is no spark.

The brushless DC motor uses a position sensor to detect the position of the rotor, and uses a power electronic switching circuit to change the current direction of the

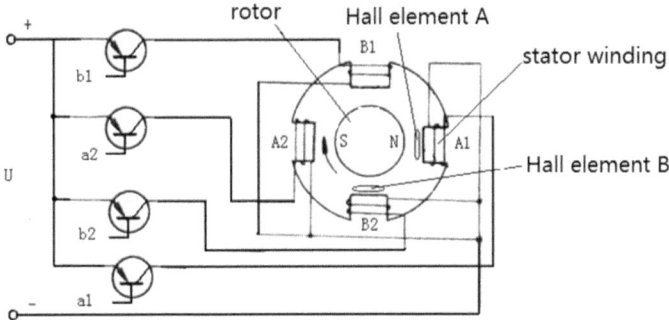

Fig. 6.20 DC brushless motor

stator winding, it can also achieve the same current direction of the winding under the fixed magnetic pole, thereby forming a rotational motion, as shown in Fig. 6.20.

There is one N pole and one S pole on the rotor. When the Hall element A detects that the N pole is under the stator winding A1, it controls b2 to energize the stator winding B2. The N pole of the motor generates attraction, and the rotor rotates 90° clockwise. When the Hall element B detects that the N pole is under the stator winding B2, control a2 to energize the stator winding A2. After the stator winding A2 is energized, it will appear as an S pole on the side close to the rotor, attracts the N pole on the rotor, and the rotor rotates 90° clockwise. If this continues, the brushless DC motor will rotate clockwise. Changing the order of electrification of the stator winding can change the rotation direction of the brushless DC motor. Use the sensor to detect the position of the rotor rotation before changing the energization of the stator coil, so it will not appear out of step phenomenon, and it is a synchronous working mode.

In order to improve the utilization rate of the stator windings, the three stator windings are connected in a three-phase symmetrical star connection, it is similar to the connection method of the AC motor. In such a DC brushless motor, the forward and reverse directions of the stator windings can be used It can form NS changing magnetic field, use Hall element to detect the rotor position, and also obtain the rotational motion of the permanent magnet rotor. As shown in Fig. 6.21. Since this working method is very similar to the working method of the frequency converter, it is sometimes called a DC frequency conversion, but its frequency change is not active, it is controlled by a position sensor.

The common brushless DC motor is a permanent magnet brushless DC motor, which can adopt a structure with a permanent magnet rotor inside or a permanent rotor outside. The structure of a brushless DC motor with an internal permanent magnet rotor is shown in Fig. 6.22, the rotor is a permanent magnet rotor, and the stator winding can change the direction of the current, that is, the direction of the magnetic field. The sensor detects the position of the permanent magnet rotor, and the magnetic field direction of the stator winding is changed by the power electronic switching circuit, forming a continuous rotating traction on the rotor.

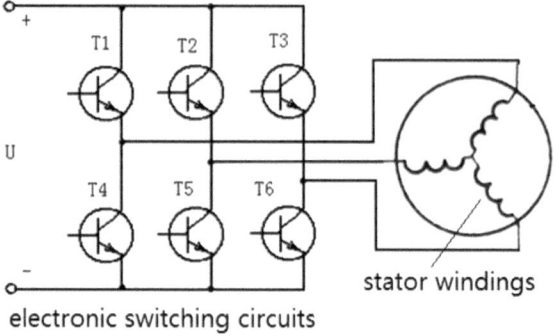

electronic switching circuits

Fig. 6.21 DC frequency conversion

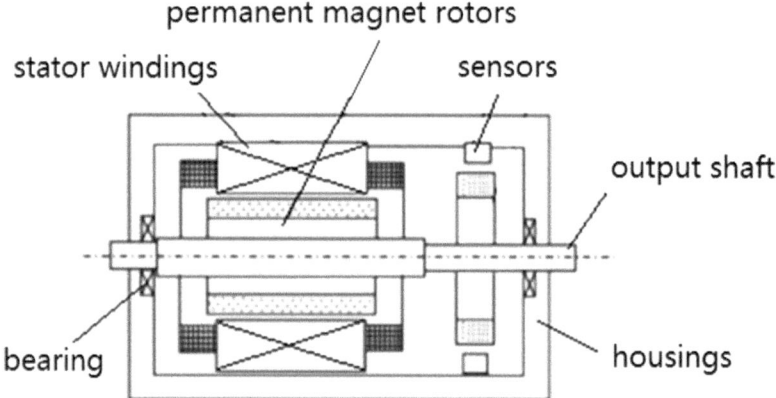

Fig. 6.22 Permanent magnet brushless DC motor

The permanent magnet materials of permanent magnet brushless DC motors are mostly rare earth permanent magnet materials. The structure of the permanent magnet rotor includes surface magnetic poles, embedded magnetic poles and ring magnetic poles, etc., as shown in Fig. 6.23.

The position sensor can be a Hall element that detects the magnetic field, or a proximity switch that measures the metal boss on the rotor, or a photoelectric switch that detects a gap, a rotary encoder, or a resolver. The position sensor detects the position of the rotor, and then Control the current direction in the stator winding so that the magnetic field of the stator always forms a rotating force on the rotor.

The brushless DC motor with an external permanent magnet rotor fixes the stator winding inside, and the external permanent magnet rotor rotates. The sensor detects the position of the outer rotor and accordingly changes the direction of the current on the internal stator winding, that is, changes the direction of the magnetic field. The magnetic rotor forms a continuous rotating traction torque, as shown in Fig. 6.24.

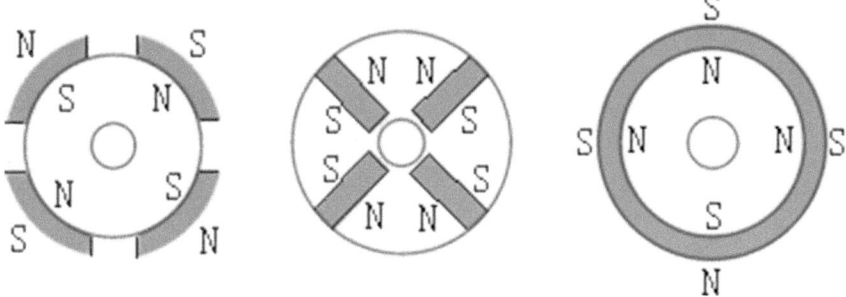

Fig. 6.23 Rotor forms of surface poles, embedded poles and ring poles

Fig. 6.24 Brushless DC
motor with external
permanent magnet rotor

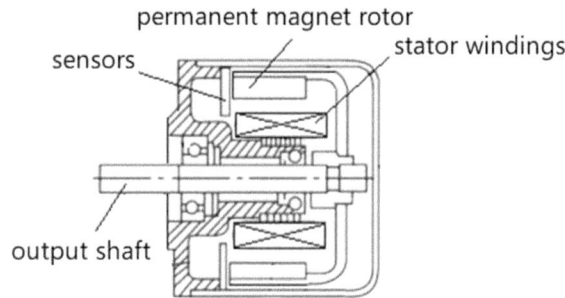

Brushless DC motors can also be made into thin disk structures. The circular stator windings and the circular permanent magnet rotor are placed opposite each other. Sensors detect the position of the permanent magnet rotor and change the direction of the magnetic field on the stator windings accordingly. The same principle can be used for permanent magnet rotors. The magnetic rotor forms a continuous rotating traction torque, as shown in Fig. 6.25.

There are many types of brushless DC motors, and there are also many shapes. Figure 6.26 shows the appearances of several brushless DC motors.

Fig. 6.25 Disk type
brushless DC motor

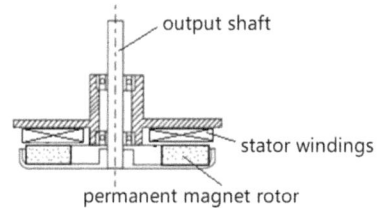

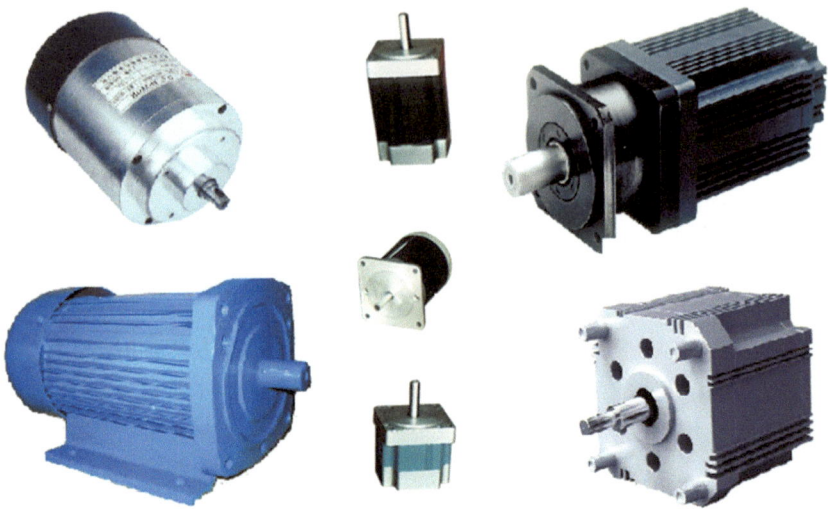

Fig. 6.26 Brushless DC motor

6.5   Stepping Motors

The stepping motor rotates step by step. Every time a pulse signal is input, the stepping motor will advance one step, so it is sometimes called a pulse motor. The stepping motor has no cumulative error, and several windings of the stepping motor are energized in a certain sequence. Direct current can form a rotating magnetic field on the stator of the stepping motor, and the rotor rotates under the action of electromagnetic force. The higher the frequency of sequential power supply, the faster the rotation speed of the magnetic field of the stepping motor. The function of the stepper motor driver is to convert the input electrical pulse signal into a corresponding stepper motor winding energization sequence, thereby changing the rotation angle of the stepper motor.

Stepper motors are divided into permanent magnet type, reactive type and hybrid type according to the excitation method. Reactive and hybrid stepping motors are divided into 3 phases, 4 phases, 5 phases, 6 phases, and 8 phases according to the number of phases of the stator windings.

Taking the permanent magnet stepping motor as an example, its stator is composed of 2-phase or multi-phase windings. After each phase winding is fed with direct current, K magnetic poles are formed in the circumferential direction. The rotor is a star-shaped permanent magnet composed of multiple magnetic poles. Along the circumferential direction, NS phases are arranged alternately, and the number of poles of the rotor is also equal to K. Take a 2-phase permanent magnet stepping motor as an example, as shown in Fig. 6.27.

In Fig. 6.27, the A-phase stator winding has four magnetic poles 1, 3, 5, and 7, and the winding directions (or wiring directions) of the four magnetic pole coils are

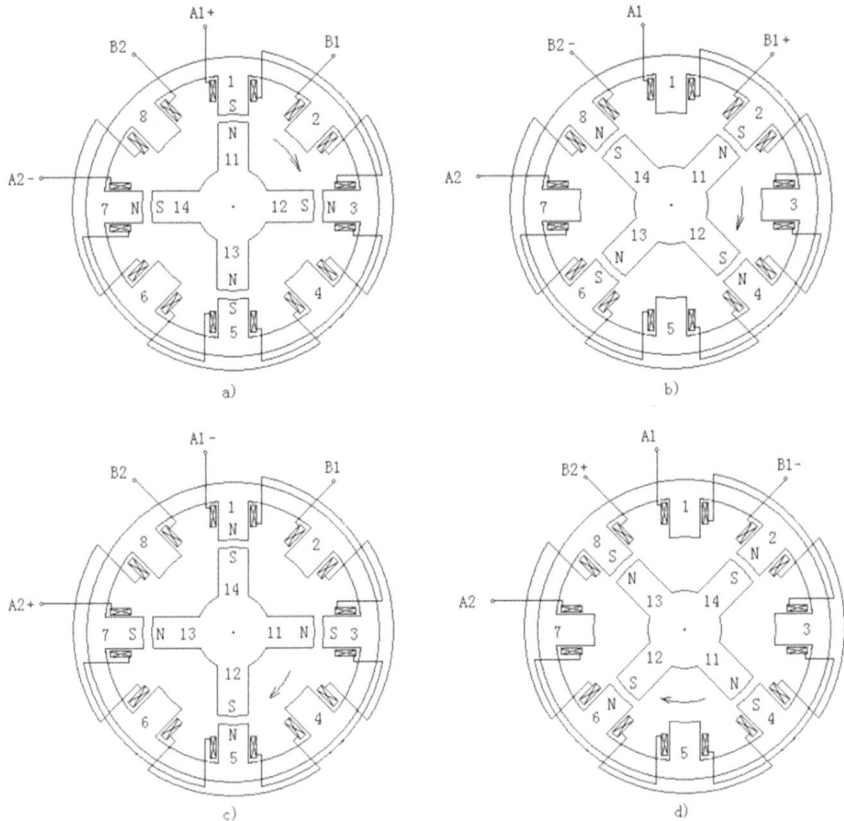

Fig. 6.27 2-phase permanent magnet stepper motor

different to form four magnetic poles arranged alternately in NS; the B-phase stator winding has 4 magnetic poles 2, 4, 6, 8, different winding directions (or wiring directions) of the magnetic pole coils can also form 4 magnetic poles arranged alternately in NS; the rotor has 4 fixed magnetic poles 11, 12, 13, 14, and The NS poles are arranged alternately.

The movement of the rotor is divided into 4 steps, and then it runs repeatedly:

Step 1: A1 of phase A is connected to the + of the DC power supply, and A2 is connected to the—of the DC power supply. The polarity of the four magnetic poles of the A phase and the polarity and position of the rotor are shown in Fig. 6.27a. Due to the attraction of the magnetic poles and repulsion, the rotor remains in this position.

Step 2: Phase A is powered off, B1 of phase B is connected to + of DC power supply, B2 is connected to—of DC power supply. The polarity of the four magnetic poles of phase B and the polarity and position of the rotor are shown in Fig. 6.27b. Due to the attraction and repulsion of the magnetic poles, the rotor rotates 45° clockwise and remains in this position.

Step 3: Phase B is powered off, A1 of phase A is connected to—of the DC power supply, A2 is connected to + of the DC power supply. The polarity of the 4 magnetic poles of phase A and the polarity and position of the rotor are shown in Fig. 6.27c. Due to the attraction and repulsion of the magnetic poles, the rotor rotates 45° clockwise and remains in this position.

Step 4: The negative power supply of phase A is powered off, B1 of phase B is connected to—of DC power supply, B2 is connected to + of DC power supply. The polarity of the four magnetic poles of phase B and the polarity and position of the rotor are shown in Fig. 6.27d. Due to the attraction and repulsion of the magnetic poles, the rotor rotates 45° clockwise and remains in this position.

Step 5: Repeat the first step, the B-phase negative power supply is powered off, the A1 of the A-phase is connected to the + of the DC power supply, and A2 is connected to the—of the DC power supply, and the rotor rotates 45° clockwise and maintain this position.

In this way, the stepping motor starts to rotate. Since the original energization sequence is repeated after 4 energization sequences, it is called 4-beat operation mode; since only one phase winding is energized each time, it is also called single-phase operation mode; The motor has a step angle (angle of rotation per step) of 45°.

The power-on sequence of phase A and phase B power supply is: A, B, (–A), (–B), A…, and the voltage waveform of each phase is similar to that of alternating current, as shown in Fig. 6.28.

In fact, for the 2-phase stepping motor, it is also possible to use a single-phase and two-phase mixed power-on mode. 8-beat operation mode: A, AB, B, B(–A), (–A),

Fig. 6.28 Voltage waveforms of A and B phases

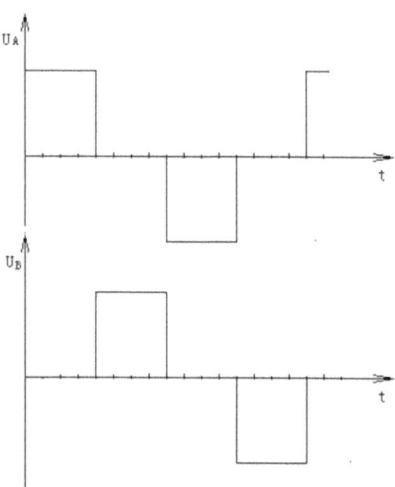

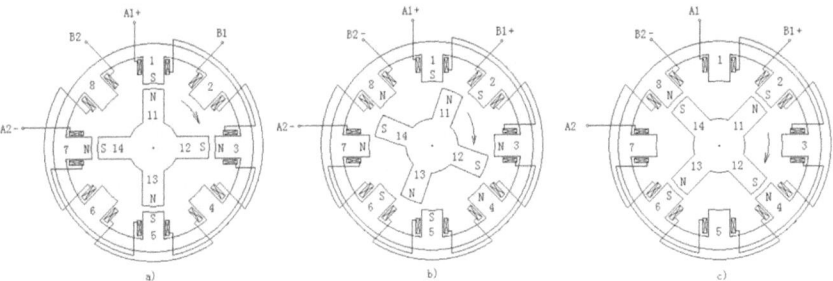

Fig. 6.29 Rotation of the stepping motor

(–A) (–B), (–B), (–B) A. The step angle of this energization method is half of the original one, which is 22.5°.

Take A, AB, B three-step power-on sequence as an example to illustrate the rotation angle and direction of the stepping motor, as shown in Fig. 6.29.

Step 1: A1 of phase A is connected to the positive of the DC power supply, and A2 is connected to the negative of the DC power supply. The polarity of the four magnetic poles of phase A and the polarity and position of the rotor are shown in Fig. 6.29a. Due to the attraction and repulsion of the magnetic poles, the rotor remains in this position.

Step 2: A1 of phase A is connected to + of DC power supply, A2 is connected to— of DC power supply. B1 of phase B is connected to + of DC power supply, B2 is connected to—of DC power supply. The polarity of the 8 magnetic poles of phase A and phase B And the polarity and position of the rotor are shown in Fig. 6.29b. Due to the attraction and repulsion of the magnetic poles, the rotor rotates 22.5° clockwise and remains in this position.

Step 3: Phase A is powered off, B1 of B phase is connected to + of DC power supply, B2 is connected to—of DC power supply. The polarity of the 4 magnetic poles of phase B and the polarity and position of the rotor are shown in Fig. 6.29c. shows that due to the attraction and repulsion of the magnetic poles, the rotor rotates 22.5° clockwise and remains in this position.

The rotor rotates 22.5° clockwise per beat, so the same stepper motor has different power-on methods, and its step angle is also different. The step angle of many stepper motors is expressed in the form of y°/0.5y° for this reason.

The principle of a single-three-beat reactive (reluctance type) stepper motor is shown in Fig. 6.30. The rotor is a silicon steel sheet. The stator winding energizes, the rotor produces magnetism. When the A5-A2 windings are energized, the rotor turns to the smallest reluctance position. The rotor 1–3 is aligned with the A5-A2 winding, when the B5-B2 winding is energized, the rotor turns to the position where the magnetic resistance is the smallest, the rotor turns 30°. The rotor 2–4 is aligned with the B5-B2 winding, C5-C2 When the winding is energized, the rotor turns to

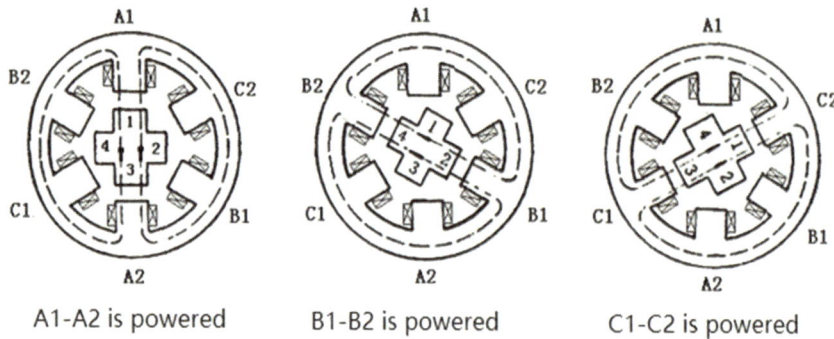

A1-A2 is powered   B1-B2 is powered   C1-C2 is powered

Fig. 6.30 Single three-beat reactive stepping motor

the position with the least reluctance, and the rotor rotates through 30°. The 1–3 of the rotor aligns with the C5-C2 winding, so reciprocating, the rotor rotates.

The shape of the stepping motor is shown in Fig. 6.31.

Stepper motors can realize open-loop positioning control without encoder feedback, no cumulative error, simple structure, and will not burn the motor if it is blocked, but the torque is small at high speeds. Stepper motors are used in CNC machine tools, valve control, automatic winding machines, it is widely used in fields such as medical equipment, bank terminals, computer peripherals, cameras and quartz clocks.

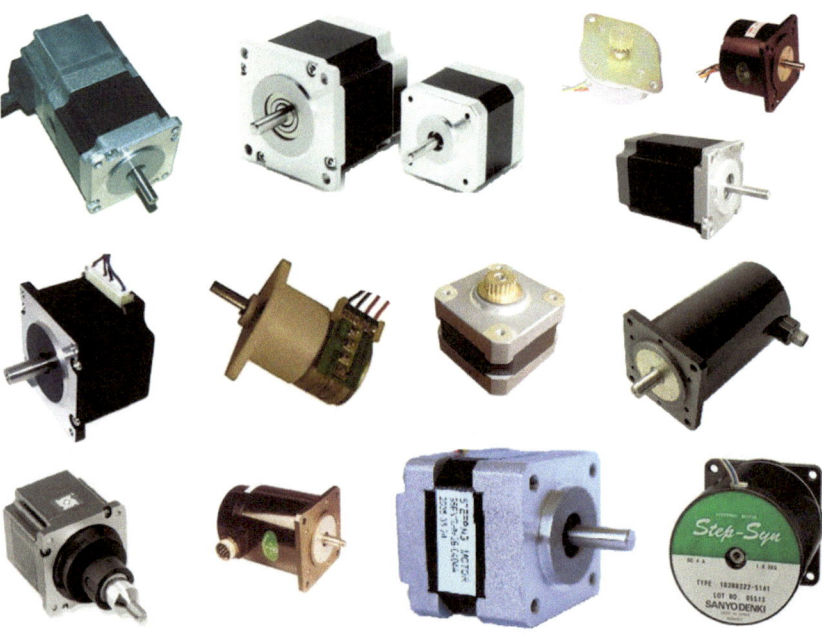

Fig. 6.31 Stepping motor

6.6 Servo Motor

Servo motors include AC servo motors and DC servo motors. The principle of servo motors is basically the same as that of DC motors and AC motors. Servo in English means "slave" in Greek. Servo motors are mainly for fast and high-precision positioning control, the servo motor can withstand high overload torque. In order to achieve these purposes, the structure of the servo motor has been specially designed. In order to obtain high starting torque, it is made of non-permanent magnet materials. The rotor of the AC servo motor has a large impedance. In order to obtain a fast motor, most of them have a slender structure with low inertia. In order to obtain an accurate position signal, the rotor generally has an encoder or a resolver for measuring the angular position. Some servo motors also need an external encoder, and the position signal of the encoder is fed back to the servo drive to achieve the angular position required by the command. The drive that controls the servo motor can receive position signals (such as pulses and rotation directions) and speed control signals (Mostly positive and negative voltage signals). In most cases, the servo drive can only achieve its high performance by driving the matching servo motor, which is very different from the frequency converter. The frequency converter is more versatile, it can drive any three-phase AC motor that does not exceed the rated current of the frequency converter. The main purpose of the frequency converter is to regulate the speed of the motor, and is not required that the rotor of the motor must have an encoder.

At present, with the rapid development of AC frequency conversion technology, some inverters already have servo functions, and the control accuracy is not significantly different from that of traditional AC servos. Therefore, there is a trend of gradual integration of inverters and AC servos.

Due to the rapid development of power electronics technology and control technology, AC servo motors have gradually replaced DC servo motors and are called the mainstream of servo motors.

AC servo motor is divided into AC permanent magnet synchronous servo motor and AC asynchronous servo motor. The rotor of AC permanent magnet synchronous servo motor is composed of permanent magnets, and the stator coil forms a rotating magnetic field. As long as the load does not exceed the synchronous torque, the permanent magnet rotor will rotates synchronously with the rotating magnetic field, which is basically similar to the AC permanent magnet synchronous motor mentioned in the previous section. For the AC asynchronous servo motor whose rotor is a hollow cup or squirrel-cage structure, its principle is similar to that of a single-phase split-phase motor. Its stator winding is composed of an excitation winding and a control winding with two phases placed at a 90° difference in space. The AC phase difference between the excitation winding and the control winding has a certain angle, so that an elliptical rotating magnetic field is generated on the stator, and the rotor cuts the magnetic field line and rotates under the traction of the electromagnetic force. Changing the power frequency of the excitation winding and the control winding can change the speed of the servo motor, changing the power supply voltage of the control

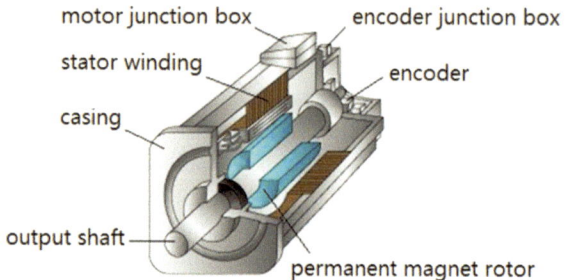

Fig. 6.32 Permanent magnet AC servo motor

Fig. 6.33 Servo motor

winding can also change the output speed of the servo motor. When the voltage of the control winding is reversed, the servo motor will rotate in the opposite direction.

The structure of the permanent magnet AC servo motor is shown in Fig. 6.32.

The price of the servo motor is relatively high, and the power is not too large. At present, it is mainly used in occasions such as large speed range, accurate positioning, fast tracking, low speed and high torque, such as precision machine tools, packaging machines, printing machinery, manipulators and other fields. In these systems, the application of AC permanent magnet synchronous servo motor is more common.

The shape of the servo motor is shown in Fig. 6.33.

6.7 Linear Motors

The principle of the linear motor is equivalent to cutting the rotor and stator of the AC or DC motor, the rotor and the stator are unfolded on a plane, and the rotor moves linearly along the direction of expansion. It can also be considered that the linear motor is an AC or DC motor with an infinite diameter. The outer surface of the rotor and the inner surface of the stator become planes, and the rotor moves in a straight line along the same direction as the stator arrangement direction.

Taking the linear stepper motor as an example, its principle is the same as that of the rotary stepper motor. Figure 6.34 is a schematic diagram of a 5-phase linear stepper motor. The mover of the linear motor consists of 5 n-shaped iron cores, the adjacent n-shaped iron core and the stator teeth are staggered by 1/5 tooth pitch. There are two oppositely connected coils on the two poles of each n-shaped iron core. The magnetic field formed by the two coils makes the n-shaped iron core one pole N and one pole S, and the magnetic flux does not enter other n-shaped iron cores. When phase A is energized, the n-shaped iron core on the A-phase coil is aligned with the teeth on the stator core, then phase A is de-energized, and phase B is energized. The mover moves 1/5 pitch to the left. Similarly, when phase B is de-energized and phase C is energized, the mover moves 1/5 pitch to the left. This is the 5-phase 5-beat operation mode. If the power-on mode is A -AB-B-BC-..., 5-phase 10-step operation mode, the mover moves 1/10 of the pitch in each step.

For linear motors with other principles, the analysis method is the same as above. The stator and rotor are also cut apart, and a linear motion magnetic field is applied on the winding side (or called primary), so that the mover (or called secondary) is forced to move along a straight line. As shown in Fig. 6.35.

The mover can be a permanent magnet or a coil, and the stator can be a permanent magnet or a coil. The linear motor with a permanent magnet as the mover is shown in Fig. 6.36. The base is provided with a linear guide rail and a stator coil. A sliding table is installed on the linear guide rail. A mover magnet is installed below the sliding table. The mover magnet is opposite to the stator coil. The stator coil is connected to the power supply to form a linear motion magnetic field, and the mover magnet will produce a linear motion.

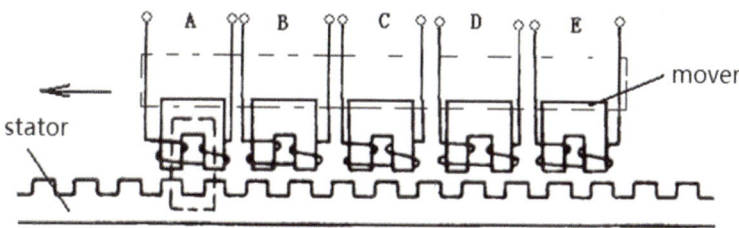

Fig. 6.34 Principle of 5-phase linear stepper motor

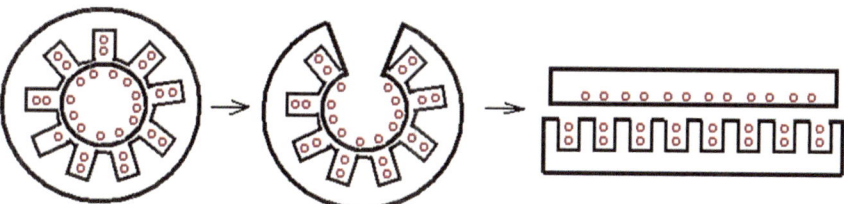

Fig. 6.35 Stator and rotor cut

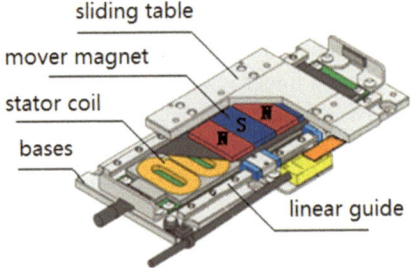

Fig. 6.36 Linear motor with permanent magnets

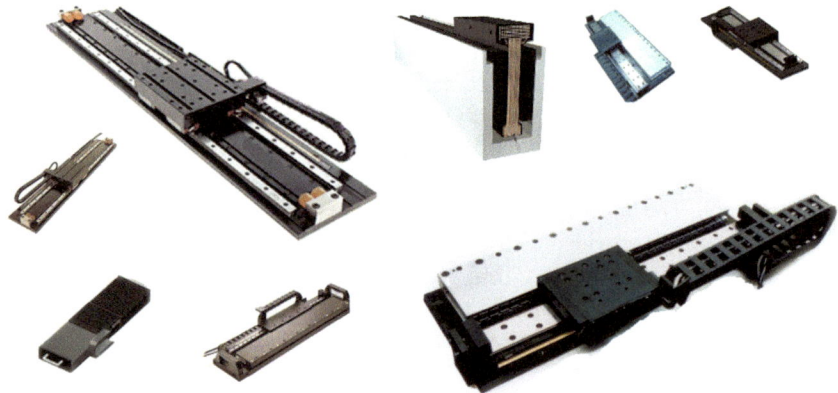

Fig. 6.37 Linear motor

When the automation equipment needs to do linear motion, the linear motor can omit the conversion mechanism such as the screw. The linear servo motor has small inertia, high speed, high acceleration, unlimited length, and high positioning accuracy. Linear motors are used in CNC machine tools, electronic device manufacturing, electronic patch equipment, manipulators, hardware processing, solar cell manufacturing, maglev trains, aircraft ejection, elevators and other fields have many applications. The shape of the linear motor is shown in Fig. 6.37.

6.8 Switched Reluctance Motors

The switched reluctance motor is made according to the principle of minimum reluctance. It is a new type of speed-regulating motor. Its structure is similar to that of a reactive stepping motor. Every time the energizing combination of the winding changes, the rotor takes a step. The operation of the switched reluctance motor also needs to use sensors to detect the position of the rotor. Using technology that can

Fig. 6.38 4-phase switched
reluctance motor

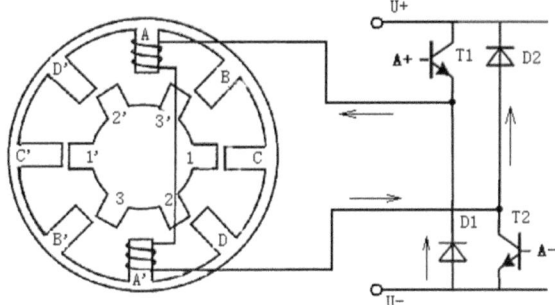

accurately predict the rotor position, it is also possible without position sensors. According to the position of the rotor, the sequence and time of energizing the stator windings are determined. Therefore, the operation mode is similar to that of a brushless DC motor, and the rotor position detector is installed in the motor.

The poles of the stator and the rotor of the switched reluctance motor are salient pole structures, and the poles are protruding, and the stator and the rotor are made of laminated silicon steel sheets. There are no windings and permanent magnets on the rotor, and there are windings on the stator. Two 180° opposite windings are connected together as "one phase". Take a 4-phase switched reluctance motor (8/6) with 8 windings and 6 poles as an example. The structure is shown in Fig. 6.38.

In Fig. 6.38, 1–1' of the rotor is aligned with the CC' winding of the stator, and then the CC' winding is de-energized, and the D-D' winding is energized. According to the "minimum reluctance principle", the magnetic flux always take the path with the least reluctance. The D-D' winding will attract 2–2' of the rotor to coincide with itself, the rotor runs counterclockwise, when the poles of the winding and the poles of the rotor coincide, the inductance on the winding is the largest. But the winding no longer has traction torque on the rotor. Detect the position of the rotor, and then energize the A-A' winding in a timely manner, and then energize the B-B', following the sequence of D-A-B-C-D, the rotor will rotate counterclockwise. If the energization sequence Change it to B-A-D-C, then the rotor rotates clockwise.

Take the A-phase winding as an example, when A + and A- are high, T1 and T2 are turned on, and the A-phase winding is energized. When A + and A- are low, T1 and T2 are turned off, and the current in the A-phase winding passes through D1 and D2 regenerates and feeds back to the power supply U, the system efficiency is higher. It also speeds up the turn-off speed of the A-phase winding. In the H-bridge circuit, the upper and lower transistors are turned on at the same time and there is a fatal problem of short circuit.

The switched reluctance motor has simple structure, low cost, large starting torque, low starting current, high-speed operation, one-phase winding failure can still reduce power operation, frequent start and stop or forward and reverse, and high efficiency. The internal structure of the switched reluctance motor is shown in Fig. 6.39.

The appearance of the switched reluctance motor is shown in Fig. 6.40.

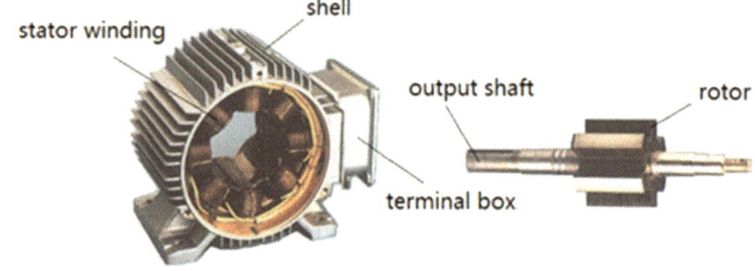

Fig. 6.39   Internal structure of switched reluctance motor

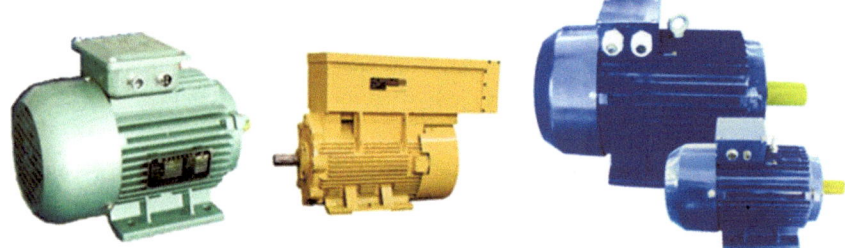

Fig. 6.40   Switched reluctance motor

6.9   Power Supply Voltage and Operating Voltage of Electrical Device

The power supply voltage in the United States is not the same as the design working voltage of electrical device. The power supply voltage is the voltage that is provided by the power supply company. Due to line loss, when the power is delivered to the device, the voltage will drop. The electrical device must be able to work within a certain voltage range. For example, for an AC120V power supply, the designed operating voltage of electrical device may be AC110-120 V.

6.9.1   Power Supply and Structure in the United States

The three-phase power supplies in the United States are: AC208V, AC240V and AC480V.

There are delta connection and wye connection for three-phase power supply, as shown in Figs. 6.41, 6.42 and 6.43. The delta configuration also derives different power supply system due to the different layouts of the ground wire and the neutral wire.

Fig. 6.41 Four-wire Wye
Configuration

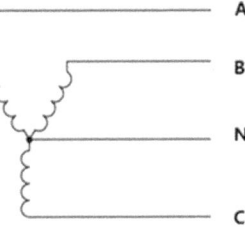

Fig. 6.42 Three-Wire Delta
Configuration (Ground)

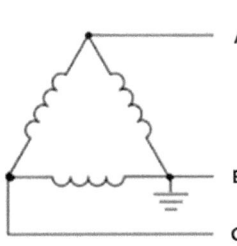

Fig. 6.43 Four-Wire Delta
Configuration (Neutral)

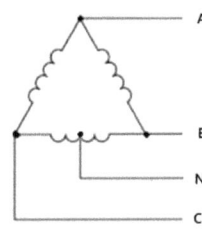

Due to the different voltages and structures of the three-phase power supplies, there are many types of power supply voltages derived from them. The most common single-phase power supply is AC120V.

The three-phase AC208V star can generate AC120V and AC208V power supply voltages, and the single-phase AC120V power supply is commonly used in residential buildings, as shown in Fig. 6.44.

Fig. 6.44 AC120V and
AC208V

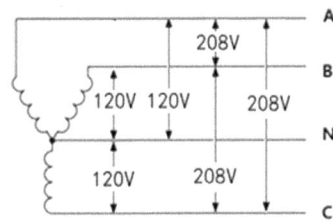

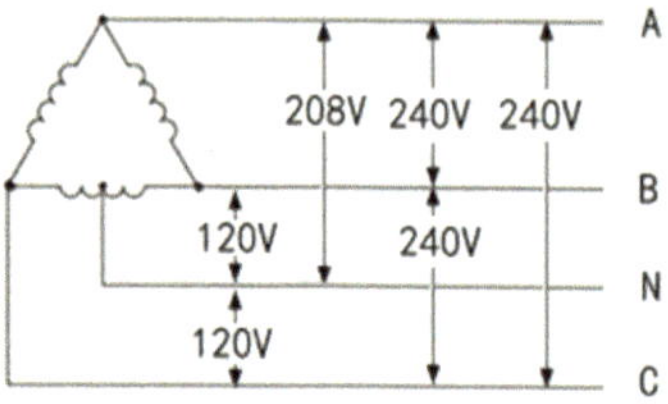

Fig. 6.45 AC120, AC240V
and AC208V

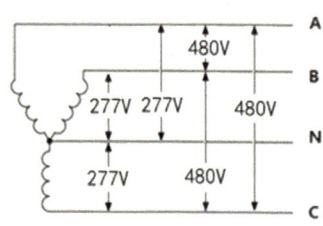

Fig. 6.46 AC277V and
AC480V

Fig. 6.47 AC120V/15A
wall socket

The three-phase AC240V delta connection can also derive the power supply voltage of AC120V and AC208V, as shown in Fig. 6.45. In the same way, the three-phase AC480V delta connection can derive the power supply voltage of AC240V and AC415V.

Three-phase AC480V wye connection can derive single-phase AC277V power supply voltage, as shown in Fig. 6.46.

6.9.2 Design Working Voltage of Electrical Device

Considering that the power transmission between the power supply and the electrical device must have line loss, the design operating voltage of the electrical device is lower than the voltage of the power supply, and the tolerance factor must also be considered.

In the three-phase AC480V power supply system, for example, the rated voltage of the electrical device is designed according to AC460V, considering the tolerance, the working voltage range is AC440-480; in the single-phase AC120V power supply system, the operating voltage range of the electrical device is AC110-120 V.

6.9.3 Power Sockets

The most commonly used wall sockets are AC120V/15A sockets, and the shapes of the two sockets are shown in Fig. 6.47. In Fig. 6.47, the hole on the left (long) of the socket is N (Neutral), the hole on the right (short) is L (Line), and the semicircle hole on the bottom is G (Ground).

Three-phase sockets generally have 3 holes or 4 holes, and if the system ground and equipment ground are separated, there are 5 holes.

Due to the influence of many factors such as current size, voltage level, number of phases, whether there is a ground wire, whether there is a neutral wire, and the configuration of the power supply, there are many types of sockets, which will not be listed here.

6.9.4 Power Supply and Structure in China

In China, the most common low-voltage power supply voltages are: single-phase AC220V and three-phase AC380V. Three-phase AC power mostly uses wye connection and three-phase four-wire power supply, as shown in Fig. 6.48.

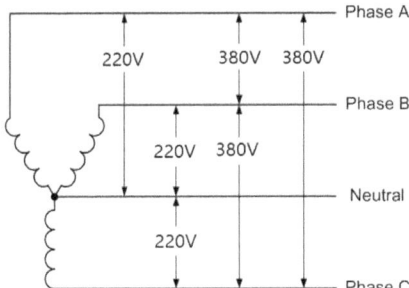

Fig. 6.48 Single-phase AC220V and three-phase AC380V

The neutral line is usually connected to the earth, and the grounded neutral point is called the zero point, and the grounded neutral line is called the zero line.

The voltage between terminal wires is called line voltage, AC380V. The voltage between the terminal line and the neutral line is called the phase voltage, AC220V. They have the following relationship.

$$\sqrt{3} * 220 = 380 \tag{6.9}$$

Chapter 7
Speed Regulation Method in Energy Saving System

In the energy-saving control system, in many occasions, it is necessary to use the method of speed regulation to distribute the load of the equipment and to start and stop the equipment. These speed regulation methods are both mechanical and electrical. Due to the rapid development of high-voltage and high-current electronic devices, the application of frequency converter speed regulation of AC motors is becoming more and more common.

7.1 Electromagnetic Slip Clutch

The magnetic powder clutch mentioned in the last chapter can realize the speed of the load on the output shaft side by controlling the magnitude of the input voltage. A similar method also has the electromagnetic slip clutch to be mentioned below.

The basic principle of the electromagnetic slip clutch is shown in Fig. 7.1. The motor (1) rotates at a constant speed. The armature (2) composed of the motor (1) and the cast steel cylinder is rigidly connected through the rotating shaft. The motor (1) drives the armature (2) rotate, the excitation winding (3) on the magnetic pole (4) has a DC voltage U_f through the slip ring brush. The current in the excitation winding (3) causes the magnetic poles (4) to establish a magnetic field. The rotating armature (2) cuts the magnetic field and induces an electromotive force. The induced electromotive force generates eddy currents in the winding. Eddy currents are generated between the armature (2) and the magnetic poles (4), and the eddy currents interact with the magnetic field to generate electromagnetic force. The direction of the electromagnetic force hinders the relative movement between the armature (2) and the magnetic poles (4). According to the characteristics of the action force and reaction force, the magnetic pole (4) rotates with the armature (2), so that the motor (1) and the load (6) are in a "connected" state. When the DC voltage U_f is zero, the electromagnetic force

F. Yao and Y. Yao, *Efficient Energy-Saving Control and Optimization for Multi-Unit Systems*, https://doi.org/10.1007/978-981-97-4492-3_7

Fig. 7.1 Electromagnetic
slip clutch

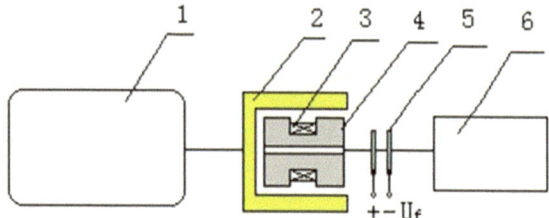

in the armature (2) disappears. The magnetic pole (4) does not follow the rotation of the motor (1), and the motor (1) and the load (6) are in a state of "disengagement".

Changing the excitation voltage U_f can change the magnitude of the eddy current in the armature (2), which also changes the magnitude of the electromagnetic force in the armature (2) and the rotational speed of the magnetic pole (4). The speed of the load (6) changes accordingly. This speed regulation method has simple structure, reliable operation and convenient control.

This kind of speed regulating device is sometimes called electromagnetic slip speed governor, and the combination of speed regulating device and electric motor is called electromagnetic speed regulating motor. In this speed regulation method, the maximum speed on the load side is lower than the rated speed on the motor side. The general speed range is 10–80%. The disadvantage of this method is that there are eddy currents in the armature. The heavier the load, the greater the electromagnetic force required and the greater the eddy current, so considerable heat will be generated in the armature. At low speeds, the transmission efficiency is very low. The torque is T_M, then the efficiency η_m of the electromagnetic clutch is as Eq. (7.1).

$$\eta_m = \frac{9550 T_M n}{9550 T_M n_0} = \frac{n}{n_0} \tag{7.1}$$

7.2   Hydraulic Coupling

In the early days, it was inconvenient to adjust the speed of some engines, and the power electronic devices were very expensive. A mechanical speed control device was connected in series between the engine and the load to adjust the speed of the load. The hydraulic coupling is one of such speed control devices.

The principle of the hydraulic coupling is shown in Fig. 7.2. The motor (1) rotates at a constant speed. The motor (1) is rigidly connected to the turbine (2) through the rotating shaft. The motor (1) drives the turbine (2) to rotate, and the turbine (2) inside Filled with a certain volume of liquid (3), the liquid (3) can be vegetable oil or water. The rotation of the turbine (2) makes the liquid (3) in it thrown out from the outer edge due to centrifugal force, and the liquid (3) enters the turbine (4). The impingement turbine (4) rotates in the same direction as the turbine (2), the turbine

Fig. 7.2 Hydraulic coupling

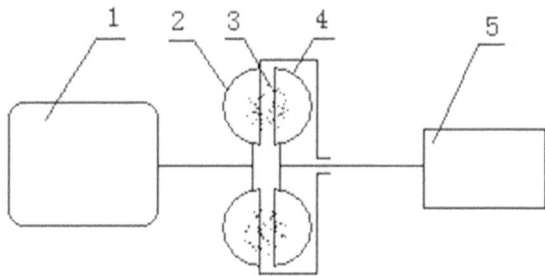

(4) is coaxially connected with the load (5), and the load (5) rotates accordingly. Adjusting the capacity of the liquid (3) can change the turbine (2) The force acting on the turbine (4), while changing the rotational speed.

This speed regulation method uses liquid to transmit kinetic energy and pressure energy, and the control method is simple and convenient. The disadvantage is that due to the speed regulation by changing the liquid volume, the speed response is slow, the internal liquid generates considerable heat, and at low speeds, the transmission efficiency is very low. The maximum speed of the load side achieved by this speed regulation method is smaller than the speed of the motor side. Assuming that the load torque is T_M, the efficiency η_m of the hydraulic coupling is as in Eq. (7.2). The range of speed regulation is about 20–97%, and the load cannot run at 100% the rated speed of the motor.

$$\eta_m = \frac{9550T_M n}{9550T_M n_0} = \frac{n}{n_0} \tag{7.2}$$

7.3 Fluid Viscous Clutch

The working principle of the liquid viscous clutch is shown in Fig. 7.3. The motor (1) rotates at a constant speed, the motor (1) is connected with the active friction plate (2) through the rotating shaft, and the motor (1) drives the active The friction plate (2) rotates, and the oil medium (3) is filled between the active friction plate (2) and the driven friction plate (4). Due to the action of friction, the rotation of the active friction plate (2) makes the oil medium (3) Rotation in the same direction also occurs. Due to the transmission of friction force, the rotation of the oil medium (3) drives the driven friction plate (4) to rotate in the same direction. The moving friction plate (4) is coaxially connected with the load (5), and the load (5) rotates, the driven friction plate (4) can move left and right through the hydraulic oil. The closer the distance between the driven friction plate (4) and the active friction plate (2) is, the smaller the gap is, the driven friction plate (4) and the oil medium (3) between the active friction plate (2). The greater the friction force, the higher the speed of the

Fig. 7.3 Fluid-viscous
clutch

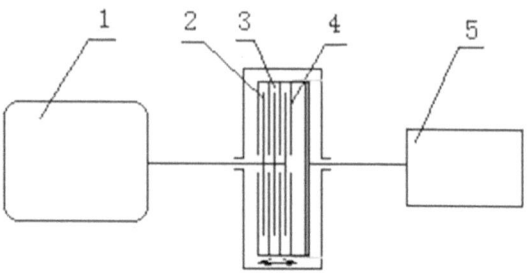

driven friction plate (4), and the higher the speed of the load (5); on the contrary, the
farther the distance between the plate (4) and the active friction plate (2), that is, the
larger the gap, the smaller the friction force transmitted by the oil medium (3). The
lower the speed of the driven friction plate (4), the lower the speed of the load (5).
In this way, the adjustment of the load speed is realized.

This speed regulation method uses the viscosity of the fluid to transfer energy.
The control is simple and convenient. The speed regulation range is about 20–100%.
When the distance between the driven friction plate and the active friction plate is
zero, the load side of this speed regulation method is the highest. The rotational
speed may be equal to the speed on the motor side. The disadvantage is that the
speed response is slow due to the use of changing the mechanical clearance for
speed regulation, and the internal medium oil generates considerable heat, and at
low speeds, the transmission efficiency is very low. Assuming the load torque is T_M,
the hydraulic viscous speed regulation clutch efficiency η_m is as follows.

$$\eta_m = \frac{9550T_M n}{9550T_M n_0} = \frac{n}{n_0} \tag{7.3}$$

7.4 Mechanical Governor

There are many methods of mechanical speed regulation, but it is not the focus of
this book. Figure 7.4 is a method of speed regulation by using belt variable diameter,
and Fig. 7.5 is a method of speed regulation by using turntable variable diameter.
There are also many ways to achieve mechanical speed regulation, which will not be
described in detail here. The stepless speed changer of many small electric motors
uses these speed regulation methods. The speed of the motor is constant, there is a
speed change device on the output shaft of the motor, and the speed change of the
output shaft can be changed by adjusting the handwheel or lever on the speed change
device.

In Fig. 7.4, the motor (1) drags the trapezoidal wheel (2) that can change the width
of the T shape. The trapezoidal wheel (2) drives the trapezoidal wheel (4) that can

Fig. 7.4 Belt variable
diameter speed regulation
method

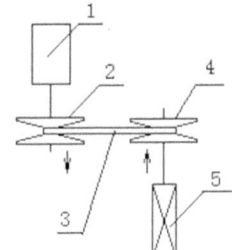

Fig. 7.5 Turntable variable
diameter speed regulation
method

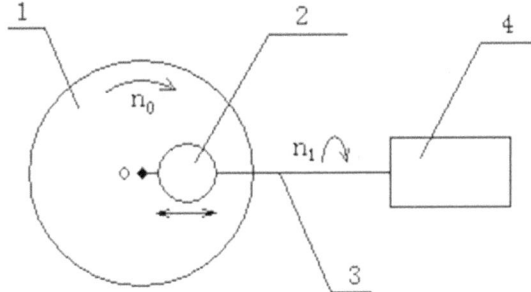

change the width of the T shape through the T belt (3). The wheel (4) drives the
load (5) to rotate. Synchronously adjusts the width of the T-shaped grooves in the
trapezoidal wheel (2) and the trapezoidal wheel (4), so that the transmission ratio
between the trapezoidal wheel (2) and the trapezoidal wheel (4) can be changed. At
the same time the speed adjustment of the load (5) is also realized. As shown by the
arrow in the figure, make the T-shaped groove of the trapezoidal wheel (2) wider. The
position of the T-shaped belt (3) in the trapezoidal wheel (2) will fall and be closer
to the axis, which is equivalent to a smaller diameter of the effective transmission
wheel. For the same motor rotation speed n_0, the linear velocity of the T-shaped
belt (3) becomes lower. In order not to change the axis positions of the trapezoidal
wheel (2) and the trapezoidal wheel (4), make the T-shaped groove of the trapezoidal
wheel (4) narrowing. The position of the T-shaped belt (3) in the trapezoidal wheel
(4) is raised, away from the axis, which is equivalent to the increase of the effective
transmission wheel diameter, so the speed on the load (5) is finally reduced. If it
needs to be raised the rotation speed of the load (5), then the adjustment direction is
opposite.

In Fig. 7.5, the motor drives a turntable (1) with a large friction coefficient to
rotate at speed n0. The spherical runner (2) is installed on the rotating shaft (3) and
can move left and right. The surface of the spherical friction wheel (2) is in contact
with the surface of the turntable (1) is in direct contact. Due to the effect of friction,
the spherical friction wheel (2) rotates. The rotating speed of the spherical friction
wheel (2) depends on the linear velocity of the contact point with the turntable (1) and
the spherical friction wheel (2). The spherical friction wheel (2) drives the load (4)

to rotate, the linear velocity of the turntable (1) is larger on the outside and smaller on the inside. When the spherical wheel (2) moves outward, the rotational speed of the spherical friction wheel (2) increases, when the spherical friction wheel (2) moves inward, the rotational speed of the spherical friction wheel (2) decreases, thus changing the rotational speed n_1 of the load (4).

The speed range of the mechanical speed control method can be very wide, and the speed of the load can exceed the speed of the motor. The transmission efficiency of these methods depends on the transmission form and structure.

7.5 Stepper Motor and Stepper Motor Driver

Stepper motors and stepper motor drivers are widely used in some situations that do not require too high precision and too high dynamic performance. It can directly realize synchronous control and positioning control without feedback signals like encoders. The main parameters of the motor are step angle, working torque, holding torque, positioning torque, no-load starting frequency, maximum operating speed, control mode, power supply voltage, etc.

The minimum step angle of the stepper motor determines the open-loop control accuracy of the stepper motor, so the step angle is one of the most important parameters of the stepper motor. Since the control method of the stepping motor has the number of phases, the stepping motor and the stepping motor driver should be used together. The wiring of a general stepping motor is shown in Fig. 7.6.

In Fig. 7.6, the power input of the stepper motor driver is either AC power supply (AC 60 V, AC 100 V, AC 220 V, etc.) or DC power supply (DC 24 V, DC 12 V, DC 36 V, etc.), and the wiring of the stepper motor can be It is the A +, A-, B +, B- mode in the figure, and it can also be U, V, W or A, B, C, D, E and other wiring modes. The pulse command input is used to control the number of steps of the stepping motor. The direction control input is used to control the rotation direction of the stepper motor (forward and reverse). The offline signal input makes the stepper motor in a free state. The appearance of the stepping motor and the stepping motor driver is shown in Fig. 7.7.

Fig. 7.6 Wiring of the stepping motor

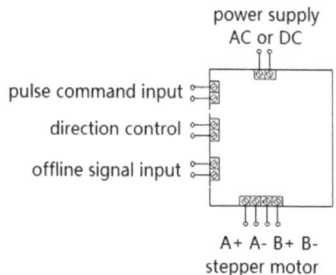

stepper motor   stepper driver

Fig. 7.7 Appearance of stepping motor and stepping motor driver

7.6 AC Servo Motor Driver

The use of AC servo motors and servo drives is very similar to frequency converters, and many of the parameter settings are also very similar. The main difference between servo drives and ordinary frequency converters is that they have higher control precision and faster control speed. It has the input port of the motor encoder (currently some frequency converters also provide the encoder input port), and the servo drive generally has a pair of encoder pulse input port and encoder pulse output port for interconnection between the servo drives. Which can be conveniently Realize the precise proportional synchronization of the speeds of the servo drives. So there is a synchronization factor on the servo drive, one is used to input the pulse number of the tracked axis, and one is used to input the corresponding tracking pulse number of the servo motor. The servo drive It can be used for positioning control, and the number and direction of pulses taken by the motor are determined by the input signal. The servo driver can control the speed and start and stop through the panel, and can also be controlled by communication. The wiring of the general servo driver is shown in Fig. 7.8.

In Fig. 7.8, R, S, T are connected to three-phase AC power supply (some are also connected to single-phase power supply AC 220 V), U, V, W are connected to AC

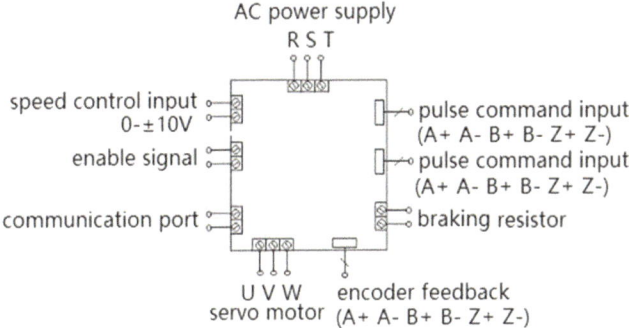

Fig. 7.8 Main wiring of the servo drive

Servo Motor   Servo Driver   Linear Servo Motor

Fig. 7.9 The appearance of the servo driver

servo motor. The speed control input (0- ± 10 V) terminal is used to control the speed of the servo motor. The enable signal controls whether the servo driver is working, and the pulse command input terminal is connected to the encoder of the previous motor or drive shaft to be tracked. It can be connected to the encoder command output of other controllers. The pulse command output is used to send the position of the servo motor to the next-level servo driver as a tracking command. The encoder feedback is connected to the encoder on the servo motor. The communication port is used to communicate with other controllers. Controllers, such as PLC, are used for data transmission and control. The servo drive is mainly selected according to output torque, maximum speed, encoder resolution, power supply, installation method, etc. The main parameters that need to be set for the servo driver are operation control mode (speed mode, position mode or torque mode), control method, maximum torque, maximum speed, etc.

When the workpiece movement requires a large acceleration (2 ~ 10 g) and precision movement without mechanical clearance, it is necessary to use a small linear motor and a servo driver. The appearance of the servo driver is shown in Fig. 7.9.

7.7   Speed Regulation Method of DC Motor

Before the large-scale application of frequency converters, DC motors and DC motor governors have always been the protagonists of speed control in the field of motor drive, and their application time can be traced back to a long time ago. The expression of the speed n_1 of the DC motor is:

$$n_1 = K(U - Ir - 2\Delta U)/\varphi_1 \tag{7.4}$$

In the Eq. (7.4), n1 is the output speed of the DC motor, K is a constant, U is the armature voltage, I is the armature current, r is the internal resistance of the armature. ΔU is the voltage drop on a brush, φ_1 is the excitation magnetic flux generated by the excitation coil.

It can be seen from Eq. (7.4) that there are two main speed regulation methods of DC motors: one is to change the voltage on the armature winding, and the other is to change the excitation current on the excitation winding. In general, the stepless speed regulation of the motor is realized by changing the armature voltage below the rated power of the motor, and the constant power speed regulation is realized by weakening the excitation when the motor speed exceeds the rated power. The main parameters of the DC speed controller are as follows: rated voltage, rated current, rated power, rated speed, speed up and down time of the controlled motor, regulator parameters of the speed control loop, PI adjustment parameters of the current control loop, maximum allowable current value, encoder parameters, etc. The parameters of the DC motor governor can be modified according to the manufacturer's instructions through the buttons on the control panel and the display screen. The wiring of the DC governor is shown in Fig. 7.10.

In Fig. 7.10, R, S, T are connected to three-phase AC power supply (some are also connected to single-phase power supply AC 220 V), U + and U- are connected to the armature winding, Uf + and Uf- are connected to the excitation winding of the motor, and the speed of the motor And the start and stop can be controlled by the panel or by external analog signal (0 ~ ± 10 V) and switch signal. In closed-loop control, the speed and position signal are input through the encoder.

The appearance of the DC speed controller is shown in Fig. 7.11.

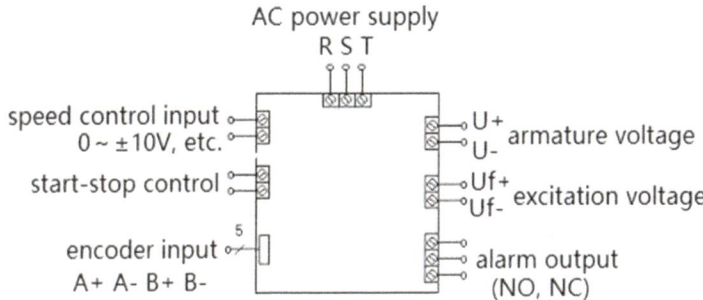

Fig. 7.10 Main wiring of the DC speed controller

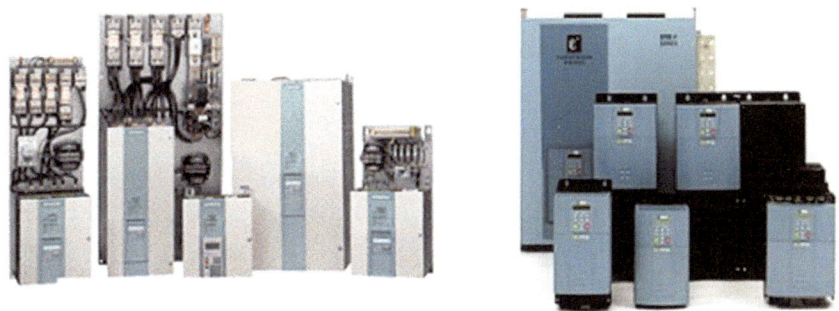

Fig. 7.11 Appearances of several DC speed regulators

7.8　Rotational Speed of AC Motors

AC motors are the most widely used power equipment in the industrial field.

From the previous chapter, the basic principle of the three-phase AC motor we know that the rotational speed n_0 (also called synchronous speed) of the stator magnetic field of the three-phase AC motor is expressed as:

$$n_0 = \frac{60 \times f}{p} \tag{7.5}$$

In the Eq. (7.5), f is the frequency of the power supply of the three-phase AC motor, and p is the number of pole pairs of the three-phase AC motor.

The expression of the rotor output speed n of the three-phase AC motor is:

$$n = (1 - s) \times n_0 = (1 - s) \times \frac{60 \times f}{p} \tag{7.6}$$

In the equation, n is the three-phase AC motor rotor speed rpm (revolutions per minute); s is the slip rate, and s represents the difference between the output rotation speed of the three-phase AC motor rotor and the magnetic field rotation speed on the stator. The slip rate of the three-phase synchronous AC motor is s = 0, that is, the rotation speed of the rotor output is equal to the rotation speed of the magnetic field on the stator. The slip rate of the three-phase asynchronous AC motor is s > 0, and the expression of the slip rate s is:

$$s = \frac{n_0 - n}{n_0} \tag{7.7}$$

7.9　Efficiency of AC Motors

The efficiency η expression of a three-phase AC motor is:

$$\eta = \frac{P_2}{P_1} = \frac{P_2}{P_2 + p_{cu1} + p_{Fe1} + p_{cu2} + p_{Fe2} + p_f + p_{ad}} \% \tag{7.8}$$

In the Eq. (7.8), P_1 is the total input power of the three-phase AC motor, P_2 is the mechanical output power of the motor, p_{cu1} is the copper loss caused by the current flowing through the copper wire resistance in the stator, and p_{Fe1} is the stator iron core turn-on the variable magnetic field leads to the iron loss caused by the existence of eddy current, etc., p_{cu2} is the copper loss of the rotor, p_{Fe2} is the iron loss on the rotor. p_f is the mechanical loss caused by the friction of the rotor bearing, etc., and p_{ad} is the additional loss caused by the transverse current in the rotor. Since the rotor rotates with the rotating magnetic field, the frequency of the alternating magnetic field in the rotor is zero during synchronous operation, and only 6–3 Hz during asynchronous rated operation. So, the iron loss p_{Fe2} on the rotor is generally very small, and circuit analysis is often ignored.

The electromagnetic power transmitted from the stator to the rotor of a three-phase AC motor is P_M. P_M is equal to the total input power P_1 minus the stator iron loss p_{Fe1} and the stator copper loss p_{cu1}. The electromagnetic torque T_M of the motor is generated by the interaction between the rotor current and the stator rotating magnetic field. The stator rotating magnetic field speed n_0, such as Eq. (7.9).

$$P_M = P_1 - p_{cu1} - p_{Fe1} = 9550T_M n_0 \qquad (7.9)$$

The electromagnetic torque T_M is transmitted to the motor rotor, the speed of the motor rotor is n. The total mechanical power on the motor rotor is P_m, P_m is equal to the electromagnetic power P_M minus the copper loss p_{cu2} and iron loss p_{Fe2} on the rotor, p_{Fe2} is negligible, as Eq. (7.10). The power transmission efficiency η_m between rotor and stator is in the following Eq. (7.10.1)

$$P_m = P_M = p_{cu2} = 9550T_M n \qquad (7.10)$$

$$\eta_m = \frac{P_m}{P_M} = \frac{n}{n_0} \qquad (7.10.1)$$

The copper loss p_{cu2} on the rotor is equal to the heat loss of the rotor current i_2 on the rotor internal resistance r_2, and it is also equal to the product of the slip rate s and the electromagnetic power P_M.

The mechanical output power P_2 of the motor is equal to the total mechanical power P_m on the rotor minus the mechanical loss p_f on the rotor and the additional loss, as shown in Eq. (7.11).

$$p_{cu2} = 3i_2^2 r_2 = P_M P_m P_M \left(\frac{P_M - P_m}{P_M} \right) P_M \left(\frac{n_0 - n}{n_0} \right) = sP_M \qquad (7.11)$$

$$P_2 = P_m - p_f - p_{ad} \qquad (7.12)$$

The efficiency η of the three-phase AC motor can also be regarded as the product of the stator efficiency η_1 and the rotor efficiency η_2, such as Eq. (7.13).

$$\eta = \frac{P_2}{P_1} = \frac{P_M}{P_1} \times \frac{P_2}{P_M} = \eta_1 \times \eta_2 \qquad (7.13)$$

The iron loss of the motor is basically unchanged, so it is also called constant loss. The copper loss of the motor increases with the increase of current (that is, the load rate increases), so it is also called variable loss. The three-phase AC motor stator and when the constant loss of the rotor is equal to the variable loss, the efficiency

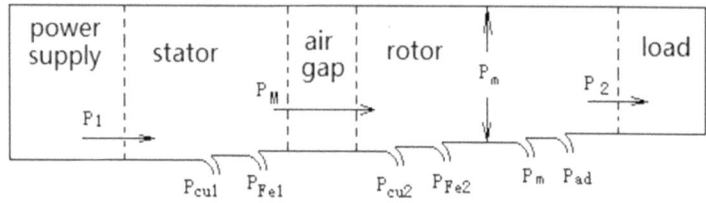

Fig. 7.12 Energy distribution of the motor

η reaches the maximum η_M. In order to use the same metal material to drag a large load as much as possible, when the three-phase AC motor is manufactured, the rated working state is generally designed to be greater than the maximum efficiency when the load rate is above. The load rate at the highest efficiency is generally 70–90%.

The relationship between the power consumption of each part of the three-phase AC motor and the total input power P_1 and mechanical output power P_2 is shown in Fig. 7.12.

7.10 Speed Regulation Method of AC Motor

From Eq. (7.6), it can be seen that there are no more than three ways to adjust the speed of a three-phase AC motor. One is to change the frequency f of the three-phase power supply, the other is to change the number of pole pairs p of the three-phase AC motor, and the third is to change the slip ratio s.

7.10.1 The Speed Regulation Method of Changing the Number of Pairs of Poles

This is the simplest method for adjusting the speed of three-phase AC motors. It is widely used on the spindles of machine tools such as boring machines and grinding machines and centrifuges. This speed adjustment method requires that the structure of the three-phase AC motor itself must be able to change poles. This speed regulation method changes the number of poles of the three-phase AC motor by changing the wiring mode of the stator coil inside the motor through the AC contactor or manual switch. There is no other intermediate link, so there is no efficiency loss of the intermediate link. The energy efficiency of this speed regulation method is the highest, of course, this does not refer to the efficiency of the pole-changing motor itself.

For example, when a three-phase AC motor is converted from a 4-pole (2 pole pair) wiring mode to a 2-pole (1 pole pair) wiring mode, the output speed of the three-phase AC motor can be doubled. When switching from 2-pole to 4-pole wiring, the speed is reduced by half. However, because the number of pole pairs p has only a few levels, this speed regulation method has steps, and cannot realize the continuous regulation of the speed of the three-phase AC motor.

The number of pole pairs of the cage rotor can be automatically changed through electromagnetic induction with the change of the number of pole pairs of the stator, while it is difficult for the wound rotor to change the number of pole pairs by changing the wiring. Therefore, the pole-changing speed regulation method is mainly used for cage-type three-phase AC motors.

7.10.2   Nine Speed Regulation Methods to Change the Slip S

There are many speed regulation methods to change the slip s, and this section gives 9 of them.

7.10.2.1   Rotor Series Resistance Speed Regulation of Wound Rotor Motor

The rotor structure of the wound rotor three-phase AC motor is shown in Fig. 7.13. In addition to the terminals of the stator winding, this motor also leads the three rotor windings out of the motor through slip rings and brushes. The leads are generally installed on the motor. At the shaft end, change the resistance value of the external resistor on each phase winding of the rotor to realize the adjustment of the motor speed.

The slip rate s of the wound rotor after the series resistance, the rated slip rate s_e before the series resistance, the self-resistance r_2 of each phase of the rotor, the series resistance R of each phase of the rotor. The electromagnetic torque T before the series resistance, after the series resistance the relationship between the electromagnetic torque T' is as follows:

$$R = \left(\frac{sT}{s_e T\prime} - 1 \right) \times r_2 \qquad (7.14)$$

Fig. 7.13 Rotor series resistance speed regulation

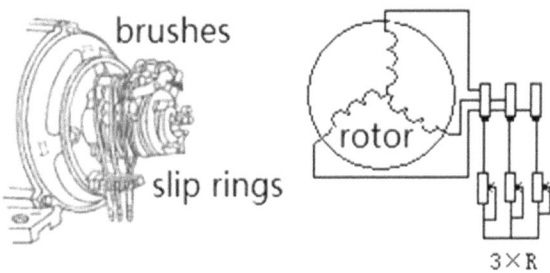

$3 \times R$

Fig. 7.14 Speed regulation
by step-by-step switching of
series connected resistors

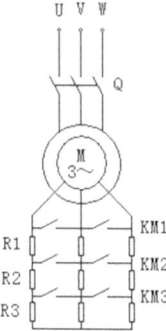

The slip rate s of the wire-wound rotor after series resistors is obtained by the Eq. (7.7), and the rated slip rate s_e before series resistors is obtained by the following equation:

$$s_e = \frac{n_0 - n_e}{n_0} \tag{7.15}$$

According to Eq. (7.14), (7.15) and (7.7), the relationship between the speed n and the resistance R of each phase of the rotor connected in series is as follows:

$$n = \frac{n_0(T - T\prime) + n_e T\prime - \frac{R}{r_2}(n_0 - n_e)T\prime}{T} \tag{7.16}$$

If the resistance of each phase of the rotor connected in series is switched in stages, as shown in Fig. 7.14. The control of the closure of KM3, KM2, and KM1 contacts in the figure is equivalent to connecting different resistors in series in the rotor. This method the speed adjustment is graded, and the series resistors are generally composed of multiple high-power metal resistors.

In order to realize the smooth adjustment of the speed, it is necessary to uniformly change the resistance value of the series resistance of each phase of the rotor. A liquid resistance governor uses sodium bicarbonate aqueous solution as the resistance liquid, and changes the interval between the moving plate and the static plate in the liquid. The size and length of the surface can evenly adjust the resistance value, so that the speed can be adjusted steplessly, as shown in Fig. 7.15, the motor drives the screw to rotate, the screw drives the screw nut to move up and down. The screw nut drives the moving plate to the electric resistance liquid moves up and down, and when the moving plate moves downward, the distance between the static plate and the moving plate decreases, and the liquid resistance decreases. When the moving plate moves upward, the distance between the static plate and the moving plate increases. As the distance becomes larger, the resistance of the liquid increases, so that the continuous adjustment of the motor speed can be realized.

Fig. 7.15 Liquid resistance
stepless speed regulator

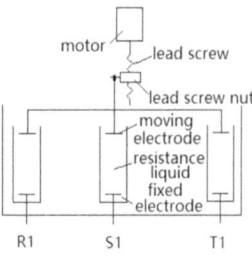

Fig. 7.16 Continuous
adjustment of series
resistance with thyristor

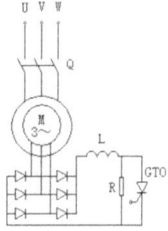

Another way to continuously change the series resistance of the rotor is to lead the wound rotor out and connect a three-phase bridge rectifier, connect a fixed resistor R at the DC terminal of the rectifier. A thyristor (GTO) or high-power transistor in parallel at both ends of the fixed resistor. The structure is shown in Fig. 7.16.

When the GTO is turned on, it is equivalent to the external resistance is zero. When the GTO is turned off, it is equivalent to the external resistance is R, and the ratio of the on-time T_1 to the off-time $(T-T_1)$ of the GTO in a cycle T is controlled. You can change the equivalent series resistance R_d, such as Eq. (7.17).

$$R_d = \left(\frac{T - T_1}{T} \right) R \tag{7.17}$$

The rotor circuit series resistance speed regulation method is simple, convenient, and easy to implement, but when the output speed of this method is low, the slip rate s is large. According to the Eq. (7.11), the copper loss P_{cu2} on the rotor will be proportional to the slip rate s bigger, less efficient. Assuming that the electromagnetic torque of the stator to the rotor is T_M, the power transfer efficiency η_m between the rotor and the stator is as in Eq. (7.18). Note that it does not refer to the total efficiency of the speed regulating motor, and the overall efficiency also needs to consider the stator efficiency and the rotor mechanical efficiency.

$$\eta_m = \frac{9550 T_M n}{9550 T_M n_0} = \frac{n}{n_0} \tag{7.18}$$

This type of governor is limited to the use of wound rotor motors with brushes for speed regulation. Since the speed regulation is achieved by increasing the slip rate, the lower the speed, the lower the efficiency. It should avoid working at low speed, and generally control the speed within the range of 50–100%. However, if this kind of governor works near the rated speed, the operating efficiency is also very high. Because after all, its highest efficiency is almost 100%, which cannot be achieved by frequency converters anyway. At light load and no load, changing the resistance of the rotor in series does not change much, so this method is suitable for heavy load speed regulation. The cold state value and hot state value of the resistance will change, this method is not suitable for occasions requiring fast response and precise speed regulation.

7.10.2.2 Cascade Speed Regulation Method of Wound Rotor Motor

A resistor R is connected in series in each phase circuit of the wound rotor to regulate the speed of the AC motor, as shown in Fig. 7.17. In fact, the speed regulation is achieved by reducing the current of the rotor circuit. The electromotive force induced by the rotor of the three-phase AC motor is E_2, internal resistance r_2, inductive reactance X_2, current i_2, series resistance R, and the heat loss i_2^2R on the external resistance are wasted in vain.

People seek other methods to adjust the current of the rotor loop to avoid the generation of i_2^2R. According to the knowledge of electricity, there are many ways to change the current in a loop. A three-phase bridge rectifier is connected externally to the lead-out end of the wound rotor. Charging a DC power supply E_3 at the terminal can also change the current value of the rotor so that the current of the rotor is i_2, as shown in Fig. 7.18.

In Fig. 7.18, the DC power supply E_3 can be used to excite the DC motor, and the DC motor is coaxially connected with the speed-regulated AC motor, as shown in Fig. 7.19. Changing the excitation and polarity of the DC motor can change E_3 Value, thereby changing the output speed of the rotor of the three-phase AC motor. Increasing the excitation E_a of the DC motor, the counter electromotive force E_3 of the DC motor increases, the armature current of the DC motor decreases, and the current of the rectifier circuit decreases. The current i_2 of the AC motor rotor decreases, the electromagnetic torque of the AC motor decreases, and the speed n

Fig. 7.17 Connecting resistors in series reduces rotor current

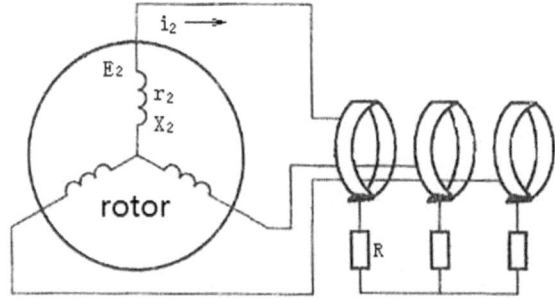

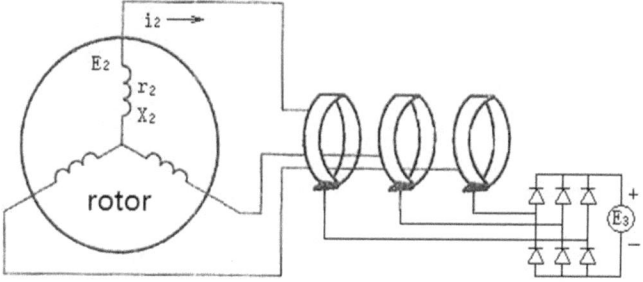

Fig. 7.18 DC power supply connected in series to reduce rotor current

Fig. 7.19 Principle of DC
motor feedback cascade
speed regulation

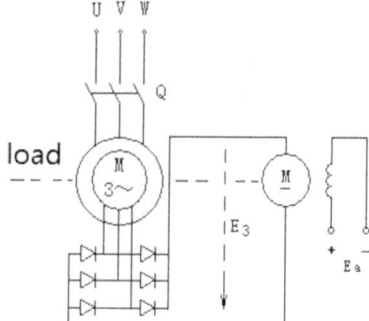

decreases. Similarly, if the excitation E_a of the DC motor is reduced, the speed n
increases. The output power of the DC motor works by dragging the load, so that the
slip power loss is recovered. Figure 7.19 is a schematic diagram of the DC motor
feedback cascade speed regulation principle.

Replace the above DC motor with an inverter, the voltage E_3 is converted into AC
through the inverter and fed back to the AC grid, so that the slip power loss can also
be recovered, as shown in Fig. 7.20. The electromotive force is connected in series
to the rotor In the loop, this is the principle of thyristor cascade speed regulation.
This speed regulation method can realize the output regulation of the motor speed
lower than the synchronous speed. This speed regulation method has high efficiency
and is suitable for occasions where the speed regulation range is not large. Because
the inverter part of this method is only responsible for converting the slip power, so
the power of the equipment is low, and the cost is lower than the method of directly
adjusting the frequency of the stator by the frequency converter. Power factors are
generally low.

In order to improve the power factor of the cascade speed regulation, a turn-
off thyristor is incorporated in the DC circuit, and the motor speed is adjusted by
controlling the ratio of the thyristor on and off (that is, the duty cycle). The lead angle
of the inverter is adjusted, fixed and take the minimum value, so that the reactive
power required by the inverter from the grid side can be reduced, and the power

Fig. 7.20 The principle of
thyristor cascade speed
regulation

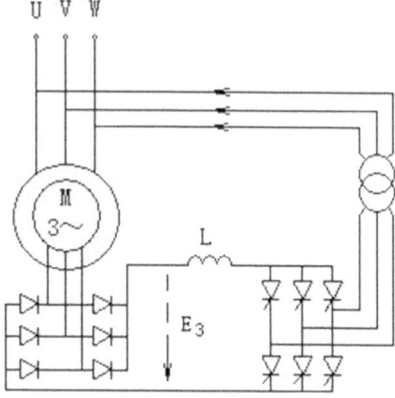

Fig. 7.21 Thyristor cascade
speed regulation system with
chopper

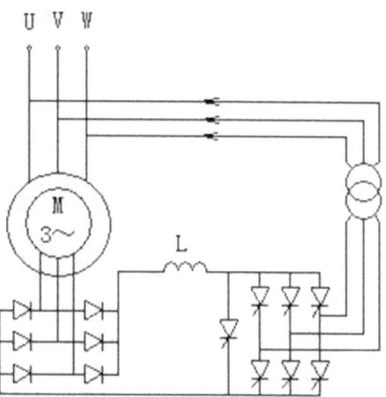

factor of the speed control system can be improved, as shown in Fig. 7.21, which is
the schematic diagram of the thyristor cascade speed control system with chopper.

The cascade speed regulation method adopts the rectification and thyristor inverter
feedback mode, and the grid interference is relatively large. Due to the feedback of
slip power, the operating efficiency is very high. At the rated speed, the efficiency is
close to 100%, which is higher than the highest efficiency of the inverter. The speed
response of this method is also faster, and the speed regulation range is generally
50–100%. The power device of this speed regulation method only needs to meet the
power requirements of the slip power part, so the total power capacity of the device
is small and the cost is low, but the motor must be a brush-wound motor.

7.10.2.3 Double-Fed Speed Regulation Method of Wound Rotor Motor

Connecting the AC power supply E_1 in series with each phase of the rotor winding can also change the current value of the rotor so that the current of the rotor is i_2, as shown in Fig. 7.22.

Use a low-power bidirectional frequency converter to supply power to the three-phase rotor coil of a three-phase AC wound motor, and change the size and phase of the rotor winding current, as shown in Fig. 7.23.

In Fig. 7.23, when the thyristor connected to the rotor winding acts as a rectifier bridge and the thyristor connected to the grid transformer acts as an inverter, it is equivalent to the above cascade speed regulation and feeds power to the grid; when the thyristor connected to the rotor winding acts as an inverter, the thyristor connected to the grid transformer rectifies, the grid supplies power to the rotor. This speed regulation method is called double-fed motor speed regulation system. In this speed regulation system, the energy of the grid is rectified and reversed to the rotor winding. The frequency, phase, and amplitude of the power supply can be adjusted. According to the superposition relationship between the positive and negative of the power supply frequency and the stator frequency, the output speed of the motor is

Fig. 7.22 AC power connected in series to the rotor winding

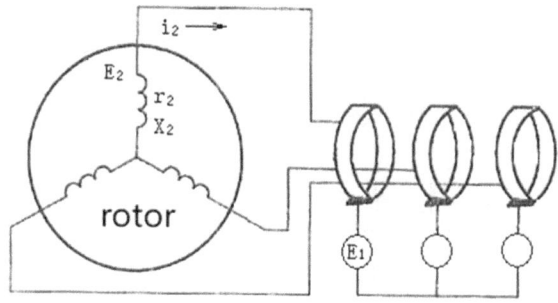

Fig. 7.23 Double-fed motor speed control system

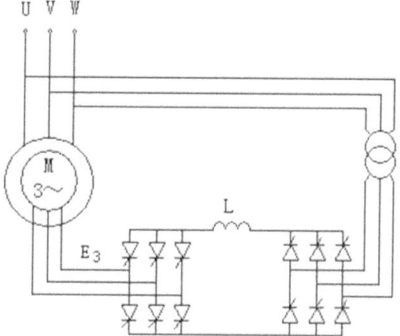

higher or lower than the synchronous speed. Because this method can realize super-synchronous operation, so this system is also called super synchronous cascade speed regulation system.

7.10.2.4 Speed Regulation Method of Brushless Doubly-Fed Motor

If the rotor winding is replaced by a cage structure, two sets of windings with different pole pairs are installed on the stator. One set is the power winding, the number of pole pairs is p_1, connected to the three-phase industrial frequency power supply, the frequency is f_1. The other is the control winding, the number of pole pairs is p_2, connected to the inverter whose frequency f_2 can be adjusted, it is required that $p_1 >$ ($p_2 + 1$), as shown in Fig. 7.24.

When the number of pole pairs is fixed, a set of rotating magnetic fields with fixed speed will be formed after the power winding is energized, and the speed of the rotating magnetic field of the control winding can be adjusted. The two groups of rotating magnetic fields work together to form a rotating magnetic field that can change the speed. This synthetic magnetic field when it acts on the rotor, the output speed can be adjusted. In short, the frequency f_2 of the power supply on the control winding can be changed by the frequency converter, and the speed n of the motor can be adjusted. This system is called brushless double-fed speed control system.

The speed n of the brushless double-fed adjustable-speed three-phase AC motor is:

$$n = \frac{60 \times (f_1 \pm f_2)}{p_1 + p_2} \tag{7.19}$$

Fig. 7.24 Brushless double-fed speed control system

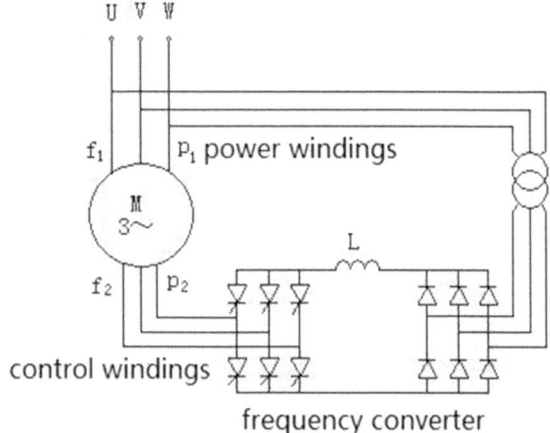

7.10.2.5   Using a Chopper Tube to Realize the Speed Regulation Method of Stator Winding Voltage Regulation

Disassemble the three stator windings of the three-phase AC motor, and then connect it to a three-phase rectifier bridge, and connect a chopper GTO to the DC side of the rectifier bridge, as shown in Fig. 7.25. When the GTO is turned on, it is equivalent to the full voltage of the three-phase stator winding. When the GTO is turned off, it is equivalent to the voltage of the three-phase stator winding is zero. The duty cycle and switching frequency of the GTO on and off can change the average working voltage of the motor stator.

This is a speed regulation method proposed by the author in the last century. Because this method is extremely simple, it has attracted the attention of many people in the industry after the article was published. Due to job changes and property rights restrictions, this method has not been studied.

This method is suitable for the speed regulation of the motor driven by the pump fan. The motor is connected in Y shape. The working pressure of the pump fan station is close to the rated head (rated pressure), so the range of motor speed adjustment is also close to the rated speed. In this case, this method's efficiency is also very high. When the speed is close to the rated speed, the operating efficiency is close to 100%, and the efficiency is low at low speed.

In order to utilize the current at the moment when the stator winding is turned off, when the GTO is turned off, the current will charge the capacitor C through the reactor; DC motor D outputs power to the load, as shown in Fig. 7.26.

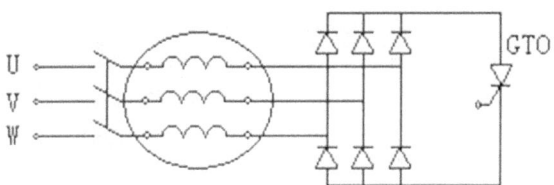

Fig. 7.25   Single thyristor speed regulation method

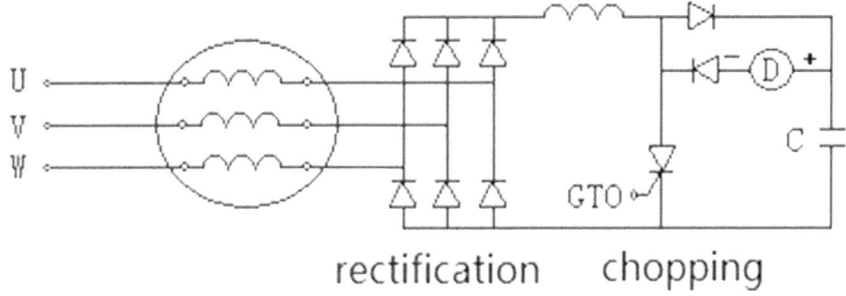

Fig. 7.26   DC motor feedback energy

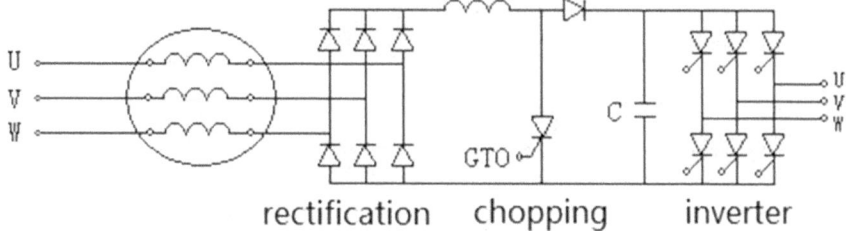

rectification   chopping   inverter

Fig. 7.27 Energy feedback grid

If the DC motor D in Fig. 7.26 is replaced by an inverter, part of the energy can also be fed back to the grid, as shown in Fig. 7.27.

7.10.2.6   Speed Regulation Method Using Thyristor to Regulate Stator Winding Voltage

Connect three sets of thyristors in series to the power wiring of the stator winding, as shown in Fig. 7.28. Adjust the effective voltage applied to the stator winding of the three-phase AC motor by changing the firing angle of the thyristors, thereby changing the output speed of the three-phase AC motor. This is similar to the principle of dimming table lamps in our home, but the dimming table lamps use a single-phase power supply, and the brightness adjustment is realized by adjusting the trigger angle of the thyristor and changing the effective voltage on the bulb.

The speed regulation method of adjusting the stator winding voltage is neither constant torque speed regulation nor constant power speed regulation, and is suitable for water pump fan loads whose torque decreases with the speed. When the speed is close to the rated speed, the operating efficiency is close to 100%, and the efficiency becomes low at low speed. This method adopts thyristor trigger adjustment, and the grid interference is large, but the speed response is fast. The speed adjustment range is generally 80–100%, cage type Both AC motors and wound rotor motors can be used.

Fig. 7.28 Thyristor voltage regulation and speed regulation method

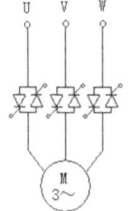

Fig. 7.29 Series reactance
voltage regulation and speed
regulation method

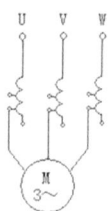

7.10.2.7 Speed Regulation Method of Stator Winding Connected in Series with Saturated Reactor to Adjust Voltage

This method is similar to the speed regulation method of a household electric fan, changing the saturated reactance connected in series to the stator winding, and due to the voltage division effect of the reactance, the voltage value applied to the stator winding can be changed, thereby controlling the speed of the three-phase AC motor, as shown in Fig. 7.29. When the speed is close to the rated speed, the operating efficiency is close to 100%, and the efficiency becomes lower at low speeds. This method has almost no interference to the power grid.

7.10.2.8 Speed Regulation Method Using Three-Phase Autotransformer to Regulate Stator Winding Voltage

Use the three-phase autotransformer to adjust the working voltage applied to the stator winding, so as to adjust the speed of the three-phase AC motor, as shown in Fig. 7.30.

This speed regulation method, when the speed is close to the rated speed, the operating efficiency is close to 100%, and the efficiency becomes lower at low speed, but this method has almost no interference to the power grid.

Fig. 7.30 Autotransformer
voltage regulation and speed
regulation method

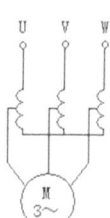

Fig. 7.31 Method of voltage
regulation and speed
regulation with series
resistors

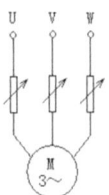

7.10.2.9   Speed Regulation Method Using Stator Winding Series
Resistance to Adjust Voltage

Use the voltage division function of the resistor to adjust the voltage applied to the
stator winding, so as to adjust the speed of the three-phase AC motor, as shown in
Fig. 7.31. If the power of the motor is large and the heat generated by the resistor
is large, this kind of series resistance voltage division speed regulation method
consumes a lot of energy. This speed regulation method, when the speed is close
to the rated speed, the operating efficiency is close to 100%, and the efficiency
becomes lower at low speed, but this method has almost no interference to the power
grid.

7.10.3   Speed Regulation Method of Changing the Frequency

In the early stage, the frequency conversion technology was limited by the limita-
tions of power electronic technology devices. With the rapid development and price
reduction of power devices and computing devices, the frequency conversion AC
speed regulation technology and products developed rapidly. Frequency conversion
technology can be applied to both asynchronous AC motors and synchronous AC
motors, it can drive squirrel-cage AC motors and wound AC motors also. Frequency
converters are used to provide variable-frequency power to three-phase AC motors.
The stepless speed regulation of the AC motor is realized, and the high-efficiency
operating area is relatively wide when operating in a full range. The rated operating
efficiency is about 94–98%. When the rated frequency is output and there is a certain
load, it is about 2–6% more wasteful than direct power frequency operation.

7.10.3.1   Voltage Type Inverter

At present, a large number of low-voltage inverters are voltage-type inverters. This
is the most widely used inverter in the industrial field, so it is also called a general-
purpose inverter. Its structure is shown in Fig. 7.32. It adopts AC–DC–AC structure,
mostly used in low-voltage frequency converters. The three-phase AC power supply
RST is connected to a three-phase rectifier bridge composed of diodes, and the AC

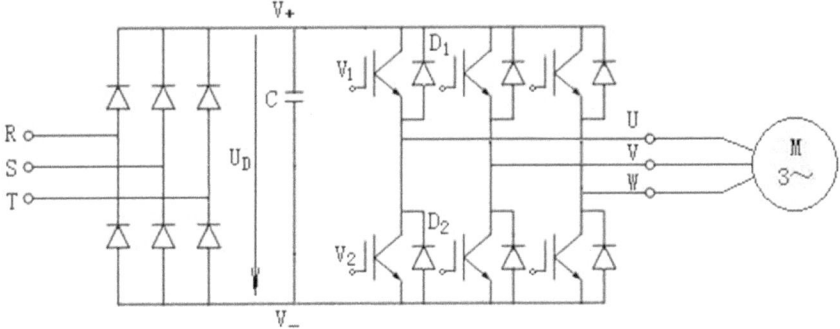

Fig. 7.32 Voltage type inverter

power is first converted into DC power, level V_+, level V_-. The DC power passes through a large capacity Capacitor C, capacitor C stores electric energy and filters it, keeps the DC voltage UD basically unchanged, which is equivalent to a voltage source (so it is called voltage type). Then the DC power passes through the inverter to become a three-phase power supply that can change the frequency and voltage. When V_1 is turned on and V_2 is turned off, the U phase is connected to V_+, and when V_1 is turned off and V_2 is turned on, the U phase is connected to V_-. The output voltage of U phase is a rectangular wave, it has two voltage levels. The situation of V and W phases is the same. So, such a frequency converter is also called a two-level frequency converter. Due to the influence of the inductance in the motor, the rising speed of the current lags behind the voltage. When the U-phase output V_+ appears and the U-phase current is negative, D_1 is turned on, and the current flows back to the DC side, and D_2 has the same effect. RST three-phase AC power supply, each phase power supply has 2 peaks, 1 positive peak and 1 negative peak in 1 sine wave cycle, 3 phases have 6 peaks in total, and the difference between the peaks is 60°. After passing through the three-phase rectifier bridge After rectification, it becomes 6 positive DC peaks, so it is also called 6-pulse rectification.

At present, most of the inverters of inverters with this structure are composed of IGBTs (insulated gate bipolar transistors). Since the sine wave output by the inverter is formed by changing the pulse width of a rectangular square wave, the harmonic component is large. When the motor is far away from the frequency converter, the distributed capacitance between the line and the ground becomes larger, and the high-order harmonics easily flow into the ground through the distributed capacitance, forming a leakage current. Affecting the nearby video signal, and tripping the leakage switch. Some measures need to be taken to solve this problem, and we will talk about these methods later. Generally, the speed range of the frequency converter is very wide, about 5–100%, the speed response is fast, and more precise speed control can be realized.

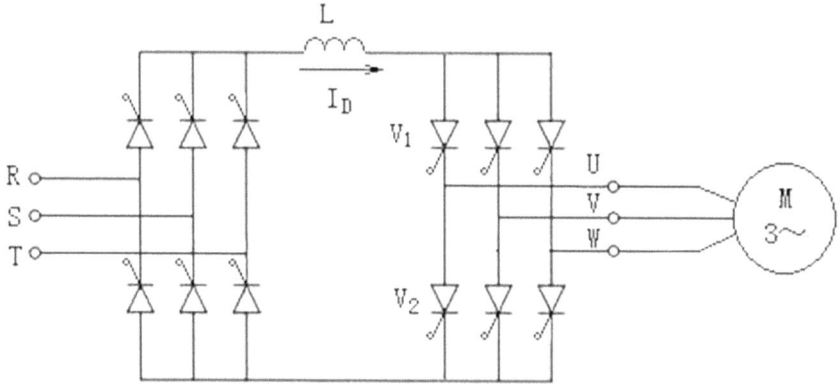

Fig. 7.33 Current-type inverter

7.10.3.2   Current-Mode Frequency Converter

The structure of the current-type inverter is shown in Fig. 7.33. The three-phase AC power supply is connected to a fully-controlled three-phase bridge, and the AC power becomes DC. The DC current flows through the large-capacity reactor L, and the reactor L stores magnetic field energy and filters the current, keep the DC current I_D flowing through the reactor L unchanged, which is equivalent to a current source (so it is called current type). Then the DC is converted into a three-phase AC current with variable frequency through the full-controlled bridge to drive the three-phase AC motor.

The two links of rectification and inverter from power input to motor output of this kind of frequency converter are symmetrical, so by changing the trigger angle of the controllable device, it can run in reverse. The electric energy generated by the motor in the power generation state can be used as the power supply. The original inverter bridge is controlled as a rectifier bridge, and the original rectifier bridge becomes an inverter bridge, which feeds the electric energy generated by the motor back to the grid to avoid waste of electric energy.

The GTO operating frequency of this kind of frequency converter should not be too high. After the variable frequency power supply is output to the motor, the noise of the motor will be relatively large, so it is rarely used in low-power three-phase AC motors. Due to the high withstand voltage and high current characteristics of turning off the GTO, this kind of frequency converter is mostly used in the occasion of driving high-voltage and high-power three-phase AC motors.

7.10.3.3   Three-Level Inverter

The structure of this inverter is shown in Fig. 7.34. The three-phase AC power supply RST is divided into two groups of power outputs that are isolated from each other

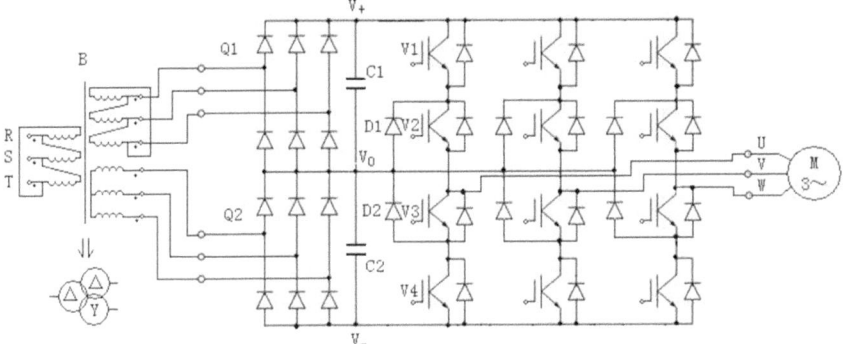

Fig. 7.34 Three-level inverter

and have a certain electrical angle difference through the phase-shifting transformer B. The two sets of power outputs are respectively connected to two groups of three-phase rectifier bridges, Q_1 and Q_2. The positive terminal of rectifier bridge Q_2 is connected to the negative terminal of Q_1 to form 0 level V_0, the positive terminal of rectifier bridge Q_1 forms + level V_+, and the negative terminal of rectifier bridge Q_2 forms -level V_-. So that the AC becomes a direct current with an intermediate 0 level, and the direct current is respectively filtered by two sets of large-capacity capacitors C_1 and C_2 to keep the direct current voltage basically unchanged. Then the direct current is converted into a variable frequency alternating current by the inverter to control the three-phase AC motor.

When transistor V_1 is turned on and V_2 is turned on, the U-phase level is V_+; when transistor V_1 is turned off and V_2 is turned on, D_1 is turned on, and the U-phase level is V_0; when transistor V_3 is turned on, and V_4 is turned on, U-phase The level is V_-, when the transistor V_4 is turned off and V_3 is turned on, D_2 is turned on, and the U-phase level is V_0. Because the DC side and the frequency conversion output side of this inverter have three levels of V_+, V_0, and V_-, so this kind of inverter is called a three-level inverter. The inverter with this structure is also a voltage type inverter. It is currently mainly used in rolling mills, locomotive traction, hoists and other fields. The output waveform of the three-level inverter is closer to the sine wave, so the harmonic component on the output side of the three-level inverter is smaller than that of the two-level inverter. The phase-shifting transformer adopts a set of $\triangle$ primary side, two sets of Y and $\triangle$ secondary sides, the phase difference of the two sets of secondary side power supplies is 30°, and the 12 DC peaks (12 pulse waves or pulse), the difference between the peaks is 30°, which is more uniform and smoother than the six DC peaks of the single rectifier bridge, so that the current wave on the grid side is closer to the sine wave and the harmonic pollution is smaller.

7.10.3.4   Multilevel Frequency Converter

At present, the medium and high voltage inverters widely used in the industry are mainly multi-level inverters. The power input side of this inverter uses a phase-shifting transformer to convert the high voltage of the power grid into multiple groups of low voltage and low voltage groups that are isolated from each other. The number and the voltage value of each group are directly related to the working voltage level of the driven three-phase AC motor. In order to make the current on the power supply side closer to a sine wave, different phase shift angles are used for each group of low voltages. The magnitude of the phase shift angle is directly related to the working voltage level of the AC motor.

The purpose of phase separation is to stagger the voltage peaks (or valleys) of each group of low-voltage outputs as much as possible, so that the rectified output DC waveform of each group of low-voltage power supplies after phase shifting has more dispersed and more uniform peaks and valleys, and the peaks and valleys of the current, so that the current waveform integrated is closer to a sine wave, and the harmonic interference to the grid is smaller.

Send each low-voltage three-phase AC RST to the rectifier bridge of a single power module. A single power module consists of 6 diodes to form a full-wave rectifier bridge. The H-type single-phase inverter bridge, the structure of a single power module is shown in Fig. 7.35. Multiple low-voltage power modules are connected in series to form a higher output voltage.

In Fig. 7.35, V_1 and V_4 are turned on, V_2 and V_3 are turned off, then the voltage V_{UV} between U and V is equal to $+ U_D$, V_3 and V_2 are turned on, V_1 and V_4 are turned off, then the voltage V_{UV} between U and V Equal to $-U_D$, V_1 and V_3 are turned on, V_2 and V_4 are turned off, then the voltage V_{UV} between U and V is equal to 0 V, V_2 and V_4 are turned on, V_1 and V_3 are turned off, then the voltage V_{UV} between U and V is also equal to 0 V. When the switch K is closed, V_{UV} is equal to 0 V. This function can ensure that the inverter continues to run with reduced capacity when the power module fails, which is very important for occasions with high safety requirements.

Taking a 3000 V-4160 V inverter as an example, the main structure of a multilevel inverter composed of a phase-shifting transformer and multiple power modules is shown in Fig. 7.36. The phase-shifting transformer has 12 low-voltage secondary

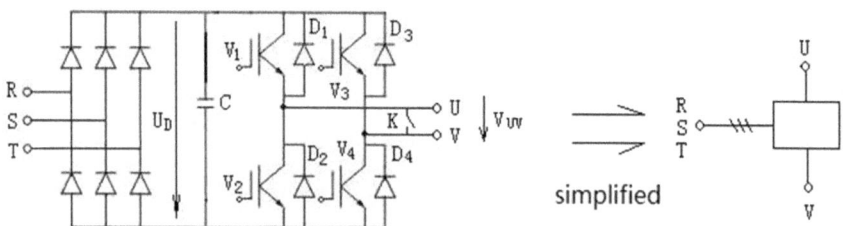

Fig. 7.35 Structure of a single power module

Fig. 7.36 Perfect Harmonic
Frequency Converter

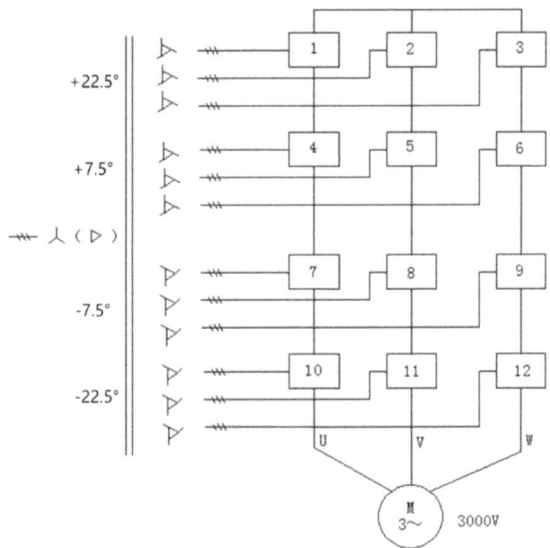

windings, which are divided into 4 groups. Each group is composed of 3 three-phase windings with the same phase. The phase angle difference between groups is 15°. One three-phase winding is selected in each group. A total of 4 three-phase windings provide power to the 4 U-phase power modules. For a 3000 V inverter, the voltage of each three-phase winding is 430 V, 4 of which form a U-phase supply voltage of 1720 V in total, and the line voltage is 2979 V. For a 4160 V inverter, the voltage of each three-phase winding is 600 V, and the V_{UV} of power modules 1, 4, 7, and 10 are connected in series to form U-phase output power. The V_{UV} of power modules 2, 5, 8, and 11 are connected in series to form V-phase output power. The V_{UV} of modules 3, 6, 9, and 12 are connected in series to form W-phase output power.

In Fig. 7.36, the U terminals of power modules 1, 2, and 3 are connected together, which is equivalent to a neutral point to form a reference voltage. According to different conduction conditions of the power modules, the voltage output to the three-phase AC motor is determined: positive and negative and voltage amplitude. The output voltage of this inverter adopts multi-level superposition, which is closer to a sine wave, the harmonic component is very small, and no output reactor is needed, so some people call this inverter a perfect harmonic-free inverter.

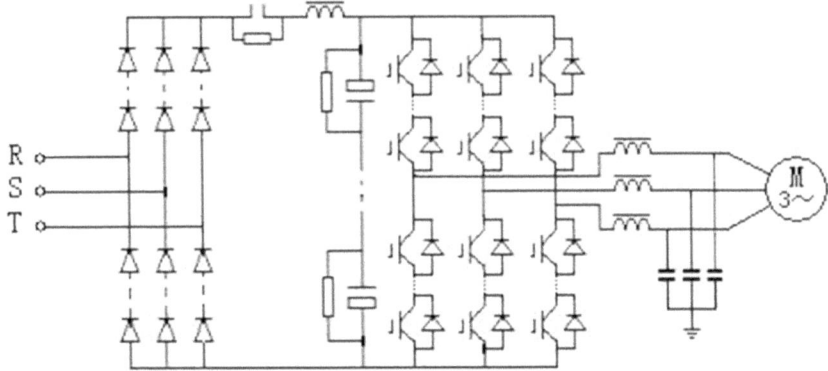

Fig. 7.37 Direct series connection of medium and high voltage inverters

This kind of frequency converter adopts the power module inverter side series connection method, which is equivalent to connecting small DC power supplies in series to form a higher voltage. The output voltage of each power module inverter is irrelevant, so each power module can be realized by mature low-voltage frequency conversion technology. The power modules are identical, with good interchangeability and convenient maintenance.

7.10.3.5   Direct Series Connection of Medium and High Voltage Inverters Without Using Input Transformer Power Devices

In fact, the three-level inverter mentioned above is also an inverter composed of power devices in series. This can be seen from the connection method of power devices in each bridge in Fig. 7.35. The output voltage level of the inverter, the higher it is, the more power devices need to be connected in series, as shown in Fig. 7.37. This kind of inverter is the same as the main circuit topology of the voltage-type inverter mentioned above. But after the devices are connected in series, in order to make each device withstand the withstand voltage in a balanced manner, to avoid damage to local devices due to excessive pressure, certain measures need to be taken.

7.10.3.6   High-Low–High Frequency Converter

Use a step-down transformer to change the high and medium voltage into low voltage, use a low-voltage frequency converter to realize variable frequency output, and then pass the three-phase power output of the variable frequency output through a step-up transformer. This inverter adopts a high-voltage-low-voltage-high-voltage structure, as shown in Fig. 7.38. The advantage of this method is that it can use mature low-voltage frequency converter technology to realize the speed control of high- and

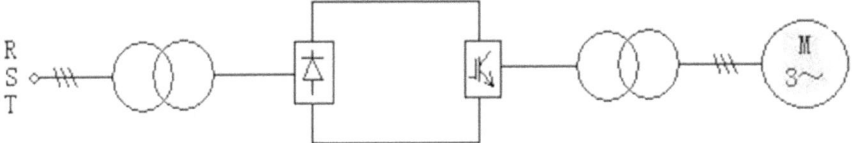

Fig. 7.38 High-low–high frequency converter

medium-voltage motors without technical obstacles. However, since this method requires two-stage voltage transformation, the operating efficiency will drop a little.

Chapter 8
Simple Usage Method of Frequency Converter and Expanding Knowledge

Since AC motors do not have brushes, the vulnerable parts of DC motors, and are easy to maintain, coupled with the rapid decline in the price of frequency converters, AC speed regulation has developed rapidly.

At present, the most widely used frequency converter is the voltage-type general-purpose frequency converter. This chapter takes this type of frequency converter as an example to describe the basic usage methods of the frequency converter. The usage methods of other types of frequency converters are slightly different, but the general situation is basically similar.

8.1 Basic Usage of Inverter

8.1.1 Selection of Inverter

1. First of all, the power of the inverter must match the rated current of the motor. Due to different power factors and different motor efficiencies, the rated current of the motor with the same power and the same voltage level has a large difference, and the inverter is affected by the current of the IGBT device. Due to class restrictions, the output current of the inverter cannot exceed the maximum allowable value, so the power of the inverter should be selected to match the rated current of the motor.
2. The nature of the load must match the type of inverter. Some manufacturers have two types of inverters: general-purpose and pump/fan. General-purpose inverters are suitable for constant torque loads, such as machine tools; pump/fan type frequency converter, only suitable for square torque load. The prices of these two kinds of frequency converters are different, and the price of general pump/fan frequency converter is lower.

© The Author(s) 2024

F. Yao and Y. Yao, *Efficient Energy-Saving Control and Optimization for Multi-Unit Systems*, https://doi.org/10.1007/978-981-97-4492-3_8

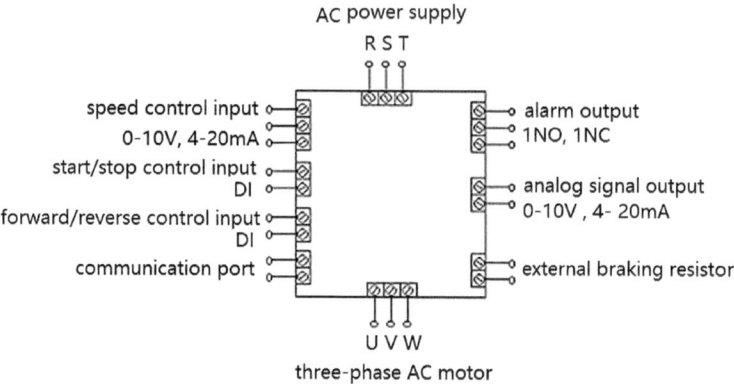

Fig. 8.1 Main wiring of the inverter

8.1.2 Main Power and Control Wiring of Inverter

The main wiring of the inverter is shown in Fig. 8.1, where R, S, T are three-phase main power supply (or single-phase AC power supply), U, V, W are connected to three-phase AC motor, and the speed control input is analog 0–10 V or 4–20 mA signal, the start/stop control input (digital input) controls the start and stop of the motor, the forward/reverse control (digital input) controls the steering of the motor, and the alarm output (digital input) is used to notify the external control device of the frequency converter In the alarm state or running state. When the motor needs to be in the power generation state (such as emergency stop, heavy objects, etc.), it needs to be connected to the braking resistor. The analog signal output is mainly used to output the frequency, current or torque parameters of the current inverter. The communication port is used for data transmission and control with the controller. In most cases, the other wires of the inverter do not need to be used.

8.1.3 Basic Parameter Setting of Frequency Converter

There are displays and buttons on the inverter panel. The display can display the output frequency, output voltage, current, setting parameters, etc. The parameter input method will be different for different manufacturers. For specific methods, please refer to the manufacturer's product manual.

There are many parameters introduced in the instruction manual of the inverter, and some inverters can even have hundreds of parameters that can be set. But in the actual application of the inverter, there are only a dozen parameters that need to be input by the user.

1. Control mode parameters:
   Frequency control mode: (1) panel control; (2) external analog signal (0–10 V or 4–20 mA) control by terminal.
   Start-stop control mode: (1) The panel controls the start/stop of the motor; (2) Use the terminal to control the start/stop of the motor by an external switch signal.
2. Parameters of the driven motor: (1) rated power; (2) rated current; (3) rated voltage; (4) rated speed; (5) number of motor poles; (6) no-load current of the motor; (7) Motor impedance; (8) Motor inductive reactance;
   If there are no these parameters in the motor instruction manual, use the factory default value of the inverter (same power as the motor), or use the motor impedance and inductance online test function provided by inverters to test.
3. Main control parameters: (1) Power supply voltage (such as AC 480 V, AC380 V); (2) Output minimum frequency (such as 0 Hz); (3) Output maximum frequency (such as 60, 50 Hz); (4) Speed up time (Such as 0.1–3600 s); (5) deceleration time (such as 0.1–3600 s); (6) torque boost mode selection; (7) V/F mode selection; V/F, vector mode, direct torque control.
4. Other parameters: Under normal circumstances, the default value can be used. If there are special requirements, please refer to the manufacturer's inverter manual.

8.1.4 Outline of Frequency Converter

The appearance of the inverter is shown in Fig. 8.2.

Fig. 8.2 Appearances of several voltage-type general-purpose inverters

8.2 Basic Usage of ABB Inverter

8.2.1 Purpose

Quickly familiarize yourself with the use of ACS510 series inverters to realize stepless speed regulation of three-phase AC motors.

8.2.2 Essentials to Master

1. Correctly connect the power supply, motor and control wiring.
2. Correctly set the voltage, current, and power of the motor in the frequency converter to be consistent with the voltage, current, and power of the actual driven motor. When the motor power adapted to the inverter does not match the actual motor power, be careful to modify the inverter parameters, otherwise the motor protection function will not work well. Correctly set the basic frequency and maximum frequency. The basic frequency is the frequency when the rated output voltage is output (generally selected as 60 or 50 Hz), and the maximum frequency is determined by the allowable operating frequency of the motor. Some Dedicated variable frequency motors run at a higher frequency (possibly much greater than the base frequency).
3. Correctly set the acceleration time and deceleration time of the motor. If the acceleration time is too small, overcurrent will occur, and if the deceleration time is too small, overvoltage will occur.
4. Correctly set the torque boost of the motor so that the motor has enough starting torque, which is very important for constant torque loads (or non-square torque loads).
5. Select the setting parameters of the panel to control the start and stop or the external terminal to control the start and stop.
6. Select the setting parameter of panel control frequency output value or external terminal control frequency output value.

8.2.3 Inverter Appearance

The appearance of the ACS510 series inverter is shown in Fig. 8.3.

Fig. 8.3 Appearance of
ACS510 inverter

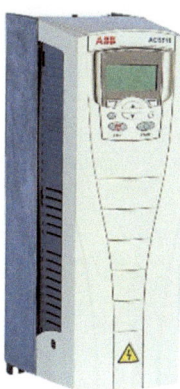

8.2.4 Inverter Model

The inverter model number is arranged as follows.

ACS510-01-09A4-4+...

AC, Standard Drive – 510 product series ─────────

Construction (region specific) ─────────
01 = Wall mounting (Setup and parts specific to IEC installation and compliance)

Output current rating ─────────
e.g. 09A4 = 9.4A, see Ratings chart for details

Voltage rating ─────────
4 = 380...480 VAC

Degree of protection Enclosure and/or other options ─────────
No specification = IP21
B055 = IP54

8.2.5 Inverter Wiring and Floating Networks

1. According to the size of the power, the ACS510 inverter has several sizes of R1–R6, and the wiring method is basically the same; there are some changes in the wiring positions of the power cables of R5 and R6. Taking the inverter with R3 shape as an example, the basic layout of the wiring terminals is shown in Fig. 8.4.

2. Floating networks (also known as IT, ungrounded, or impedance/resistance grounded networks) refers to the power supply system in which the neutral point of the power grid is not grounded or high impedance grounding. For example, the power supply network of a ship or a coal mine is a neutral point ungrounded

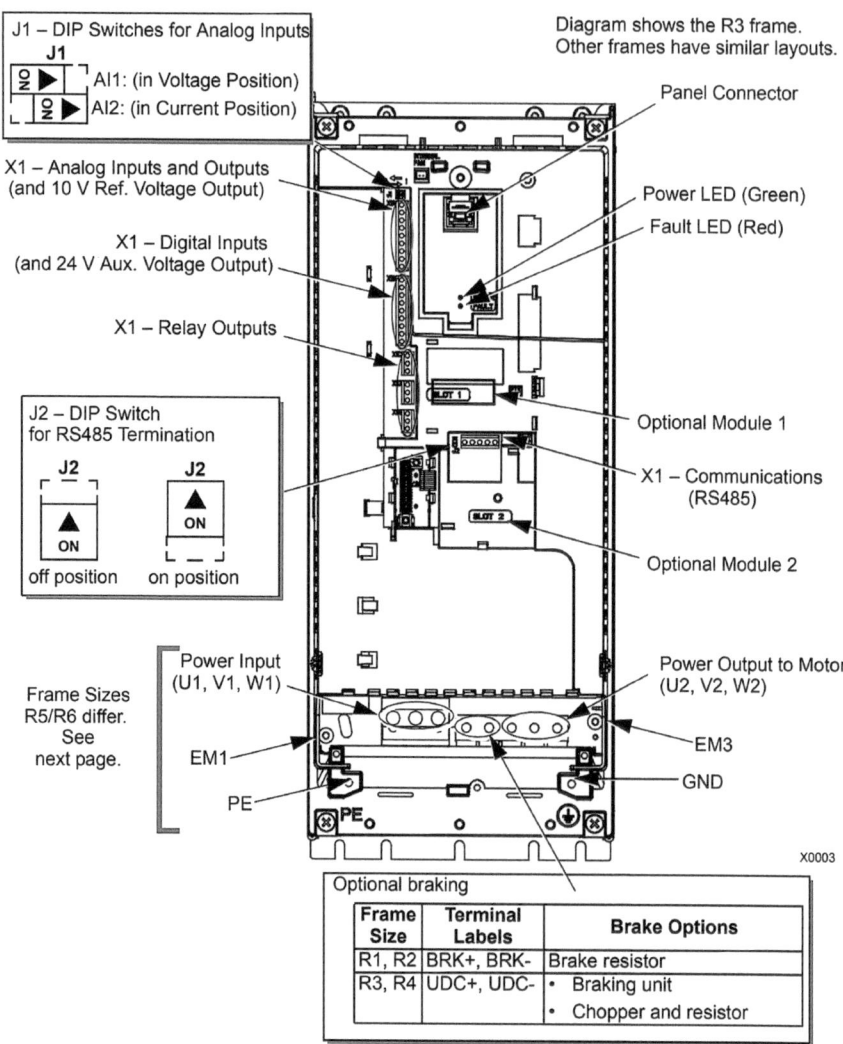

Fig. 8.4 Layout of wiring terminals of frame R3 inverter

system. The advantage of this kind of power supply system is: the impedance between the strong electric ground and the weak point ground is relatively large. It is not easily affected by the performance of the earth, and to a certain extent prevents electromagnetic interference caused by the common ground. The disadvantages are: Affected by the parasitic capacitance, the ground potential of the circuit changes, and it is also easy to cause charge accumulation and electrostatic breakdown, resulting in interference. In order to solve this problem, a high-impedance resistor is used to ground the neutral point to release the charge accumulated due to parasitic capacitance, which is the high-impedance grounded grid.

For the R1–R4 shape inverter, the filter capacitor in the internal RFI filter is grounded through the EM1 and EM3 screws in Fig. 8.4; for the R5–R6 shape inverter, the filter capacitor in the internal RFI filter is grounded through The F1 and F2 screws are grounded.

When the external power grid is a floating network, a high-impedance power grid, or an unsymmetrically grounded system, remove the EM1 and EM3 screws or the F1 and F2 screws.

3. U1, V1 and W1 are connected to the three-phase AC power supply, and U2, V2 and W2 are connected to the three-phase AC motor. The cross-sectional area of the power cable should be selected according to the maximum operating current of the motor. In order to reduce electromagnetic radiation, the motor side cable must be used Shielded cable, armored cable or metal conduit, the shielding of the cable is connected to the protective earth PE (Protective earth). Changing the phase sequence of U, V, W to the motor will change the direction of rotation of the motor.

4. When the motor is often in the state of braking and lowering heavy objects, for the frequency converter with the shape of R1 and R2, BRK+ and BRK− must be connected with a suitable braking resistor. Otherwise the frequency converter may cause an overvoltage alarm. For inverters with the shape of R3–R6, UDC+ and UDC− should be connected with appropriate braking units or braking choppers and braking resistors.

5. The functions of the control terminals are shown in Table 8.1.

Terminals 3, 6 and 9 are internally equipotential.

There are two connection methods of PNP and NPN for digital input signal, as shown in Fig. 8.5.

Cables connected to control terminals: (1) Do not share a single cable for analog signals and numerical signals; (2) Do not share a single cable for DC24V and AC115/ 230 V control lines; (3) It is best to use twisted pairs for relay control signals; (4) It is best to use twisted-pair braided shielded cables for analog signals, as shown in Fig. 8.6, the shield is grounded, and each analog signal uses a separate twisted pair; (5) The requirements for digital signal cables are looser, but it is best Also use braided shielded cable, with the shield grounded.

6. In order to quickly set parameters, ACS510 provides some conventional factory control modes. Different application types are available for users to choose in the form of application macros. Macros are a set of pre-defined parameters. Using application macros, the number of parameters that need to be defined in the process is reduced to the minimum. When different application macros are selected, the definitions of analog I/O, digital input and relay output are also different. The parameter item selected by the application macro is 9902, which has 8 application methods. Standard type, three-wire type, manual/automatic type and PFC control type (pump and fan), etc.

Table 8.1 Control terminal functions

		X1	Hardware Description
Analog I/O	1	SCR	Terminal for signal cable screen. (Connected internally to chassis ground.)
	2	AI1	Analog input channel 1, programmable. Default[2] = frequency reference. Resolution 0.1%, accuracy ±1%.
			J1:AI1 OFF: 0…10 V (R_i = 312 kΩ)
			J1:AI1 ON: 0…20 mA (R_i = 100 Ω)
	3	AGND	Analog input circuit common (connected internally to chassis gnd. through 1 MΩ).
	4	+10 V	Potentiometer reference source: 10 V ±2%, max. 10 mA (1kΩ ≤ R ≤ 10kΩ).
	5	AI2	Analog input channel 2, programmable. Default[2] = not used. Resolution 0.1%, accuracy ±1%.
			J1:AI2 OFF: 0…10 V (R_i = 312 kΩ)
			J1:AI2 ON: 0…20 mA (R_i = 100 Ω)
	6	AGND	Analog input circuit common (connected internally to chassis gnd. through 1 MΩ).
	7	AO1	Analog output, programmable. Default[2] = frequency. 0…20 mA (load < 500 Ω). Accuracy ±3%.
	8	AO2	Analog output, programmable. Default[2] = current. 0…20 mA (load < 500 Ω). Accuracy ±3%.
	9	AGND	Analog output circuit common (connected internally to chassis gnd. through 1 MΩ).
Digital Inputs[1]	10	+24 V	Auxiliary voltage output 24 V DC / 250 mA (reference to GND), short circuit protected.
	11	GND	Auxiliary voltage output common (connected internally as floating).
	12	DCOM	Digital input common. To activate a digital input, there must be ≥ +10 V (or ≤ -10 V) between that input and DCOM. The 24 V may be provided by the ACS510 (X1-10) or by an external 12…24 V source of either polarity.
	13	DI1	Digital input 1, programmable. Default[2] = start/stop.
	14	DI2	Digital input 2, programmable. Default[2] = fwd/rev.
	15	DI3	Digital input 3, programmable. Default[2] = constant speed sel (code).
	16	DI4	Digital input 4, programmable. Default[2] = constant speed sel (code).
	17	DI5	Digital input 5, programmable. Default[2] = ramp pair selection (code).
	18	DI6	Digital input 6, programmable. Default[2] = not used.
Relay Outputs	19	RO1C	Relay output 1, programmable. Default[2] = Ready
	20	RO1A	Maximum: 250 V AC / 30 V DC, 2 A
	21	RO1B	Minimum: 500 mW (12 V, 10 mA)
	22	RO2C	Relay output 2, programmable. Default[2] = Running
	23	RO2A	Maximum: 250 V AC / 30 V DC, 2 A
	24	RO2B	Minimum: 500 mW (12 V, 10 mA)
	25	RO3C	Relay output 3, programmable. Default[2] = Fault (-1)
	26	RO3A	Maximum: 250 V AC / 30 V DC, 2 A
	27	RO3B	Minimum: 500 mW (12 V, 10 mA)

Take the standard macro as an example, set the value of parameter 9902 to 1 (ABB standard), the definitions and connections of its analog I/O, digital input and relay output are shown in Fig. 8.7.

Other application macros have different definitions of analog I/O, digital input and relay output, please refer to the manual.

7. For a simple application, only 14 terminals is ok: U1, V1 and W1 terminals are connected to AC power supply, U2, V2, W2 and PE are connected to three-phase AC motor. 2 and 3 control terminals (0- 10 V) connected to the PLC analog

PNP connection (source)   NPN connection (sink)

X1			X1		
10	+24 V		10	+24 V	
11	GND		11	GND	
12	DCOM		12	DCOM	
13	DI1		13	DI1	
14	DI2		14	DI2	
15	DI3		15	DI3	
16	DI4		16	DI4	
17	DI5		17	DI5	
18	DI6		18	DI6	

Fig. 8.5 Two connection methods of PNP and NPN

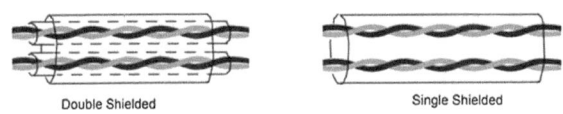

Double Shielded   Single Shielded

Fig. 8.6 Control cable

output module AO1, to control the speed of the motor; the control terminals 11 and 12 are short-circuited, and the terminals 10 and 13 are connected to the PLC digital output (relay) module DO1, to control the start/stop of the inverter.

8.2.6 Parameter Setting

1. Selecting an application macro will set all parameters to the default value of the macro, and there are not many other parameters that need to be set.
2. The parameters inside the inverter can be set through the assistant control panel or the basic control panel through the keyboard panel, and the control panel can also be used to monitor the running status of the inverter. The appearance and functions of the assistant control panel are shown in Fig. 8.8.
3. The parameters and steps to be set are as follows:

   (1) Select language, code 9901 (LANGUAGE):0 English.
   (2) Select different application macros, code 9902 (APPLIC MACRO): 1 = ABB standard, frequency or speed control; 2 = three-wire macro; 3 = alternating macro; 4 = motor potentiometer macro; 5 = manual/ Auto macro; 6 = PID control macro; 7 = PFC control macro; 15 = SPFC control macro.
   (3) According to the data on the motor nameplate: set rated (nominal) voltage 9905 (MOTOR NOM VOLT); rated current 9906 (MOTOR NOM CURR); rated frequency 9907 (MOTOR NOM FREQ); rated speed 9908 (MOTOR

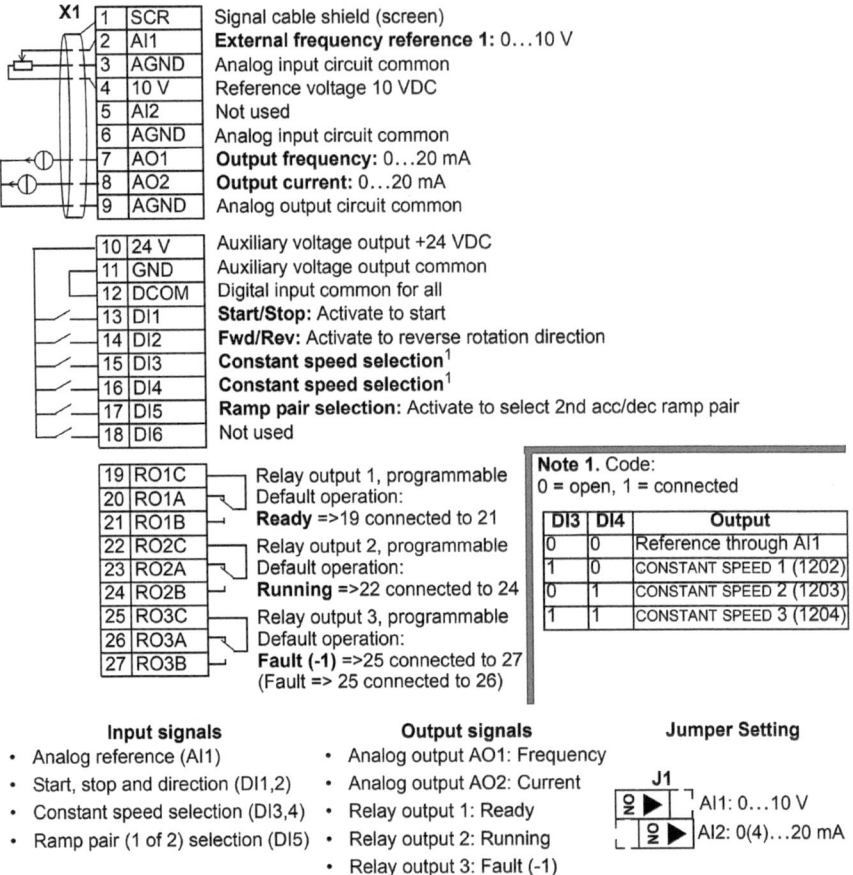

Fig. 8.7 Connection method of ACS510 standard application macro

NOM SPEED); rated power 9909 (MOTOR NOM POWER); power factor 9915 (MOTOR COSPHI): 0.01–0.97, 0 = automatic identification.

(4) Start/stop and direction: start/stop 1001 (EXT1 COMMANDS): 2 = use DI1 and DI2 to control start-stop and direction, 8 = control panel to control start-stop and direction, 10 = communication control; direction control 1003 (DIRECTION): 1 = forward, 2 = reverse.

(5) Frequency given 1103 (REF1 SELECT): 0 = control panel, 1 = AI1 given, 2 = AI2 given, 8 = communication given.

(6) the minimum frequency, code 2007 (MINIMUM FREQ): 0 Hz, the highest frequency, code 2008 (MAXIMUM FREQ): 60 Hz or 50 Hz.

(7) According to the load inertia, set the acceleration time, code 2202 (ACCELER TIME 1): the time required to reach the highest frequency from 0 Hz, and the shortest time is the time when the inverter does not stop due to overcurrent failure. The deceleration time, code 2203 (DECELER

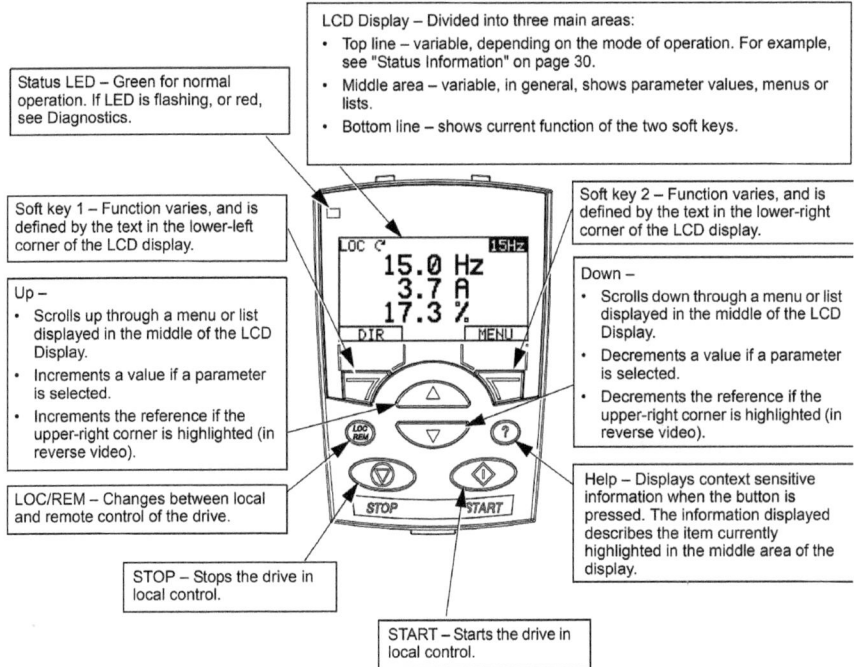

Fig. 8.8 Appearance and functions of the panel

TIME 1): the time required to reach 0 Hz from the highest frequency, the shortest time is the time when the inverter does not stop due to overvoltage fault.

(8) Dangerous frequency control 2501 (CRIT SPEED SEL): 0-off, 1-on; the lower limit 2502 (CRIT SPEED 1 LO) of critical frequency 1, the upper limit 2503 (CRIT SPEED 1 HI) of critical frequency 1.

(9) U/f ratio control, code 2605 (U/F RATIO): 1-linear, 2-square torque;

(10) Operation enable 1601 (RUN ENABLE): 0- no need for external signal permission, 1-DI1 signal control;

(11) parameters save1607 (PARAM SAVE): 0 = DONE, value changes automatically when all parameters are saved; 1 = SAVE, saves altered parameters to permanent memory.

After the above 11 steps are completed, start the inverter again, and it can run according to the selected function.

4. Special tips

(1) If you do not know the setting value of this parameter, please choose the factory default value. If you accidentally mess up the parameters in the

inverter, making the inverter unusable, the easiest way is to restore the parameters to the factory default values, and then set the necessary parameters above.

(2) If the load driven by the motor (such as a fan, long axis pump) has mechanical resonance at frequency f0 during operation, the lower limit of the dangerous speed is 2502, and the upper limit of the dangerous speed is 2503. The output frequency of the inverter skips the frequency f_0 (a certain frequency range) to avoid the resonant frequency of the machine.

8.2.7 Other Notes

1. Low-speed operation of the inverter:
   Ordinary three-phase asynchronous motors are designed according to the rated working conditions. Ordinary three-phase asynchronous motors rely on their own fans to dissipate heat. When ordinary three-phase asynchronous motors driven by frequency converters work at low speeds for a long time, the fans slow, the heat dissipation effect will be poor, and the motor may stop or burn out due to overheating. If you need to work at low speed for a long time, it is best to use a three-phase variable frequency motor. The heat dissipation of the variable frequency motor is solved by an independent power supply fan, so There is no low-speed heat dissipation problem, and the low-speed starting torque of the variable frequency motor is higher than that of ordinary motors.

2. Derating rating:
   The standard operating altitude does not exceed 1000 m. When the altitude of the place of use increases, the output current and power of the inverter need to be derated. When the altitude is 1000–2000 m, it needs to be derated by 1% for every 100 m increase.

   The ambient temperature does not exceed 40 °C, when the temperature of the place of use increases, the output current and power of the inverter need to be derated for use. When the temperature is 40–50 °C, it needs to be derated by 1% for every 1 °C increase.

   The increase of the switching frequency will also reduce the frequency converter. When the switching frequency is 8 kHz, the inverter needs to derate by 20%.

8.3 The Principle of Frequency Converter (Beginners Do not Need to Master)

8.3.1 Main Circuit Structure of General Frequency Converter

The main circuit of the general-purpose frequency converter is composed of three parts: rectification part, DC part and inverter part, the structure is shown in Fig. 8.9, adopting AC–DC–AC structure.

The rectification part consists of 6 diodes to form a three-phase rectification bridge, which converts the three-phase AC power supply RST into direct current. The DC part consists of several large-capacity capacitors and equalizing resistors, the direct current passes through the filter capacitor to keep the DC voltage U_D stable. Since the current large-capacity electrolytic capacitors have low withstand voltage and the consistency of the capacitance value is poor. In order to avoid the difference of the voltage drop of each capacitor is too large, the voltage drop of one capacitor is higher than the withstand voltage value, causing the capacitor to break down. The voltage equalization effect of the resistor can be used to basically ensure that the voltage drop on each capacitor is consistent. In the figure, the capacitance values of capacitors C_1 and C_2 are equal to store electric energy and filter. Resistors R_3 and R_2 have the same resistance value and play the role of voltage equalization. The inverter part consists of 6 IGBT modules and 6 anti-parallel diodes, which convert the DC voltage U_D into a three-phase AC that can change the frequency and effective voltage.

In order to enhance the filtering effect, the capacity of the filter capacitors C_1 and C_2 is generally large. The voltage on the filter capacitor is zero when the inverter is first powered on, which will inevitably cause the charging current of the filter capacitor to be very large. Causing the instantaneous voltage of the power supply grid to generate the steep drop will cause other equipment on the same power grid to trip or malfunction and interfere with the normal operation of the power grid. In order to solve this problem, a current-limiting resistor R1 is added to limit the maximum

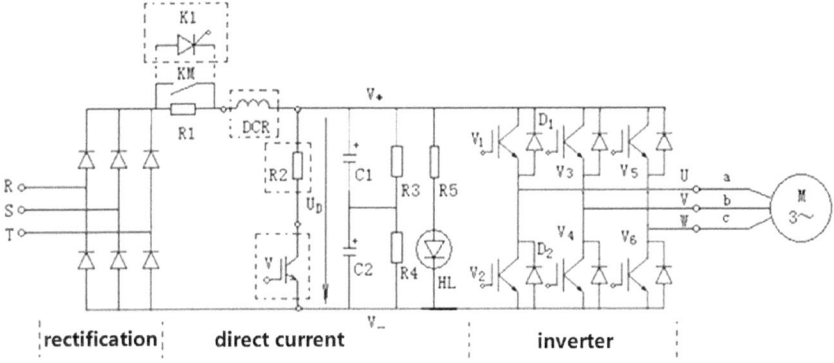

Fig. 8.9 Main circuit structure of a general-purpose inverter

charging current of the filter capacitor. After completion, in order to eliminate the voltage-drop and heat loss of R_1, use the KM contact or thyristor K_1 to bypass R_1 again.

In Fig. 8.9, the function of the DC reactor DCR is to use the restraining effect of the inductance on the current, smooth the input current of the power supply. Improve the power factor of the inverter, and at the same time reduce the charging current of the filter capacitor when the inverter is initially powered on. The DC reactor is a standard accessory in some frequency converters, and some frequency converters (or large-capacity frequency converters) are optional accessories.

In the occasions that need to stop quickly or lower heavy objects, the motor is in the state of generating electricity. The electric energy generated by the motor charges the filter capacitor through the anti-parallel diode, resulting in an increase in the DC voltage inside the inverter. If the energy is not properly processed, it will as a result, the DC voltage U_D exceeds the upper limit. Considering the withstand voltage of the IGBT and filter capacitor, the inverter will generate an overvoltage alarm and stop. If it cannot be stopped, it must be dealt with. There are two ways to deal with it. One is to feed back this part of energy to the grid, and the other is to use the braking resistor to consume the energy. In Fig. 8.9, the second method used in large quantities uses the braking resistor to consume energy. R_2 is the braking resistor, and V is the IGBT that acts as a switch in the braking unit. When the DC voltage U_D exceeds the upper limit, V When it is turned on, the brake resistor R_2 will consume the excess energy of the filter capacitor. The braking resistor and braking unit are built-in standard configurations in small-capacity inverters, and are optional accessories for large-capacity inverters.

In Fig. 8.9, the function of R_5 and HL is to indicate the presence or absence of voltage on the large-capacity filter capacitor. When the inverter is powered off, since the charge on the filter capacitor is not discharged immediately, the residual voltage is enough to cause personal injury. In order to avoid danger caused by people touching the external terminal of the filter capacitor before the discharge of the capacitor is completed, HL is used to indicate the presence or absence of the voltage of the filter capacitor. Only after the HL indicator is off can touch wiring or maintenance be performed.

8.3.2 Sine Wave Pulse Width Modulation (SPWM) Mode and Implementation

In Fig. 8.9, when V1 of the inverter part is turned on and V2 is turned off, the U-phase outputs V+, and when V1 is turned off and V2 is turned on, the U-phase outputs V-. The output voltage of the U-phase has two levels rectangular wave. V and W phase are the same, due to the influence of the inductance in the motor, the current rising speed is lagging behind the voltage, when the U phase output V+, and the U phase current is negative, D1 is turned on, and the current flows back to the DC side. D2 does the same.

The voltage waveform output by the inverter is a series of rectangular pulse waveforms with equal voltage amplitude and unequal width. The frequency and pulse width of the rectangular square wave are controlled by sinusoidal pulse width modulation (SPWM), which is equivalent to a sine wave. The principle of sine wave equivalence is that the area enclosed by the rectangular wave and the sine wave in each time period is equal. The area enclosed is equal to the area enclosed by the rectangular wave in the time period and the horizontal axis of time. Changing the frequency and pulse width of the pulse wave can realize the change of the frequency and amplitude of the equivalent sine wave, which is the variable frequency variable voltage output.

The more pulses that make up a rectangular wave, the closer its equivalent waveform is to a sine wave. The number of pulses is measured by the carrier frequency in the inverter. In Fig. 8.10, the carrier frequency is 5 times the frequency of the positive wave. The early generation method of SPWM is to use a triangular wave as the carrier, and use a sine wave that can change the frequency and amplitude to modulate the signal wave, and the frequency of the triangular carrier can be adjusted. Take the star connection method of the motor as an example, the voltage of the neutral point is OV, use a comparator to compare the signal wave of each phase with the triangular carrier wave. Use the switching signal output by the comparator to control the IGBT output of the corresponding phase. So as to U-phase as an example, when the sine wave of the U-phase signal is higher than the triangular wave, the U-phase IGBT is turned on, and the U-phase outputs U_D. When the U-phase signal sine wave is lower than the triangular wave, the U-phase IGBT is turned off and outputs 0 V. Assume that the triangular carrier wave is U_t, the U-phase signal wave is U_u, the U-phase output is U_{U0}, the V-phase signal wave is U_v, the V-phase output is U_{V0}. The line voltage of U-phase and V-phase is U_{UV}, the frequency of the carrier U_t is the signal wave U_u and the signal wave 3 times the U_v frequency, as shown in Fig. 8.11a).

For the U-phase signal wave, when the amplitude of the signal wave U_{U0} is higher than the triangular carrier U_t, the U-phase outputs a DC voltage U_D. When the amplitude of the signal wave U_{U0} is lower than the triangular carrier Ut, the U-phase outputs 0 V, as shown in Fig. 8.11b) shown.

For the V-phase signal wave, when the signal wave U_{V0} amplitude is higher than the triangular carrier U_t, the V-phase outputs a DC voltage U_D. When the signal wave

Fig. 8.10 Sinusoidal pulse width modulation (SPWM) mode

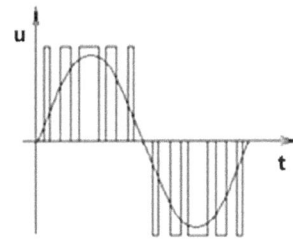

Fig. 8.11 Relationship
between signal wave and
carrier wave

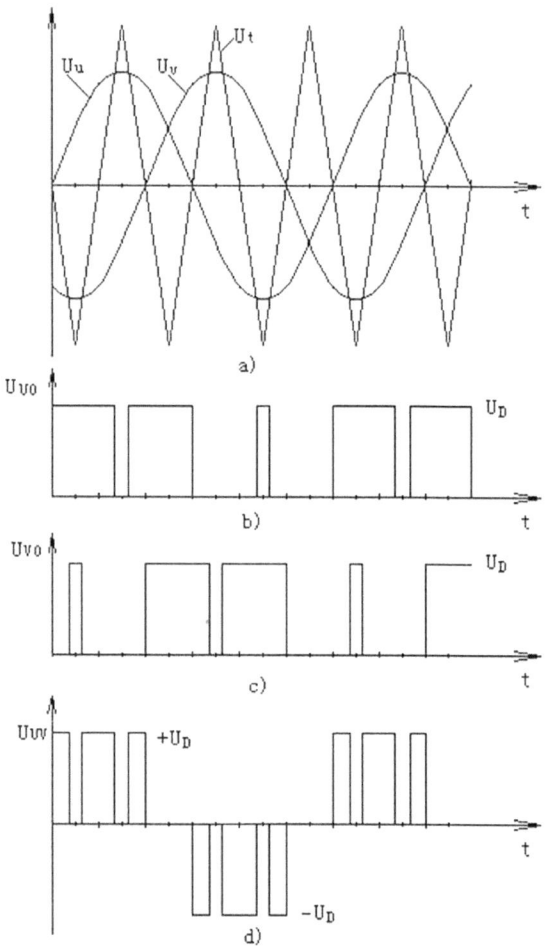

U_{V0} amplitude is lower than the triangular carrier U_t, the V-phase outputs 0 V, as shown in Fig. 8.11c) shown.

The line-to-line voltage $U_{UV} = U_{U0} - U_{V0}$ of U-phase and V-phase, the waveform of U_{UV} is shown in Fig. 8.11d), U_{UV} is the equivalent sine wave composed of rectangular waves. Because the carrier frequency is three times the frequency of the signal wave, so the wave head of each U_{UV} is composed of three rectangular waves with varying widths.

Changing the frequency of the signal wave changes the frequency of the output equivalent sine wave. Changing the amplitude of the sine wave changes the width of the rectangular wave in the equivalent sine wave, which also changes the effective voltage value, which will not be explained here. With the rapid development of digital

Fig. 8.12 Output voltage
and current waveforms of
inverter at high carrier
frequency

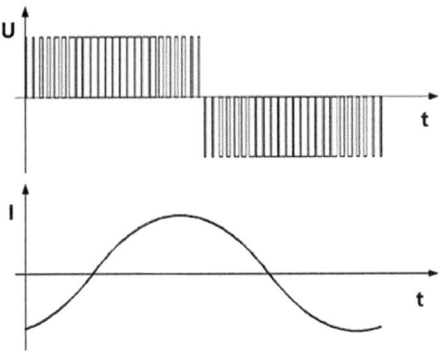

technology, the current SPWM method no longer needs these complicated transformations, and can be easily realized directly with a dedicated chip or a calculation method.

Although the output voltage waveform of the frequency converter is a series of rectangular pulse waves, due to the suppression of the current change by the inductive load of the motor, the current waveform in the motor is a fluctuating and phase-lag approximate sine wave. The higher the carrier frequency, The smaller the current fluctuation, the smoother the current waveform. When the carrier frequency is high enough (such as 12 kHz), the current waveform output from the inverter to the motor is basically a smooth sine wave, as shown in Fig. 8.12. The current waveform lags behind the voltage, etc. When the carrier frequency is high, the low-frequency torque output is also stable, and the noise of the motor is small. It should be noted that the high carrier frequency will increase the loss of the inverter itself, increase the temperature of the inverter, and cause voltage glitches (du/dt) When the carrier frequency is low, the noise of the motor is large, the loss of the motor is large, the torque is reduced, and the temperature of the motor is high.

8.3.3   V/F Control of Frequency Converter

When the output frequency f of the inverter decreases, the inductance XL of the motor also becomes smaller. If the output voltage U remains high, the current flowing into the stator winding of the motor will increase greatly, and the stator winding will be burnt out. In order to avoid this problem, occur, it is necessary to reduce the output voltage U while reducing the frequency f. Conversely, when the output frequency f of the inverter increases, the output voltage U of the inverter will increase accordingly, so that the inverter will output the maximum frequency, has a maximum output voltage, which can ensure that the motor can output rated power. In order to meet this requirement, it is necessary to keep U/f changing according to a certain law.

In order to keep the motor output constant torque at different operating frequencies (generally below the basic frequency), it is necessary to keep the air gap flux between the stator and the rotor constant. If the flux is too low, the output of the motor is insufficient, and the flux is too high, saturation occurs, and the winding will be burned out due to excessive excitation current. Therefore, keeping the air gap flux constant is one of the best operating modes for the inverter.

The effective value E of the electromotive force of each phase of the three-phase AC motor stator, such as Eq. (8.1).

$$E = K \times f \times N \times \varphi_m \tag{8.1}$$

where E is the effective value of the electromotive force induced by the air-gap flux in each phase of the stator winding, K is the coefficient. f is the frequency of the stator (Hz), and N is the turns of each phase of the stator winding in series Number, φ_m is the air gap magnetic flux of each pole.

Transform Eq. (8.1) to get φ_m as Eq. (8.2).

$$\varphi_m = \frac{E}{K \times f \times N} = \left(\frac{1}{K \times N}\right) \times \frac{E}{f} \tag{8.2}$$

It can be seen from Eq. (8.2) that for a known motor, K and N are fixed values, in order to keep φ_m constant, Eq. (8.3) needs to be maintained.

$$\frac{E}{f} = (K \times N) \times \varphi_m = constant \tag{8.3}$$

When the stator frequency f is high, the voltage drop on the stator winding impedance can be ignored, and it is approximately considered that the stator electromotive force E of each phase is equal to the stator phase voltage U of each phase. Then the Eq. (8.3) becomes:

$$\frac{E}{f} \approx \frac{U}{f} = constant \tag{8.4}$$

This is the constant control mode of the voltage-frequency ratio (U/f) of the frequency converter.

When the stator frequency f is low, the voltage drop on the stator winding impedance can no longer be ignored. In order to actually keep φ_m constant, the stator phase voltage U of each phase must be increased to compensate for the stator voltage drop and increase the starting current, as shown in the Fig. 8.13.

In Fig. 8.13, curve 1 is the U/f curve with low-frequency compensation when the motor is driving a constant torque load. Curve 2 is the U/f curve without low-frequency compensation when the motor is driving a constant torque load. Curve 3 is the U/f curve with low frequency compensation when the motor drives the water

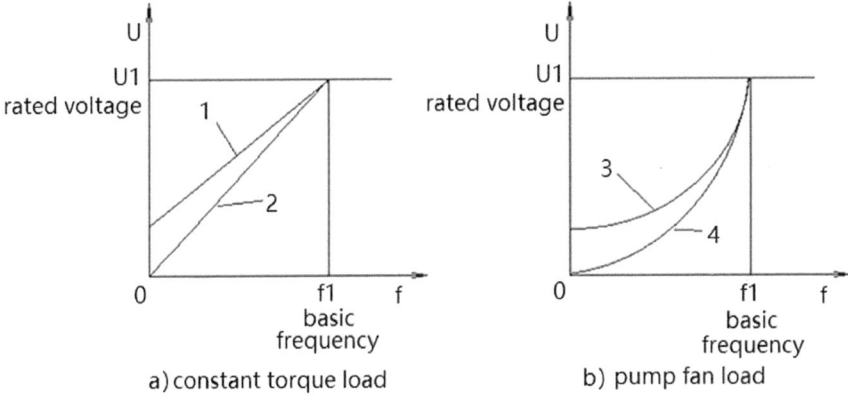

Fig. 8.13 Voltage-frequency ratio (U/f) curve

pump fan load, and curve 4 is the U/f curve without low frequency compensation when the motor drives the water pump fan load.

It can also be seen from Fig. 8.13 that since the motor cannot work at a state higher than the rated voltage, when the frequency converter is higher than the basic frequency f1, f increases, and the output voltage U remains unchanged at the rated voltage U1, so U/f curve becomes a horizontal straight line. According to the Eq. (7.9), this will cause the magnetic flux φ_m to decrease with the increase of the f, and the torque will also decrease. The motor becomes a constant power speed regulation, and a constant torque speed regulation (U/f is equal to a fixed value) as an example, draw the change curve of magnetic flux φ_m as shown in Fig. 8.14.

In Fig. 8.14, curve 2 is the U/f curve without low-frequency compensation when the motor drives a constant torque load, and curve 1 is the flux change curve corresponding to curve 2. When f is lower than the basic frequency f1, it is a constant flux φ_{me}, it corresponds to the constant torque speed regulation mode; when f is higher

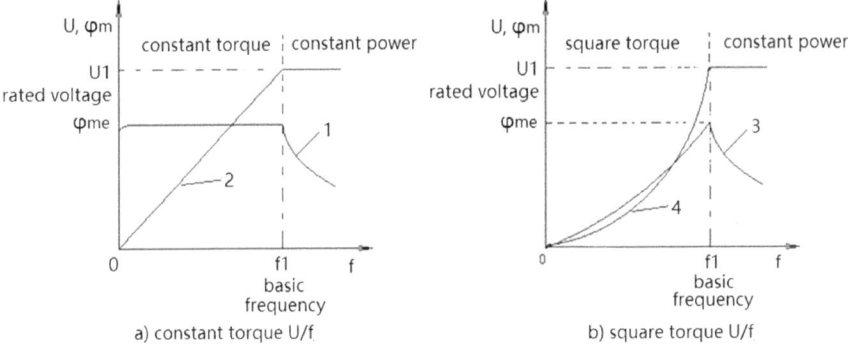

Fig. 8.14 Three types of U/f curves

than the basic frequency f1, the magnetic flux φ_m is the reciprocal change function of f, corresponding to constant power speed regulation; curve 4 is when the motor drags the square torque load, there is no The U/f curve of low-frequency compensation, curve 3 is the magnetic flux change curve corresponding to curve 4. When f is lower than the basic frequency f1, it corresponds to the square torque speed regulation mode. When f is higher than the basic frequency f1, the magnetic flux φ_m is f the reciprocal change function, corresponding to constant power speed regulation.

8.3.4 Vector Control of Inverter

The motor drives the load to move. If the no-load torque of the motor is ignored, according to Newton's second law, its motion equation is as Eq. (8.5).

$$T_M - T_L = J \frac{d\omega}{dt} \tag{8.5}$$

where T_M is the electromagnetic torque of the motor (N · m), T_L is the resistance torque (N · m) of the load on the motor side. J is the total moment of inertia of the motor and the load on the motor side (kg · m^2), ω is the rotational angular velocity of the motor (rad/s).

$$\omega = \frac{2\pi n}{60} \tag{8.6}$$

where n is the motor speed (r/min).

$$J = mr^2 = \frac{G}{g} \times \left(\frac{D}{2}\right)^2 \tag{8.7}$$

where m is the mass of the rotating body (kg), G is the weight of the rotating body (N), D is the diameter of the rotating body (m), and r is the radius of the rotating body (m).

Equation (8.5) becomes Eq. (8.8).

$$T_M - T_L = \frac{GD^2}{375} \frac{dn}{dt} \tag{8.8}$$

It can be seen from the Eq. (8.8) that the acceleration, deceleration and constant speed operation of the load driven by the motor are directly controlled by the electromagnetic torque of the motor. The change of the electromagnetic torque of the motor directly controls the running speed of the load.

$T_M > T_L$, the acceleration is greater than zero, and the load will run at an accelerated speed;

$T_M < T_L$, the acceleration is less than zero, and the load will decelerate; $T_M = T_L$, the acceleration is equal to zero, and the load will run at a constant speed.

For a DC motor, if the positive and negative poles of the excitation power supply U_f and the armature power supply U are determined, the magnetic field direction of the main pole magnetic field (the excitation magnetic field) and the direction of the electromagnetic torque of the motor are also determined. The flow direction of the excitation current I_f and I_f form the direction of the magnetic field is shown in the figure. Due to the effect of the commutator and brushes, the direction of the armature winding current i_a under the main pole magnetic field remains unchanged. As shown in Fig. 8.15, the direction of the armature current i_a is under the N pole (The upper half) flows inward, represented by ⊙.The direction of the armature current i_a flows outward under the S pole (lower part), represented by ⊕. The direction of the armature current is perpendicular to the direction of the main pole magnetic field φ, and the direction of the electromagnetic force F on the conductor can be determined by the left-hand rule. The palm is facing the N pole (the magnetic force line is from N to S), the four fingers point to the direction of the current, and the thumb is the direction of the electromagnetic force (the force of the conductor), so that the torque direction of T_M is determined, as shown in the figure.

The relationship between the electromagnetic torque T_M of the DC motor and the magnetic flux φ of the main pole magnetic field and the armature current i_a is shown in Eq. (8.9).

$$T_M = C_M \times \varphi \times i_a \qquad (8.9)$$

Fig. 8.15 Field current and armature winding current of a DC motor

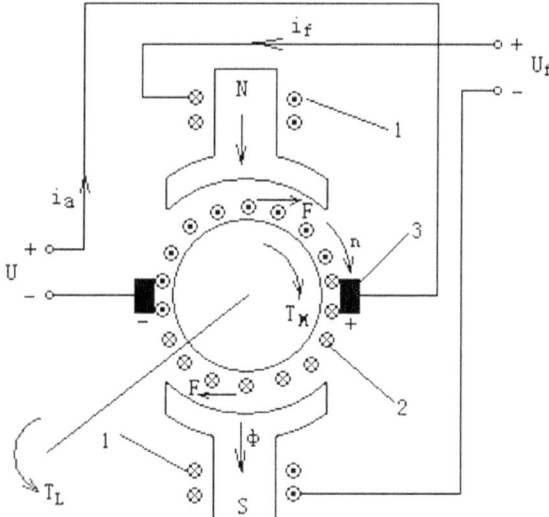

where C_M is the torque constant, the excitation current if generates the main pole magnetic field φ, and the armature magnetic field generated by the armature current i_a is perpendicular to the main pole magnetic field. Since the excitation current if and the armature current i_a can be controlled separately, the torque control of the DC motor is relatively simple. The fixed excitation current if can control the armature current i_a to control the motor torque, and the fixed armature current i_a can control the excitation current if to control the motor torque.

In the AC motor, since the stator current contains both the excitation current and the armature current, a certain conversion is required to make the torque control of the AC motor the same as the torque control of the DC motor.

Taking a 2-pole AC motor as an example, the positions of the three-phase windings of the stator are fixed, and the spatial positions differ by 120°. The three-phase AC current with a phase difference of 120° is connected, and the rotational angular velocity of the three-phase AC is ω, thus a NS with an angular velocity of ω is produced. For the rotating magnetic field (magnetomotive force) of the poles, please refer to the previous chapters for the analysis process. The two-phase stator windings with fixed spatial positions have a difference of 90° in space, and the two-phase alternating current with a difference of 90° is connected. The rotational angular velocity of the alternating current is ω, and the same It can also generate a rotating magnetic field (magnetomotive force) of an NS pole with an angular velocity of ω. As long as the voltage amplitude is selected properly, this rotating magnetic field is exactly the same as that of a three-phase AC motor. For the analysis process, please refer to the principle of the single-phase AC motor. The spatial position of the excitation winding and the armature winding of the DC motor differ by 90°, see the principle of the DC motor. The direction of the magnetic field (magnetomotive force) of the DC motor is fixed, but if we let the coordinate axis of the DC motor rotate at the same angular velocity ω of the AC motor, and keeps the magnitude of the magnetomotive force of the rotating DC motor equal to the magnitude of the magnetomotive force of the above-mentioned two-phase AC motor, then the rotational magnetomotive force of the DC motor under the rotating coordinates is exactly the same as the magnetomotive force of the two-phase AC motor. The rotating magnetic field between the three-phase AC motor, the two-phase AC motor and the rotating DC motor is completely equivalent. According to the principle of the DC motor, we control the DC The excitation current of the motor excitation winding can control the strength of the magnetic field and keep the strength of the magnetic field constant. We can control the torque of the DC motor by controlling the armature current of the armature winding. By reversely pushing back through the coordinate transformation, we can have the current and voltage values of the three-phase AC motor that achieve the same torque control effect are obtained, which is the basic idea of vector control. For the simplicity of control, the magnetic flux of the rotor excitation is guaranteed to be constant, that is, to keep E_2/f constant, and E_2 is the induced electromotive force of the rotor.

Vector control is a torque control method that imitates a DC motor. It first decomposes the speed signal of a three-phase AC motor into the interaction between the

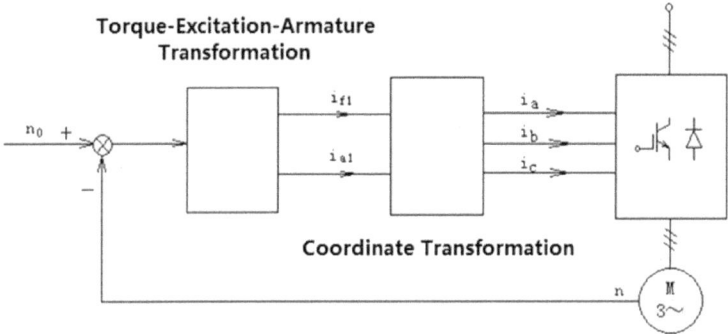

Fig. 8.16 Schematic diagram of vector control

excitation magnetic field and the armature magnetic field, calculates the DC excitation current i_{f1} and the armature current i_{a1}. Then after a series of coordinate transformations, it is converted into control signals for controlling the three-phase currents i_a, i_b and i_c to control the inverter part of the frequency converter, as shown in Fig. 8.16. After the given speed signal n_0 changes, the excitation current i_{f1} is maintained No change, only the armature current i_{a1} is adjusted, so as to adjust the speed of the three-phase AC motor like a DC motor. The low-speed performance of the vector control mode is much better than that of the U/f mode, and it can provide sufficient torque even at 0 Hz.

Vector control is the same as U/f control: vector control keeps the relationship E_2/f between rotor induced electromotive force and frequency constant, U/f control keeps the relationship between stator supply voltage and frequency U/f constant. Both of them can avoid under-excitation or over-excitation. The difference between vector control and U/f control is that vector control also adjusts the phase angle of the three-phase output current at the same time, which ensures that at low speed, only the armature current is increased to make the torque larger, while keeping the excitation current constant, avoiding overexcitation, so the vector control of the three-phase AC motor is better than the U/f control speed regulation performance.

8.3.5   Direct Torque Control of Frequency Converter

The torque of a three-phase AC motor is equal to the stator flux linkage F1 multiplied by the rotor flux linkage F2 and then multiplied by the sine of the angle between the two vectors. The flux linkage is in a derivative relationship with the voltage vector, and the two are perpendicular to each other. Direct torque control, the idea is to control the stator flux linkage in a basically constant range by controlling the voltage vector, so as to realize the direct control of the torque.

Let's talk about the concept of six voltage vectors in the inverter part of the inverter. The inverter part of the inverter is composed of six transistor bridges, which supply power to the three windings a, b, and c of the three-phase AC motor, as shown in Fig. 8.9. If 3-bit binary numbers are used to represent the three bridge arms, 1 means that the upper transistor is turned on, 0 means that the lower transistor is turned on, such as 100 means that V1, V4 and V6 are turned on, and 011 means that V2, V3 and V5 are turned on There are 6 conduction modes of 010, 101, 001, and 100, plus 0 V short-circuit power supply mode 000 and 111, and there are 8 combinations in total. 000 and 111 are only applied when the torque is greater than the load torque. Considering that the three windings a, b, c differ by 120° in spatial position, for a 2-pole AC motor with neutral star connection, 100 conduction mode, the voltage vector of the three-phase winding is shown in Fig. 8.17.

The amplitude of U_a is equal to $U_D/2$, the amplitude of U_b and U_c is also equal to $U_D/2$, the combined total voltage vector is U_1, the amplitude of U1 is equal to U_D, and the corresponding 101 synthesized voltage vector is U_2, U_2 is counterclockwise than U1 Rotate 60°, the voltage vector corresponding to 001 synthesis is U_3, U_3 rotates 60° counterclockwise than U_2, the voltage vector corresponding to 011 synthesis is U_4, U_4 rotates 60° counterclockwise than U_3, and the voltage vector corresponding to 010 synthesis is U_5, U_5 Rotate 60°counterclockwise than U_4, and the resulting voltage vector corresponding to 110 is U_6, and U_6 is rotated 60°counterclockwise than U_5. The six voltage vectors U_1, U_2, U_3, U_4, U_5, and U_6 are shown in Fig. 8.18.

Direct torque control continuously measures the torque and flux linkage, applies different voltage vectors according to the direction and size of the current flux linkage, controls the amplitude and rotation direction of the stator flux linkage. The amplitude

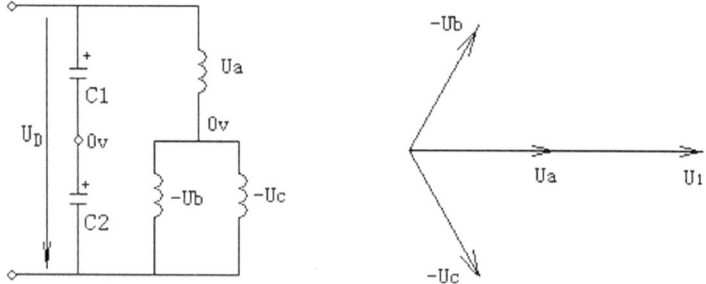

Fig. 8.17 Synthetic vector U1 of three-phase winding voltage in 100 conduction mode

Fig. 8.18 Six voltage vectors corresponding to six conduction modes

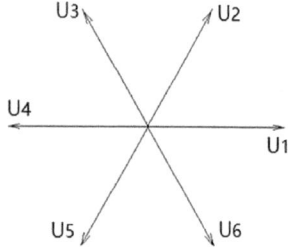

of the stator flux linkage changes within a small range and is basically constant, thereby achieving direct control of the torque. The advantage of this method is that the response speed is fast, and the disadvantage is that the switching frequency is not fixed, resulting in relatively large noise.

8.4 Expansion of Inverter Application (Beginners Do Not Need to Master)

8.4.1 Harmonics of Frequency Converter and Countermeasures

From Fig. 8.10, we can see that since the sine wave is approximately equivalent to the rectangular wave, the voltage waveform output by this type of inverter contains a large number of high-order harmonic components. When the motor is far away from the inverter, the distributed capacitance between the line and the ground becomes larger. The capacitive reactance is inversely proportional to frequency, capacitive reactance $X_c = 1/(2\pi fC)$. The higher the carrier frequency, the smaller the capacitive reactance is, and the easier it is for high-order harmonic currents to flow into the earth. The line loss will become larger. The leakage current will cause the three-phase current imbalance, trip the leakage switch, and affect the nearby video signal through the ground. At the same time, the voltage glitch will also It will damage the insulation of the motor, so some measures need to be taken to solve these problems. Harmonic hazards can be dealt with by the following methods:

1. Reduce the carrier frequency and reduce the high-order harmonic current radiation. The disadvantage of this method is that the operating noise of the motor will become larger;
2. Add an output reactor between the output side of the frequency converter and the motor, as shown in Fig. 8.19. The output reactor is made of three wires wound several times in the same direction on an iron core, so that the synthesis of the fundamental wave current magnetic field is zero, which has a strong inhibitory effect on high-frequency harmonics, and can also extend the driving distance between the inverter and the motor, from tens of meters (such as 50 m) to hundreds of meters (such as 400 m). At the same time weaken Voltage spikes (du/dt) are also beneficial for motor insulation.
3. Add output filter NF1 on the output side of the inverter, as shown in Fig. 8.20, the inductance L × 3 of the output filter is formed by winding 3-phase wires on the high-frequency magnetic core for 3–4 turns, and attention should be paid, the output filter is directional. The inductance side must be connected to the inverter, and the current limiting resistor R × 3 and capacitor C × 3 are connected to the motor. They cannot be connected in reverse, otherwise, there will be a current impact on the IGBT power module of the inverter.

Fig. 8.19 Adding an output reactor to the output side of the inverter

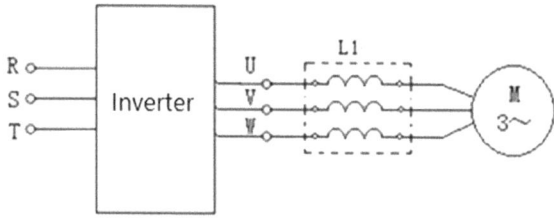

Fig. 8.20 Adding an output filter to the output side of the inverter

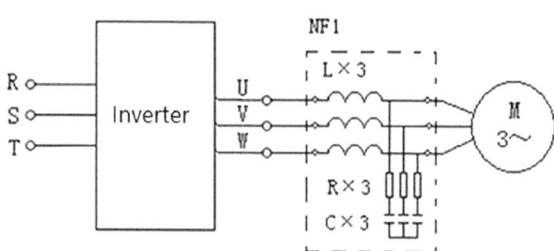

In order to reduce the interference of the frequency converter to the power supply grid and reduce the power interference to other frequency converters in the same network, the following methods can be adopted.

4. Add an input reactor on the input side of the inverter, as shown in Fig. 8.21. The input reactor is made of three wires wound several times on the same iron core in the same direction. The synthetic magnetic field of the fundamental wave current is zero. It can suppress high-frequency harmonics, and because the input reactor can suppress the current, the input reactor can also improve the power factor of the inverter.

5. Add an input filter on the input side of the inverter, as shown in Fig. 8.22. The inductance L × 3 of the output filter is formed by winding 3-phase wires on the high-frequency magnetic core for 3–4 turns. It has a suppressive effect on high-frequency harmonics.

Fig. 8.21 Adding an input reactor to the input side of the inverter

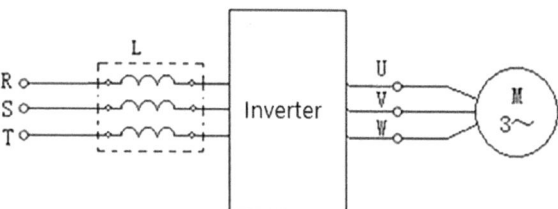

Fig. 8.22 Adding an input filter to the input side of the inverter

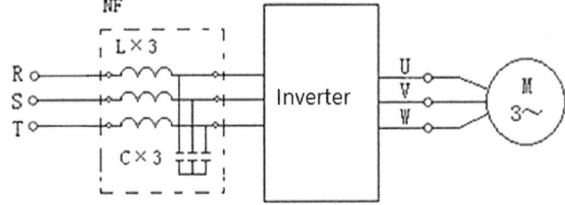

8.4.2 Estimation of Input and Output Reactors

In order not to significantly reduce the output capacity of the inverter, the voltage drop U_L on the input and output reactors of the inverter should meet the requirements of $(2-5)\%$ not greater than the rated voltage U_e, that is:

$$U_L = (2 - 5\%)\ U_e = 2\pi f L I_e \tag{8.10}$$

where L is the inductance (H) of the AC reactor, I_e is the rated current (A) of the inverter, and f is the power frequency (Hz).

8.4.3 Heat Dissipation and Reactive Power Compensation of the Frequency Converter

1. The heat dissipation problem of the inverter

The ventilation conditions of the frequency conversion cabinet where the frequency converter is placed must meet the ventilation volume and ambient temperature requirements in the manual. However, the calculation of the ventilation volume of the frequency converter is generally complicated. In order to simplify this problem, the author can use the following simple method based on years of experience. Method to deal with: In the case that there is no fan above the frequency conversion cabinet, the ventilation area of the ventilation holes (including side ventilation holes) above the frequency conversion cabinet should be designed to be larger than the area of the inverter radiator, as shown in the Eq. (8.11).

$$S1 + S2 + S3 > \alpha \times S \tag{8.11}$$

where the reserved coefficient α takes 1.2–1.5, S represents the total area of the air outlet of the inverter radiator, S1 represents the total area of the air outlet on the top cover of the inverter cabinet, S2 represents the total area of the upper cooling vents, S3 represents the total area of the side cooling vents, as shown in Fig. 8.23. If the total area of the cooling holes is not enough, you can also increase the height of the upper air outlet, or increase the fan to force suction.

Fig. 8.23 Layout of ventilation holes in the inverter cabinet

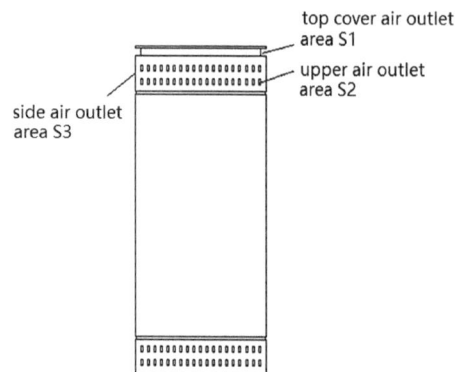

top cover air outlet area S1

upper air outlet area S2

side air outlet area S3

Due to the vague and general heat dissipation requirements of the frequency converter, the phenomenon that the frequency converter cannot work normally due to improper heat dissipation design has occurred in many projects, which should arouse sufficient attention.

2. The problem of reactive power compensation of the frequency converter

The power factor of the frequency converter is generally relatively high and no reactive power compensation is required.

Since the motor and the power supply are isolated through the frequency converter, no reactive power compensation is required on the motor side. On the power input side of the inverter, the power factor problem caused by high-order harmonics is difficult to be effective with capacitor compensation, and high-frequency fluctuations will also affect the life of the capacitor.

8.4.4   Calculation and Estimation of Braking Resistor

1. Calculation of braking resistor:

It can be seen from the Eq. (8.5) that when the motor driving the load needs to reduce the speed from the initial speed n_0 to n_1 within t_1 (s), if the resistance torque T_L of the load is not enough to meet the requirements, it needs the motor provides braking torque T_M', such as Eq. (8.12).

$$T_M' - T_L = \frac{GD^2}{375} \frac{n_1 - n_0}{t_1} \qquad (8.12)$$

Transform Eq. (8.12) to get the braking torque T_M' that the motor needs to provide, such as Eq. (8.13).

$$T'_M = T_L + \frac{CD^2}{375} \frac{n_1 - n_0}{t_1} \tag{8.13}$$

The braking torque that can be generated by the internal loss of the motor itself is about 20% of the rated torque T_e of the motor. If this part of the braking torque is still not enough, the motor needs to provide an additional resistance torque T_R. It is provided through the braking resistor R on the inverter, as shown in Eq. (8.14).

$$T_R = T'_M - 0.2T_e \tag{8.14}$$

The power consumed by the braking resistor is P (W), as shown in Eq. (8.15).

$$P = \frac{U_z^2}{R} \tag{8.15}$$

In the Eq. (4.15), U_Z is the voltage (V) on the DC bus of the frequency converter when the braking resistor is working. The power represented by the upper resistance torque T_R of the motor and the initial speed n_0 should be equal to the power P consumed by the braking resistor, such as Eq. (8.16).

$$\frac{T_R \times n_0}{9.55} = \frac{U_z^2}{R} \tag{8.16}$$

solve R, such as Eq. (8.17).

$$R(\Omega) = \frac{9.55 \times U_z^2}{T_R \times n_0} \tag{8.17}$$

Considering that there is a braking situation at the maximum speed, the initial speed n_0 in the Eq. (8.17) can be calculated by the rated speed ne. The calculation of the above Eq. (8.17) needs to know the moment of inertia converted to the motor side, sometimes this is difficult.

2. Estimation of braking resistor:

Generally, when the current flowing through the braking resistor is 50% of the rated current, the resistance torque T_{R0} is equal to the rated torque T_e, and the current Eq. (8.18) and torque Eq. (8.19) are as follows:

$$I_{R0} = 50\% I_e \tag{8.18}$$

$$T_{R0} = T \tag{8.19}$$

The braking resistor value R_0 corresponding to this point is shown in Eq. (8.20):

$$R_0 = \frac{U_Z}{I_R} = \frac{2U_Z}{I_e} \tag{8.20}$$

Generally, the resistance torque T_R is (0.8–2) times of the rated torque T_e, and there is Eq. (8.21):

$$T_R = (0.8 - 2)\, T_e \tag{8.21}$$

The torque is proportional to the current, and the current is inversely proportional to the resistance. Therefore, the value range of the braking resistor R is as follows (8.22):

$$R = \frac{R_0}{0.8 - 2} \tag{8.22}$$

The calculation of the maximum power of the braking resistor is shown in Eq. (8.15). In actual selection, the actual power of the resistor should be determined by multiplying a proportional coefficient (0.11–0.4) according to the frequency and intensity of the motor in the power generation state.

Chapter 9
Controllers Used in Energy-Saving Control Systems—PLC

PLC is short for Programmable Logic Controller. At present, PLC has two external forms: integrated (compact) and modular. The integrated type is to combine the PLC power supply, CPU processor, memory, and a certain number of I/O together to form a whole, as shown in Fig. 9.1a. This type of PLC has low cost, fixed I/O addresses, and is easy to use, the disadvantage is that the program stock is small and the functions are few. Modular PLC is composed of modules with different functions. The types of modules include: power supply module, CPU module, digital input module, digital output module, analog input module, analog output module, communication module, positioning module, counting module, etc. According to the engineering needs, a certain number of modules are combined on a base plate (or rack) to form a flexible assembled PLC, as shown in Fig. 9.1b. This kind of PLC is more expensive than the integrated PLC, but the functions and scale of control are also more powerful. This chapter will take the modular PLC as an example to explain the basic usage of PLC.

9.1 Simple Way to Get Started with Modular PLC

1. First determine the signal module (SM) you are using, and find out the address arrangement method of the digital input (DI) module, digital output (DO) module, analog input (AI) module and analog output (AO) module.

Take the S7-300 PLC with four input and output modules as an example to illustrate the address assignment of the modules, as shown in Fig. 9.2.

Where PS is a power supply module, CPU is a central processing unit (control module), SM are input and output modules.

The address of the first SM module (the left-most) closest to the left CPU;
The address of the analog input (AI) module is PIW256, PIW258, PIW260–PIW270;

© The Author(s) 2024   177
F. Yao and Y. Yao, *Efficient Energy-Saving Control and Optimization for Multi-Unit Systems*, https://doi.org/10.1007/978-981-97-4492-3_9

a)   b)

Fig. 9.1. The appearance of PLC

Fig. 9.2 Address allocation
of input and output modules

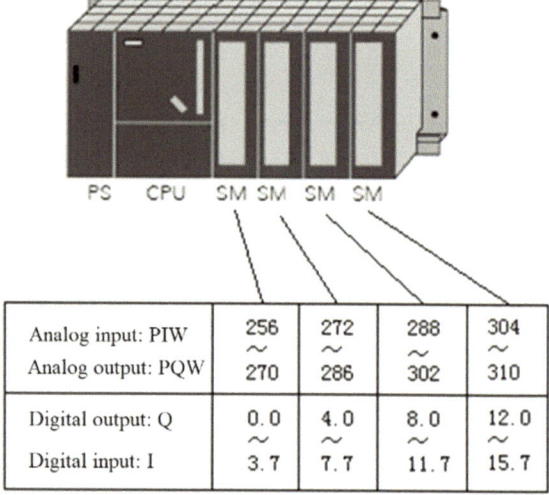

	256 ~ 270	272 ~ 286	288 ~ 302	304 ~ 310
Analog input: PIW Analog output: PQW	256 ~ 270	272 ~ 286	288 ~ 302	304 ~ 310
Digital output: Q Digital input: I	0.0 ~ 3.7	4.0 ~ 7.7	8.0 ~ 11.7	12.0 ~ 15.7

The address of the analog output (AO) module is PQW256, PQW258, PQW260–PQW270;

The address of the digital input (DI) module is I0.0, I0.1, I0.2–I3.7 (32 in total), or IB0, IB1, IB2, IB3, or IW0, IW2, or ID0.

The address of the digital output (DO) module is Q0.0, Q0.1, Q0.2–Q3.7 (32 in total), or QB0, QB1, QB2, QB3, or QW0, QW2, or QD0.

The addresses of SM modules at other locations can be deduced by analogy.

If the 16-way DI module is inserted into the first SM location, the address I0.0 represents the first digital input, IB0 represents upper 8 bits, IB1 represents lower 8 bits, IW0 represents all 16 bits.

If the 32-way DI module is inserted into the first SM location, the address I0.0 represents the first digital input, IW0 represents upper 16 bits, IW2 represents lower 16 bits, ID0 represents all 32 bits.

2. Confirm the address programming method of PLC internal memory and data block;

Take Siemens S7-300 PLC commonly used memory M and data block DB as an example.

(1) When used in bits such as M0.0, M0.1, M127.7, DB1.DBX0.0, DB1.DBX 0.1, DB10.DBX 240.0, etc., DB1 represents data block 1.

(2) When used by byte, such as MB0, MB1, MB64, DB1.DBB0, DB1.DBB1, DB15.DBB7, etc.

(3) When used by word, such as MW0, MW2, MW64, DB1.DBW0, DB1.DBW2, DB3.DBW 64, etc.

(4) When using double words, such as MD0, MD4, MD64, DB1.DBD0, DB1.DBD4, DB3.DBD 88, etc.

(5) MD0 is composed of MW0 and MW2, MW0 is composed of MB0 and MB1, MB0 is composed of M0.0, M0.1, M0.2, M0.3, M0.4, M0.5, M0.6 and M0.7.

(6) DB1.DBD0 is composed of DB1.DBW0 and DB1.DBW2, DB1.DBW0 is composed of DB1.DBB0 and DB1.DBB1, and DB1.DBB0 is composed of DB1.DBX0.0–DB1.DBX0.7.

(7) By analogy, it should be noted that there should be no repetition during use. For example, if M0.0–M0.7 is used, MB0, MB1, MW0 and MD0 should not be used for other purposes.

3. Find and modify the address of the PLC programming port and communication port: plug in the programming cable (RJ45 network port or USB port) between the PLC and the programming PC, open the PLC programming software, find the PLC. Set all the default communication ports on the PLC to the address you arranged. If multiple PLCs are connected through fieldbus or industrial Ethernet, arrange that the communication addresses of each PLC should not overlap.

4. According to the module models and positions in the actual PLC, insert modules of the same model modules in the corresponding position in the programming software, so that the address of the module in the software is determined as mentioned above.

5. Now you can start programming.

6. It should be noted that:

(1) There is a corresponding relationship between the value collected by the analog input module in the PLC and the value of the actual process parameter. For example, the pressure is 0–1 MPa, the corresponding pressure sensor outputs 0–10, and 0–10 V is input to the AI module of PLC. The corresponding value in PLC is 0–27,648, such as the PIW256 analog input value of S7-300. Some PLCs may be corresponding to 0–32,768. Beginners must pay attention to this.

(2) Similarly, there is also a corresponding relationship between the output value of AO module and the actual controlled parameters. For example, frequency converter's 0–60 Hz, the value corresponding to the AO module is 0–27,648, such as PQW256. Some PLCs may correspond to 0–32,768.

(3) When performing double-word floating-point operations inside the PLC, the PLC value must be converted into the corresponding process value before performing the normal calculation of energy-saving control. For example, if the pressure is 0.5 MPa, the value of the AI module's PIW256 is 13824. It converts to 0.5 and then performs floating point operations. For analog output, it should also be converted into the corresponding output value of PLC. For example, if the inverter outputs 30 Hz, then the AO module of PLC outputs $27,648/2 = 13,824$.

9.2   Getting Started with PLC Programming—Ladder Diagram

The commonly used programming language for PLC is ladder diagram, and the expression method of ladder diagram is similar to that of electrical diagram. In order to let beginners quickly master it, we will use ladder diagram programming method to describe the PLC programming method below. Since the ladder diagram programming methods of most PLCs follow the provisions of the international electrotechnical standards, the programming forms are basically the same.

9.2.1   "AND"

Only if the two conditions "A" and "B" are met, C will have output. The program in the PLC is shown in Fig. 9.3, and the logic relationship of "AND" is shown in Table 9.1. A and B are both normally open contacts.

Fig. 9.3 "AND"

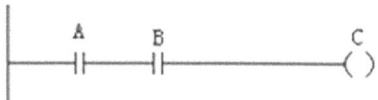

Table 9.1 "AND" logical table

A	B	C
0	0	0
0	1	0
1	0	0
1	1	1

Fig. 9.4 "Or"

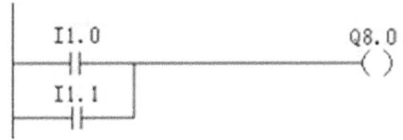

Table 9.2 "Or" truth table

I1.0	I1.1	Q8.0
0	0	0
0	1	1
1	0	1
1	1	1

9.2.2 "OR"

One condition of I1.0 "or" I1.1 is realised, then Q8.0 will output 1 (or closed). The ladder diagram of PLC is shown in Fig. 9.4, and the truth table of "or" is shown in Table 9.2. I1.0 and I1.1 are both normally open contacts. Logic tables and truth tables mean the same thing.

9.2.3 "Output"

() means output, output logic 1 (or high) if the conditions are satisfied, and output logic 0 (or low) if the conditions are not met, the ladder diagram in PLC is shown in Fig. 9.5. The logic relationship of the "output" is shown in Table 9.3.

Fig. 9.5 "Output"

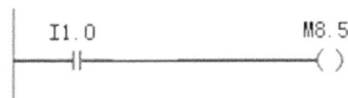

Table 9.3 "Output" logical table

I1.0	M8.5
0	0
1	1

Fig. 9.6 "Set"

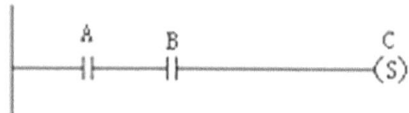

Table 9.4 "Set" logical table

A	B	C
0	0	Stay the same
0	1	Stay the same
1	0	Stay the same
1	1	1

9.2.4 "Set"

Set bit (S) indicates that it will be set to 1 when the conditions occur; when the conditions do not occur, the original state will be maintained without any action. For example, when A "and" B occur, C is set to 1. The ladder diagram of PLC is shown in Fig. 9.6, and the logic relationship of "set" is shown in Table 9.4.

The difference between the set (S) output and the immediate output () is that the immediate output () outputs low (0) when the previous conditions are not met; for the set output (S), when the previous conditions are not met, the output does not change state.

9.2.5 "Reset"

The reset output (R) resets to low when the previous conditions are met, and does not change state when the conditions are not met. For example, when I128.0 and Q12.0 meet the conditions, Q12.7 is reset to the low position, the PLC ladder diagram is shown in Fig. 9.7, and the "reset" logic table is shown in Table 9.5.

Fig. 9.7 "Reset"

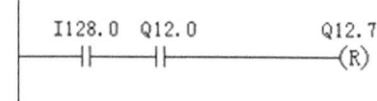

Table 9.5 Logical table of "reset"

I128.0	Q12.0	Q12.7
0	0	Stay the same
0	1	Stay the same
1	0	Stay the same
1	1	0

Fig. 9.8 Data assignment

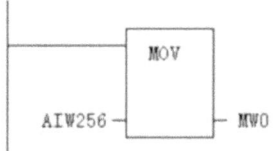

Fig. 9.9 Data plus

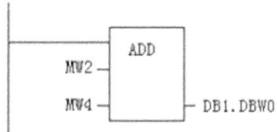

9.2.6 Data Transfer "MOV"

If you want to put the analog quantity AIW256 into the memory MW0, the PLC ladder diagram is shown in Fig. 9.8.

9.2.7 "ADD"

If you want to add the values in memories MW2 and MW4 and put the result into DBW0 of data block DB1, the ladder diagram of the PLC is as shown in Fig. 9.9. If you want to use floating point numbers for addition, you can use MD2 plus MD6. But you cannot add MD4 with MD2, because MD2 occupies 4 words, including MW4.

9.2.8 "SUB"

If MW6 is subtracted from DB1.DBW2 and the result is stored in MW8, the PLC ladder diagram is as shown in Fig. 9.10. If you want to do floating point subtraction, use MD6 to subtract DB1.DBD2.

Fig. 9.10 Data subtraction

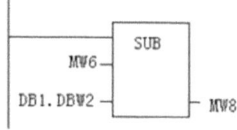

Fig. 9.11 Data
multiplication

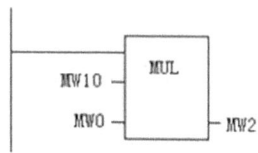

Fig. 9.12 Data division

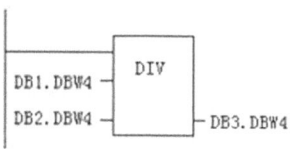

9.2.9 "MUL"

If the data MW10 is multiplied by MW0 and the result is stored in MW2, the PLC ladder diagram is shown in Fig. 9.11. If you want to perform multiplication with floating point numbers, multiply MD0 by MD10.

9.2.10 "DIV"

If the data DB1.DBW4 is divided by DB2.DBW4, and the result is put into DB3.DBW4, the ladder diagram of PLC is shown in Fig. 9.12. If you want to divide by floating-point numbers, divide DB2.DBD4 by DB1.DBD4.

9.2.11 Counter C (Counter)

If I0.0 has a rising edge pulse, the counter C2 adds 1, the counter is cleared when I0.1 is high, MW0 stores the current counter value, and the counter C2 is an up counter. Then the PLC ladder diagram is shown in Fig. 9.13.

If I0.0 has a rising edge, the counter C2 will be decremented by 1. When I0.1 is high, the counter will be cleared. When I0.2 is high, the number in MW2 will be put

Fig. 9.13 Up counter

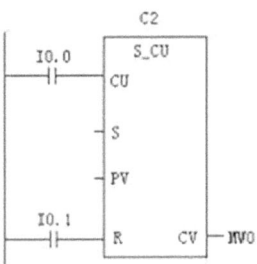

Fig. 9.14 Down counter

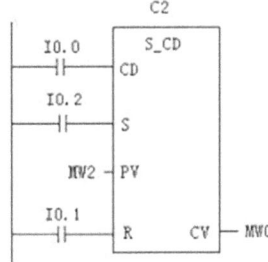

Fig. 9.15 Timer

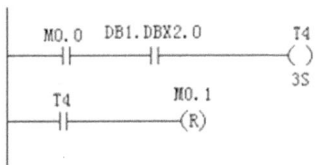

into the counter. MW0 will store the current counter value, and counter C2 will be a down counter. The PLC ladder diagram is shown in Fig. 9.14.

9.2.12   Timer T (Timer)

If both M0.0 and DB1.DBX2.0 are high, the timer T4 (3 s) starts, and when T4 (3 s) expires, M0.1 is reset. The ladder diagram of the PLC is shown in Fig. 9.15.

9.2.13   Greater Than or Equal to (≥)

If MW6 is greater than or equal to MW256, set M0.7, and the ladder diagram of PLC is shown in Fig. 9.16.

Fig. 9.16 Greater than or equal to

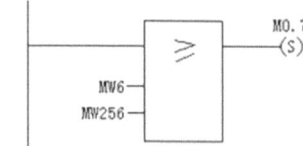

Fig. 9.17 Is equal to

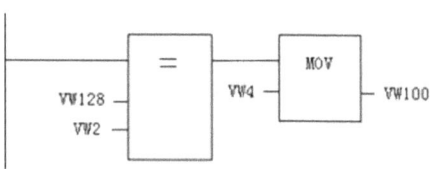

Fig. 9.18 Less than

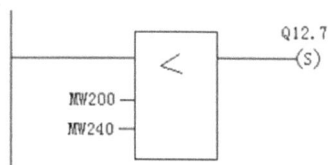

9.2.14 Equal to (=)

If the number in the internal data area VW128 is equal to VW2, put VW4 into VW100, and the PLC ladder diagram is shown in Fig. 9.17.

9.2.15 Less Than (<)

If the number in MW200 is less than the number in MW240, set Q12.7, and the ladder diagram of PLC is shown in Fig. 9.18.

9.2.16 Greater Than (>)

If MW128 is greater than MW0, subtract 1 from MW128 and put it back into MW128. The ladder diagram of PLC is shown in Fig. 9.19.

Fig. 9.19 Greater than

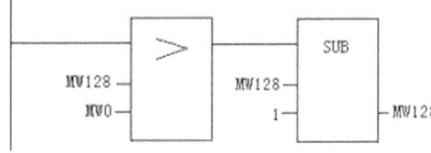

Fig. 9.20 Less than or equal to

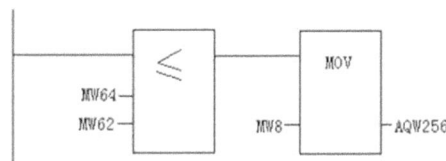

Fig. 9.21 Rising edge action

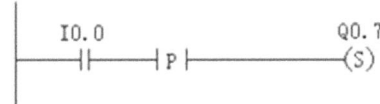

9.2.17 Less Than or Equal to (≤)

If MW64 is less than or equal to MW62, then MW8 is output to the analog quantity AQW256, and the ladder diagram of PLC is shown in Fig. 9.20.

9.2.18 Rising Edge Action (P)

If I0.0 changes from low to high (there is a rising edge), set Q0.7 to high, and the ladder diagram of PLC is shown in Fig. 9.21.

9.2.19 Falling Edge Action (N)

If M0.0 and M1.1 change from satisfying conditions to not satisfying conditions (falling edge action), then subtract 1 from MW8 and return to MW8. The ladder diagram of PLC is shown in Fig. 9.22.

Fig. 9.22 Falling edge
action

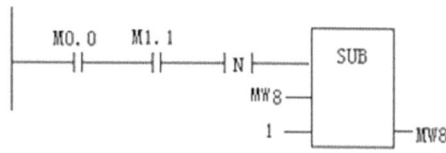

Fig. 9.23 Per second pulse
program

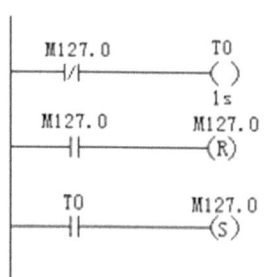

9.2.20 Per Second Pulse Program

In some PLC instructions, there is a special per second pulse bit, but there are also
no per second pulses in some PLCs. The following is a way to make M127.0 go high
once per second, and only execute it once. The length of time can be changed by
changing the timer time. The PLC ladder diagram is shown in Fig. 9.23.

The working sequence of the PLC program is executed sequentially from top to
bottom. The program is executed to the end, it returns to the top program. In the
PLC working cycle from top to bottom, M127.0 becomes low again, and the T0
timer starts counting again; after 1 s, T0 becomes high, M127.0 becomes high, and
the program executes downward until the end, and M127.0 is always high state, the
program returns to the top and then executes from top to bottom. Since M127.0 is
high, T0 stops timing. Then, repeat the above process.

9.2.21 PID Closed-Loop Control

PID closed-loop control can be performed in PLC. The ladder diagram of PLC
is shown in Fig. 9.24. When M0.0 in the figure is closed, PID control starts. The
number of PIDs varies from PLC to PLC. When applying PID in PLC, define the
input address, output address, and set value storage address, and then define the data
block address corresponding to the P, I, and D parameters. So that the operator on
the man–machine interface or the upper computer can modify the actual situation.
Then set the positive and negative effects of PID control (such as heating and cooling
control), sampling period, maximum output, minimum output and other parameters
(different PLCs will be different).

Fig. 9.24 PID closed-loop control

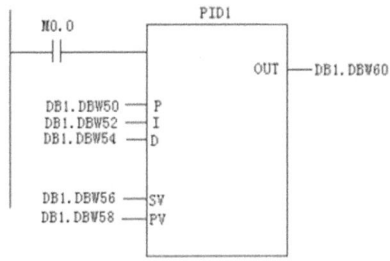

PID can be programmed by yourself and use the scheduled interrupt program to calculate the PID output value. The method is as follows:

$$u_i = u_{i-1} + \Delta u$$
$$\Delta u = P(e_i - e_{i-1}) + Ie_i + D(e_i - 2e_{i-1} + e_2) \tag{9.1}$$

Among them, u_i is the current PID output value, u_{i-1} is the last calculated PID output value, and Δu is the current PID correction value. e_i is the error this time, it is equal to the set value SV of the controlled parameter minus the actual value PV, $e_i = $ SV − PV, e_{i-1} is the error last time, e_{i-2} is the error before e_{i-1}. P is the proportional parameter, I is the integral parameter, and D is the differential parameter. The three parameters P, I, and D need to be adjusted to obtain the best control effect.

9.3 PLC Programming Software

9.3.1 Module Configuration

Take the module configuration in Fig. 9.2 as an example.

In Siemens STEP7 programming software, insert 5A power supply module "PS307 5A", CPU314 module "CPU314", 16-way digital input module "DI16 × DC24V", 16-way relay output module "DO16 × Relay output", 8-way analog input module "AI8 × 12-bit", 4-channel analog output module "AO4 × 12-bit". As shown in Fig. 9.25. The third slot is reserved for the interface module IM, which is used to connect expansion racks to increase the number of SM modules in a larger control system.

After selecting the hardware board, remember the I/O address of each board shown in Fig. 9.25.

The address of the 16-way digital input module SM321 is I0.0–I3.7, IB0–IB1, IW0.

The address of the 16-way digital output module SM322 is Q4.0–Q5.7, QB4–QB5, QW4.

The address of the 8-way analog input module SM331 is PIW288–PIW302.

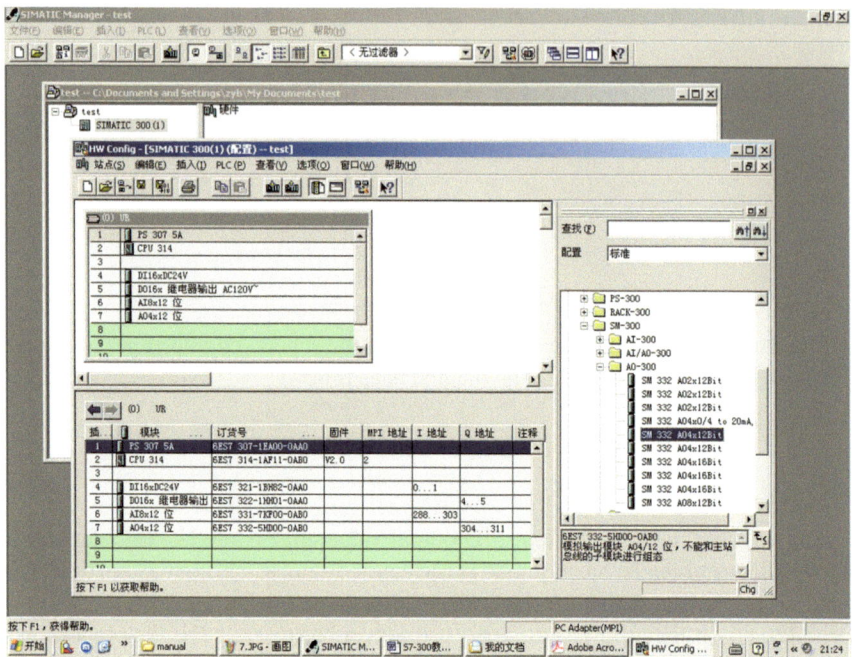

Fig. 9.25   Installing the power supply, CPU, and I/O cards

The address of the 4-way analog output module SM332 is PQW304–PQW310.

The address of default MPI communication port is 2, as shown in Fig. 9.25.

The MPI port is a communication bus defined by Siemens itself. If there are more than one S7-300 stations in the same network, the address of the MPI port of "CPU314" needs to be modified.

After completing this step, the subsequent software program will know how many modules in this PLC system has, what modules they are, and where they are.

9.3.2   Software Programming

Open the instruction tree of STEP software on the left, drag the normally open contact ⊣⊢, normally closed contact ⊣/⊢ and set ‑(s) in the "bit logic" into segment 1, and input the corresponding addresses I0.0, I0.1 and Q4.0. As shown in Fig. 9.26. I0.0 is a normally open contact. I0.1 is a normally closed contact.

The meaning of this program is: if the digital input bit I0.0 is closed, and the digital input bit I0.1 has no action, then the digital output bit Q4.0 is set to close.

STEP7 complies with the international standard IEC 61,131-3.

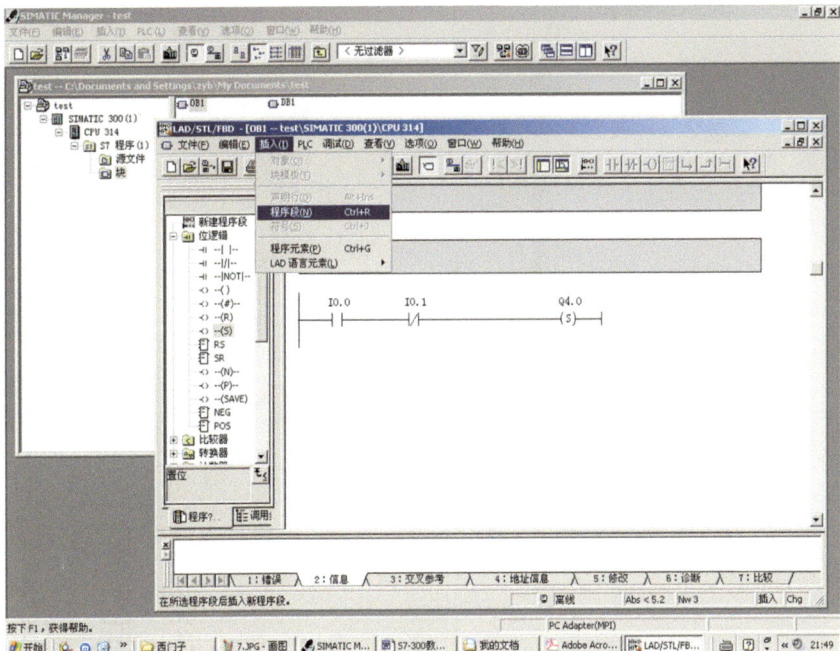

Fig. 9.26 Program segment 1

Chapter 10
Human–Machine Interface and Configuration Software

10.1 Basic Usage of the HMI

10.1.1 Main Purpose of the Human–Machine Interface

Human–machine interface (HMI, MMI) is generally used to communicate with PLC (frequency converter, PID) and other controllers. Human–machine interface is generally used to display and record the data collected or calculated by PLC and other controllers, and control the data that needs to be controlled. The set value or the switch signal of the equipment is sent to the PLC and other controllers. The man–machine interface with touch function can directly switch the buttons on the screen and input data on the LCD screen. If the human–machine interface uses membrane buttons, you need to press the buttons on the display to enter data. The screens of most man–machine interfaces are shown in Fig. 10.1.

10.1.2 Wiring of the Man–Machine Interface

The general external wiring of the HMI is shown in Fig. 10.2.

Most touch screen HMI are powered by DC24V and have two communication ports, one of which is a programming port (RJ45) for connecting to a PC with programming software installed, and the other (RS485) is a communication port for connecting to control devices (such as PLC).

© The Author(s) 2024
F. Yao and Y. Yao, *Efficient Energy-Saving Control and Optimization for Multi-Unit Systems*, https://doi.org/10.1007/978-981-97-4492-3_10

Fig. 10.1 The screen of the
man–machine interface

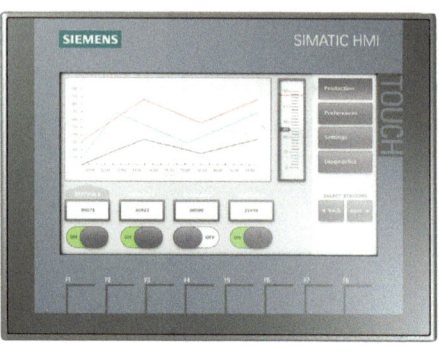

Fig. 10.2 General wiring of
the man–machine interface

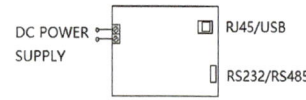

10.1.3   Communication Connection of HMI

In the PC for programming the HMI, open the HMI programming software, first select
the model of the HMI, then the PLC model. Select communication port, communication protocol, etc. This is the most important, the first step, beginners must solve
this problem.

10.1.4   Display Data

Select the display component, click the component, and select the data block, memory
or input and output register in the PLC corresponding to the display. Generally, there
is no direct decimal function in the human–machine interface, so it is generally calculated in the PLC through addition, subtraction, multiplication and division without
decimal points. Just define the position of the decimal point on the HMI. For example,
if you want to display a pressure value, the analog input of the pressure is PIW288.
Assuming that 1 MPa corresponds to the 27,648 in the PLC, then for any pressure, to
display X.X X MPa on the HMI, define 2 decimal places, do the following calculation
on the PLC.

$$(PIW288 \times 100)/27648 = DB1.DBW0 \qquad (10.1)$$

In order to prevent the overflow of numerical calculation, the above calculation
should be performed in double-byte calculation or floating-point. Multiplication is
performed first to reduce numerical calculation errors.

Fig. 10.3 Procedure for
pressure judgment

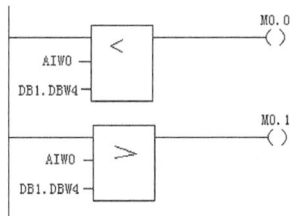

For example, the value of PIW288 is 13824, then DB1.BW0 is equal to 50, define two decimal points, 0.50 MPa is displayed on the HMI, and the value displayed on the HMI corresponds to the actual pressure value.

10.1.5 Set Data

Add a data input component, click this component, and select the data block or memory to be set. Setting parameters in the HMI is opposite to the displaying data. Assuming that the pressure in the production process is to be controlled by constant pressure, set the pressure to 0.40 MPa and store it in DB1.DBW2. That is, DB1.DBW2 = 40, and the corresponding value in the PLC is placed in DB1.DBW4, then do the following calculation in the PLC.

$$(DB1.DBW2 \times 27648)/100 = DB1.DBW4 \tag{10.2}$$

Then DB1.DBW4 is the value corresponding to 0.40 MPa in the PLC, 11,059.

The program for pressure judgment in the PLC can be written as shown in Fig. 10.3.

In Fig. 10.3, M0.0 high indicates that the actual pressure is lower than the set pressure, and M0.1 high indicates that the actual pressure is greater than the set pressure. According to the state of M0.0 and M0.1, control the analog output value (such as PQW304), to increase or decrease the speed of the corresponding inverter.

10.1.6 On/off Display

For example, we want to display I0.0 of the PLC.

Select a lamp component, click on the component, select the IW0 of the PLC, select I0.0 bit, then this display component is connected to I0.0 bit.

When I0.0 closes, the lamp component displays on. When I0.0 opens, the lamp component is off.

10.1.7   On/off Control

For example, we want to control Q4.0 of the PLC, on or off.

Select the button component, click the component, select QW4, select Q4.0, then this component is connected to Q4.0 bit. Define the button to be pressed to close and raised to release.

When the button component is pressed, Q4.0 will close (on), when the button component releases, Q4.0 will open (off).

We can also define the button to be closed when pressed and held when raised.

10.1.8   Curve Display

Add a graph component, click the component, and select the analog input, the data block or memory to be displayed. In HMI programming software of the PC, directly select the corresponding curve display component and correspond to the corresponding data block (such as DB6.DBW16), memory (such as MW8) or analog input (such as PIW288).

10.1.9   Display of Bar Graph

Add a bar graph component, click this component, and select the analog input, the data block or memory to be displayed. In the HMI programming software of the PC, directly select the corresponding bar graph display component and correspond to the corresponding data block (such as DB6.DBW16), memory (such as MW8) or analog input (such as PIW288).

10.1.10   Appearance of HMI

Human–machine interface is used more and more in the field of automation, and it exists in a large number of fields such as machine tools, automated production lines, and process control. The appearance of the human–machine interface is generally shown in Fig. 10.4.

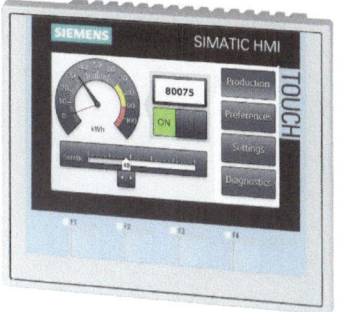 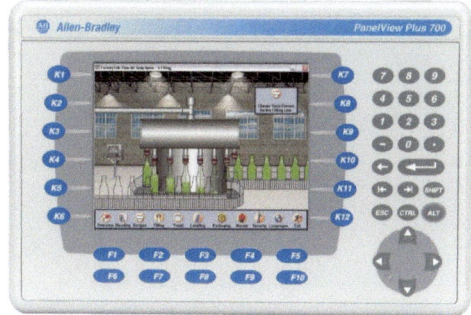

Fig. 10.4 The appearance of the HMI

10.2 Configuration Software

10.2.1 Purpose of Configuration Software

In order to make the control process intuitive and maintain a large amount of historical data, people often need to program specific software on the computer with a programming language, so that the software can communicate with the on-site equipment (such as PLC, sensors, PID, etc.). In this way, the computer can be used to display and control the production process and save the production data, but such software needs to compile different software for different projects, the workload is extremely high and time-consuming. The configuration software is produced to solve this requirement. It no longer requires programmers to understand the computer programming language. The configuration personnel only need to combine various ready-made components to complete very complex data display, control and data processing.

With the popularization of computers and the decline in prices, in order to improve the intuitiveness of the control process and the rapidity of programming. The use of configuration software on the computer to directly display, store, control and share data on the industrial control process on the network has become very popular and very simple. There are many types of general configuration software available at home and abroad. Since the configuration software is installed on the computer, its functions are much more powerful than those in the HMI in the previous section. Migration, correction of calculated values, etc. have all become very convenient.

Figure 10.5 is the application screen of the configuration software in a production process.

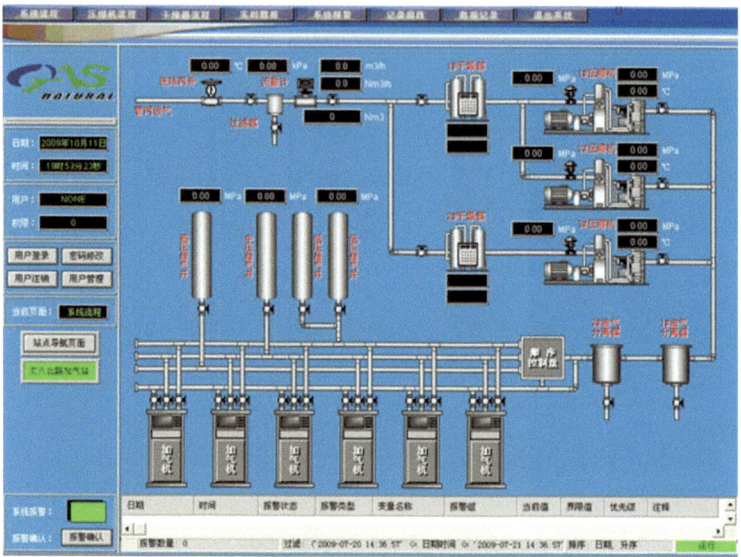

Fig. 10.5 Application screen of configuration software

10.2.2   General Usage of Configuration Software

1. First install the development version and running version (or common version for development and operation) of the configuration software on the industrial computer. If there is no driver corresponding to the monitored device in the software, you need to install the driver.
2. Open the configuration software and create a new project
3. Select the type of board (or communication port, communication format) used for communication between the computer (PC) and the following and the type of PLC (or other controller) connected below. This step is very important. Beginners must attach great importance to this link.
4. Define the memory, data block, input/output (I/O) storage area, etc. that need to be displayed, controlled, and recorded on the PLC as data labels.
5. Add a component that displays data, click on the component, and correspond to the above data label.
6. Add a data input component for setting parameters, click the component, and correspond to the data block and memory in the PLC.
7. Add a switch value display component, click on the component, and correspond to the above data label, and define which digit to display.
8. Add a control button component, click on the component, and correspond to the above data label, and define which bit is to be set or reset.
9. Add a bar graph component, click on the component, and correspond to the above data label.

10. Add a trend graph component, click on the component, and correspond to the above data label, and define the frequency of updating and recording the data and the total length of the data, etc.

11. Run the configuration software, then the PC will automatically establish the relationship between the data display, data input, switch display, switch control, data trend display, and storage of the above configuration with the downlink PLC.

10.2.3   Common Configuration Software

Common configuration software: Wincc, Intuch, Fix, etc.

10.3   Quick Start of WINCC Configuration Software

Siemens WINCC configuration software is connected with PLC and other industrial control equipment, which can realize the collection, control, display and storage of field data and equipment status, especially when it cooperates with Siemens' own PLC and control equipment, it is very convenient.

10.3.1   Purpose

Take the configuration software Wincc and "S7–300" PLC to form a control and monitoring system as an example to realize the display of a hot water flow (analog input); by changing the frequency of the inverter, control the speed of a pump's motor (analog output); A device running state (digital input); a device start/stop (digital output).

The PLC system that completes this function consists of a power supply module PS307 (2A), a CPU314, a 16-way input module "SM321-DI16XDC24V", the 16-way relay output module "SM321-DO16XRelay", an 8-way analog input module "SM331-AI8X12Bit", a 4-channel analog output module "SM332-AO4X12Bit".

After selecting the hardware board, remember the I/O address of each board, as shown in Fig. 9.25.

The address of the 16-way digital input module SM321 is I0.0~I3.7, IB0~IB1, IW0.

The address of the 16-way digital output module SM322 is Q4.0~Q5.7, QB4~QB5, QW4.

The address of the 8-way analog input module SM331 is PIW288~PIW302.

The address of the 4-way analog output module SM332 is PQW304~PQW310.

The address of default MPI communication port is 2, as shown in Fig. 9.25.

10.3.2 Basic Steps

1. The connection between the PC and the programmable controller is connected through a dedicated communication cable (PC/MPI) provided by Siemens.
2. The correspondence between input and output data between Wincc and programmable controller S7–300.
3. Wincc programming method.

10.3.3 Wincc Programming Software Operation and Communication Settings

After Wincc is installed correctly, it already has the driver of S7–300PLC. In the "Programs" under the "Start" menu, start "SIMATIC"–"Wincc"–"Windows Control Center".

Click "New Project" in the "File" menu, in the opened "Wincc Project Manager", select "Single User Project", and click "OK".

In the "Create New Project" dialog box, enter the name of the new project "TEST", select the path to store the project, and click the "Create" button.

On the opened programming configuration screen, on the left side of the screen, right-click "Variable Management", in the opened "Add New Driver" dialog box, select "SIMATIC S7 Protocol Suite.chn", and click "Open", as shown in Fig. 10.6 shown.

Click the newly loaded "SIMATIC S7 Protocol Suite.chn" under "Variable Management" on the left side of the screen, right-click "MPI", and select "New Driver Connection". In the opened "Connection Properties" dialog box, since the default MPI address of 300PLC is 2. In order to be simple and unchanged, write "MPI2" in the name column. Sometimes you can also write the name of the workshop or station where the PLC is located for easy memory, as shown in Fig. 10.7. The MPI bus is S7–300 comes with a standard bus, which can connect 127 PLCs or PCs and other equipment, and the cost is low. If users need more powerful communication, they can also choose "PROFIBUS" and other protocols.

In the "Connection Properties" dialog box, click "Properties", and in the opened "Connection Properties-MPI" dialog box, select "Station Address" as 2, which means the S7–300PLC of MPI address being 2 is connected. Click "OK", as shown in Fig. 10.8.

In this way, a PLC with MPI address 2 is installed under the "MPI" bus, and the name of the PLC station is "MPI2", as shown in Fig. 10.9. If more PLCs need to be installed under the MPI bus, then continue to right-click "MPI", select "New Driver Connection", and add a new PLC. But the "name" of each PLC and the MPI "address" cannot be the same, and the MPI bus can connect up to 127 PLCs.

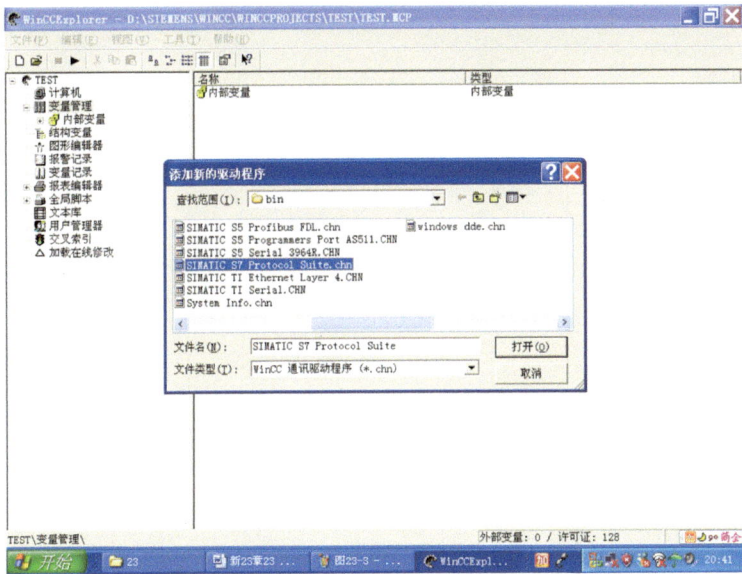

Fig. 10.6 Adding a new driver

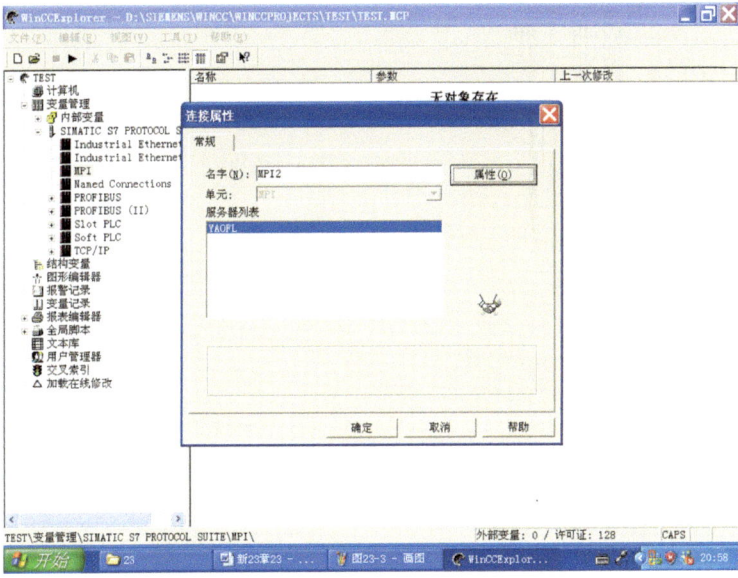

Fig. 10.7 Naming the new connection

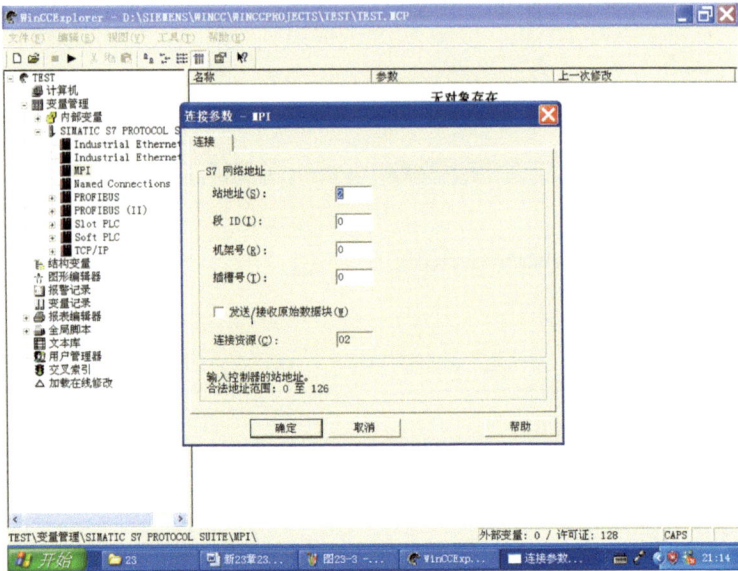

Fig. 10.8   Define the MPI address of the PLC connected to WINCC

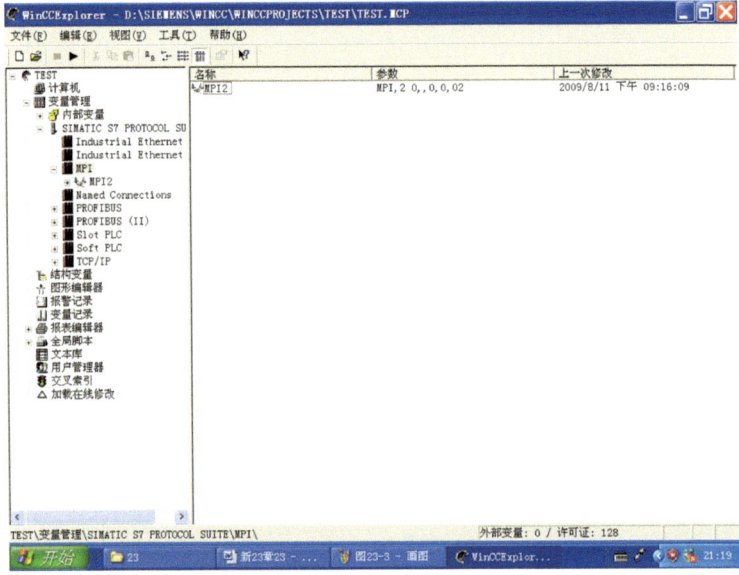

Fig. 10.9   Naming the name of the PLC station

10.3.4 Add "Variable" Connected with PLC

Add variables that need to be monitored or controlled in the PLC. Right-click "MPI2", select "Add New Variable", and in the "General" tab of the "Variable Properties" dialog box, enter the new variable "Name" as "Flow 1". In the PLC, the first analog input signal PIW288 of the analog input module needs to be sent to DB1.DBW0 with the MOV instruction. For the specific method, please refer to the previous chapter of PLC. Select "Data Type" as "unsigned 16 digits", click the "Select" button on the left of the "Address" to open the "Address Properties" dialog box, if you choose that the variable comes from DB1.DBW0 of the PLC, the "Data" is "DB", and the "DB Number" is 1. Fill in "0" for "DBW", as shown in Fig. 10.10.

In the "Linear Calibration" item, determine the corresponding relationship between the data of the variable in the PLC and the value displayed in the Wincc monitoring software. "Process value range" indicates the value of the variable signal in the PLC, 4–20 mA in S7–300 The analog input corresponds to "0–27,648". Value1 of "Process Value Range" selects 0, and Value2 of "Process Value Range" is 27648. "Variable Value Range" indicates the display value of the variable in Wincc, assuming the flow full scale of the meter is 2000m3/h. Then the Value1 of the "variable value range" is 0, and the Value2 of the "variable value range" is 2000, as shown in Fig. 10.11.

Add a new variable "Flow 1" under "MPI2". Right-click "MPI2", "Add New Variable", in the "General" tab of the "Variable Properties" dialog box. Enter the "Name"

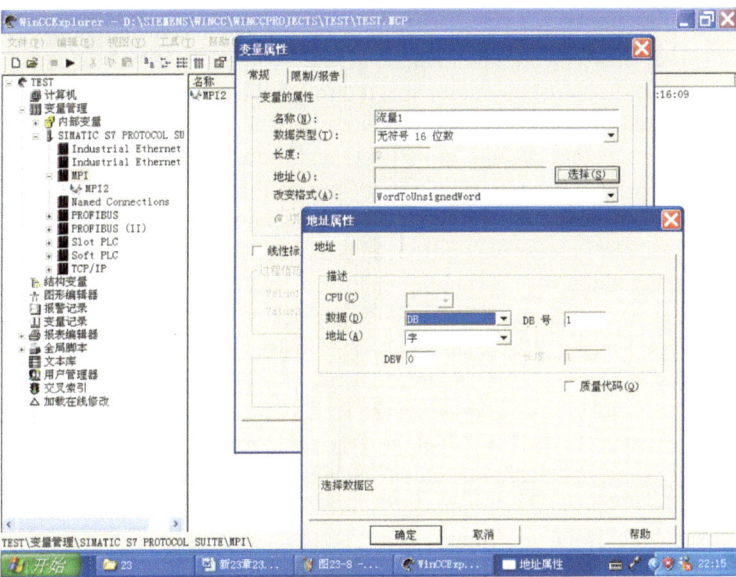

Fig. 10.10 Add a new variable "flow 1"

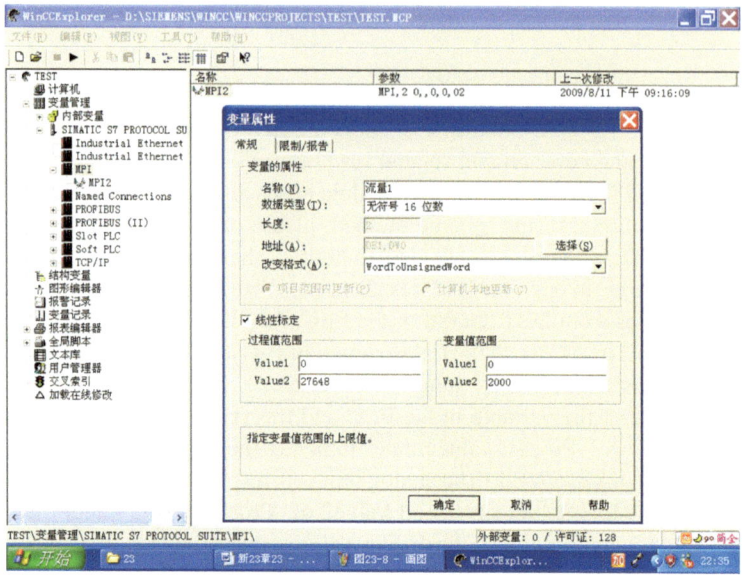

Fig. 10.11 Define variable properties

of the new variable as "Speed Output", and select the "Data Type" as "Unsigned 16-bit". Click the "Select" button on the left side of "Address" to open the "Address Properties" dialog box. If you select this variable to output to PLC's DB1.DBW2, then "Data" is "DB", "DB No." is 1. " Fill in "2" for the "DBW" item. As shown in Fig. 10.12. Use the MOV command in the PLC to send DB1.DBW2 to the PQW304 analog output, and control the motor speed through the frequency converter.

In the "Linear Calibration" item, determine the corresponding relationship between the data of the variable in the PLC and the value displayed in the Wincc monitoring software. "Process value range" means that the corresponding value of the variable 0–10 V in the PLC is "0–27,648", select 0 for Value1 of "Process Value Range", Value2 of "Process Value Range" is 27648. "Variable Value Range" indicates the display value of the variable in Wincc, assuming that the rated speed of the motor is 1470 rpm, then "Variable Value Value1 of "Range" is 0, and Value2 of "Variable Value Range" is 1470, as shown in Fig. 10.13.

In this way, a new variable "speed control" is added under "MPI2". Enter the running status indicator, the address in the PLC is I0.0. Right click "MPI2", "Add New Variable", in the "General" tab of the "Variable Properties" dialog box, enter the new variable "Name" as "Running Status". Select "Binary Variable" for "Data Type", click the "Select" button on the left of "Address" to open the "Address Properties" dialog box, select "Data" as "Input", and "Address" as "Bit". The "I" item is 0, the "bit" is 0, click "OK". As shown in Fig. 10.14. So that the first input signal of the PLC digital input card can be collected. In this example, since the digital input card is in the first block position, it is I0 and the first input bit 0 is used.

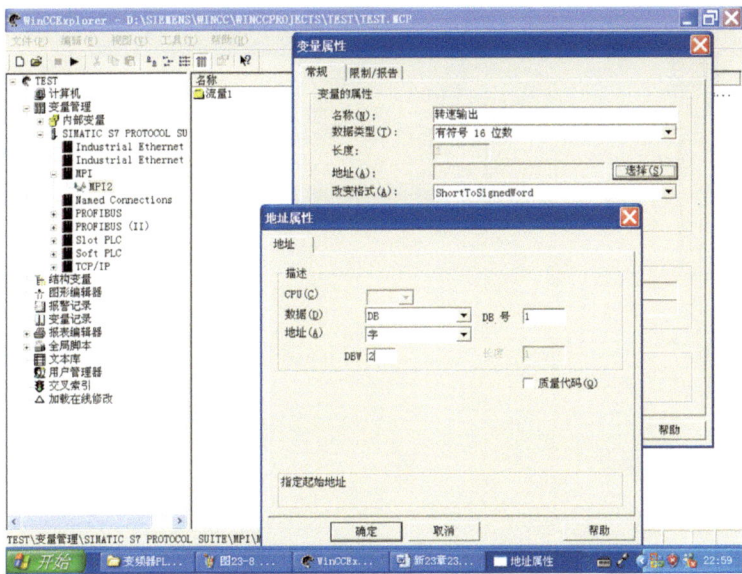

Fig. 10.12 Define a new variable "speed output"

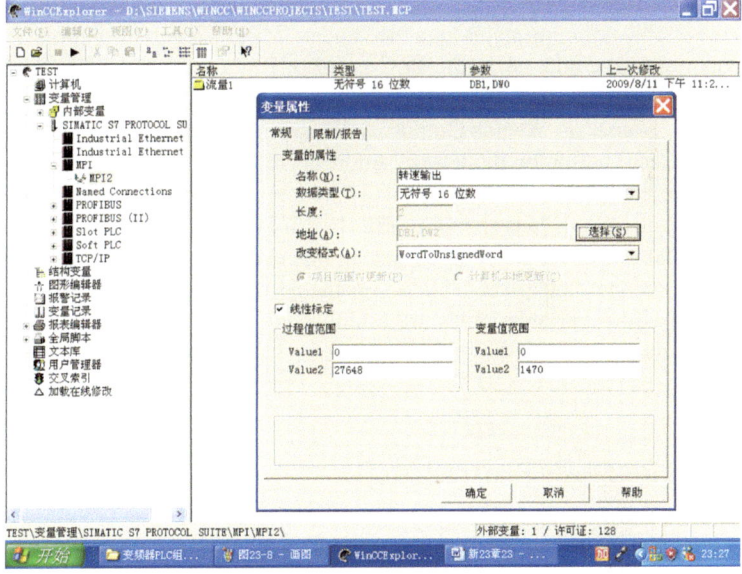

Fig. 10.13 Defining variable properties

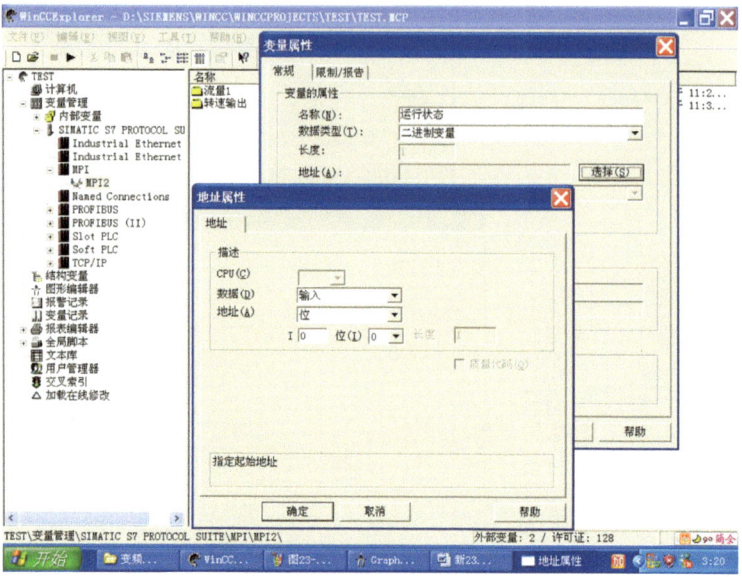

Fig. 10.14 Defining a new variable "running state"

A new variable "running status" is added under "MPI2". Insert the button variable, corresponding to the address in the PLC is Q4.0. Right-click "MPI2", "Add New Variable", in the "General" tab of the "Variable Properties" dialog box, enter the new variable "Name" as "Switch Output", select "Binary Variable" for "Data Type". Click the "Select" button on the left of "Address" to open the "Address Properties" dialog box, select "Data" as "Output", and "Address" as "Bit", The "Q" item is 4, the "bit" is 0, click "OK", as shown in Fig. 10.15. So that the first output signal of the PLC digital output card can be output. In this example, since the digital output card is in the second block, it is Q4 and uses the first output bit 0.

In this way, four variable labels are established: flow 1, speed output, running status and switch output, as shown in Fig. 10.16.

10.3.5 Adding a New Screen

Click "Graphics Editor" on the left side of the screen to create a "New Screen". The default file name of the new screen is "NewPd10.Pdl". Right-click and rename the default name to "Control Experiment. Pdl", as shown in Fig. 10.17.

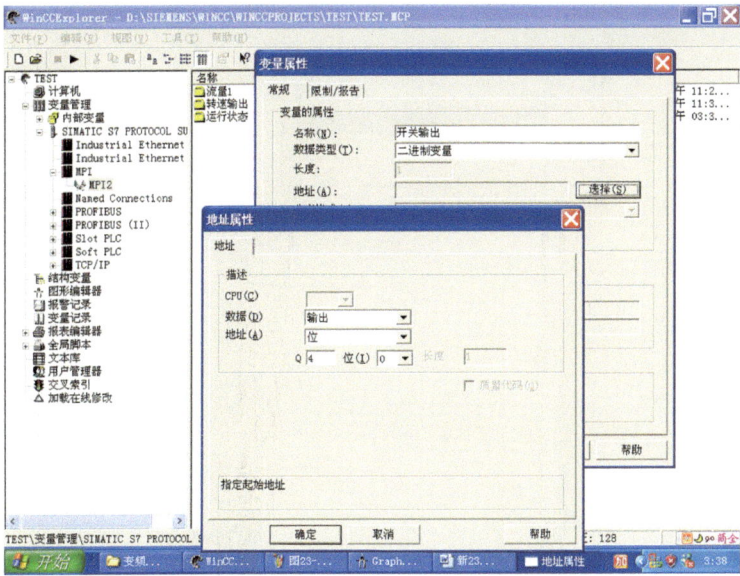

Fig. 10.15 Define a new variable "switch output"

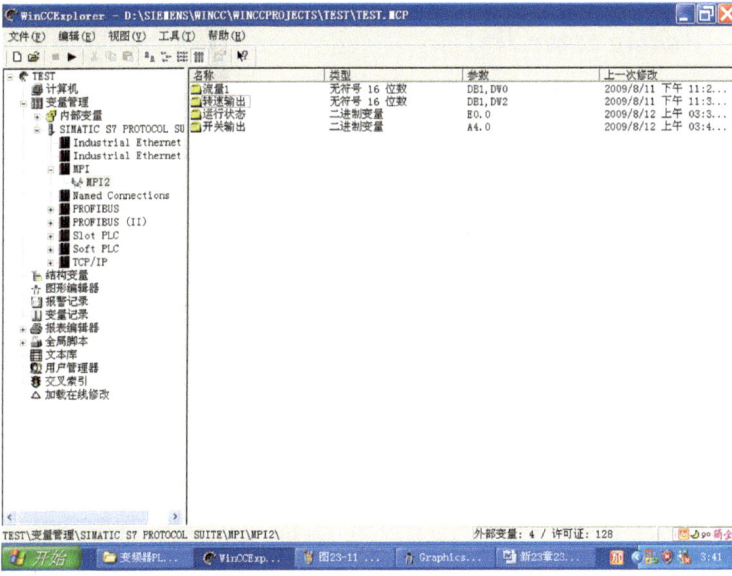

Fig. 10.16 Complete 4 variable labels

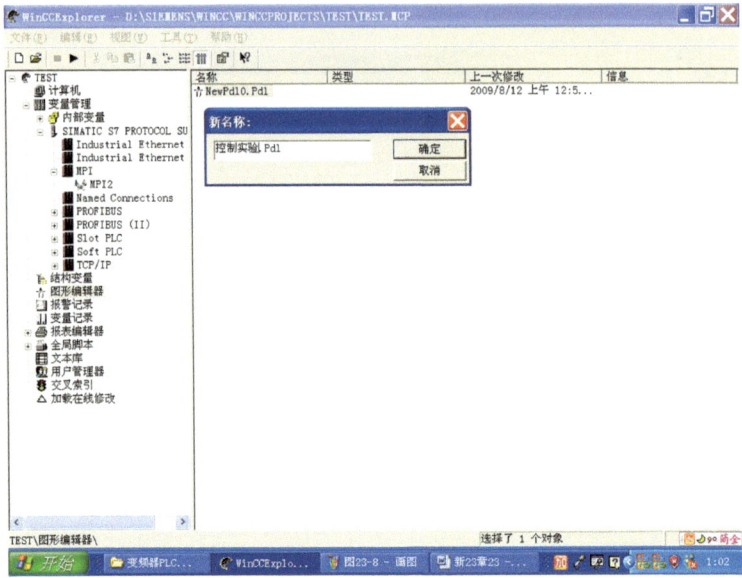

Fig. 10.17 New screen

10.3.6   Adding Static Text

Double-click "Control Experiment.Pdl" to edit and configure the screen. In the "Object Palette" on the left side of the "Graphic Editor" screen, click "Static Text", click the position where the text is to be placed, and then, in the text prompt "???". Enter the first text to be displayed: "Flow rate (m3/h):". At the top of the screen, select "Arial" for "Font", and 18 for "Font Size", adjust the text position again. As shown in Fig. 10.18.

Similarly, put 4 static texts, "Speed Control (1470r/min)", "Jog Button" and "Running Status", as shown in Fig. 10.19.

10.3.7   Analog Value and Data Display

Insert the display of the variable "Flow 1". Click the "Input/Output Field" under the "Smart Object" in the "Object Palette" on the right side of the screen, and place the logo prompt at the position where it needs to be placed. In the "I/O Field Configuration" In the dialog box, select "Type" as "Input", the "Update" speed of the variable as 2 s. Select 12 for "Font Size", select black for "Color", click the "Variable Selection" button, and open "Variable-Project" dialog box, select the variable "Flow 1" to be displayed, and then click "OK", as shown in Fig. 10.20.

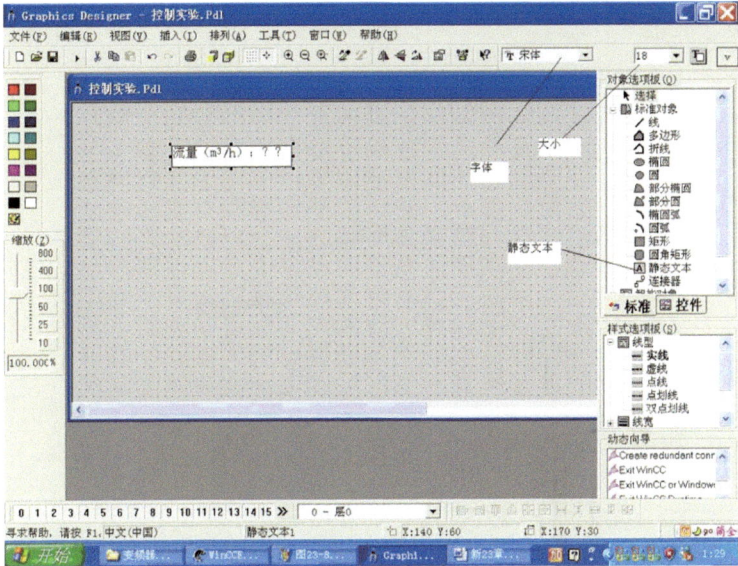

Fig. 10.18 Add text "flow (m3/h):"

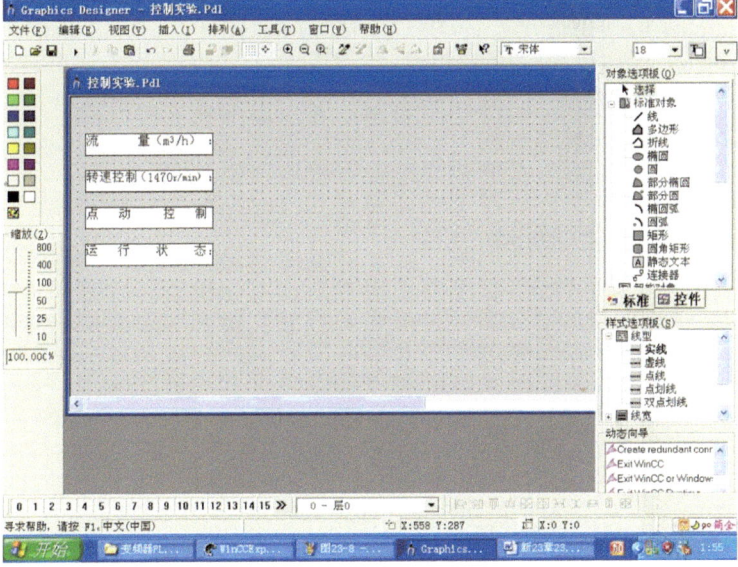

Fig. 10.19 Add 4 texts

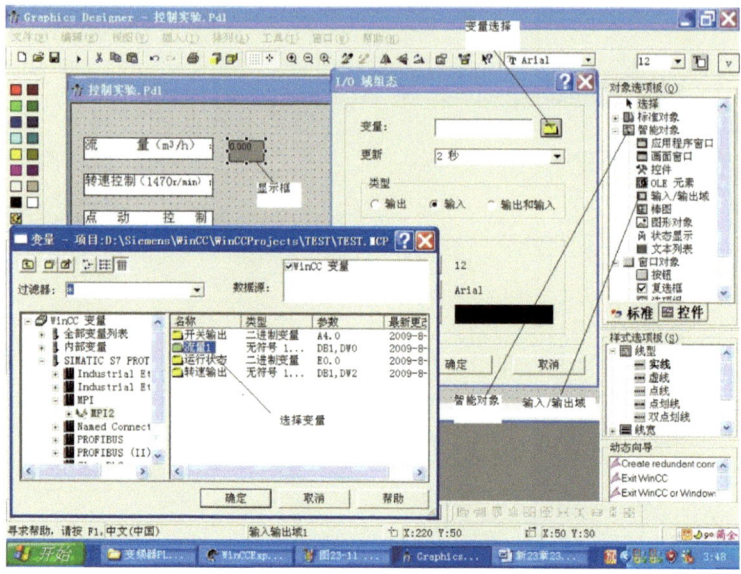

Fig. 10.20 Flow display

Right-click the placed "Display Box", select "Properties", open the "Object Properties" dialog box, in the "Properties" tab. Click "Input Output", in the "Output Format" item, select the 4-digit integer format "9999", click "OK", as shown in Fig. 10.21.

10.3.8 Output Analog and Data

Insert the variable "speed control" output, click the "input/output field" under the "smart object" in the "object palette" on the right side of the screen. Put the mark prompt at the position where it needs to be placed, in the "I/O field configuration" dialog box, select "Type" as "Output". So that after Wincc runs, the operator can click the variable with the mouse to input the control data, the "update" speed of the variable is 2 s, click the "variable selection" button, open the "Variable-Project" dialog box, select the variable "Speed Control" that needs to be controlled, and then click "OK", as shown in Fig. 10.22.

Right-click the placed "I/O" display box, select "Properties", open the "Object Properties" dialog box, in the "Properties" tab, click "Input/Output", in the "Output Format" dialog box, you can also According to the process requirements, in the "Input a Format" column, create a display format by yourself, such as 4 integers and 1 decimal, and the display style is "9999.9", click "OK", as shown in Fig. 10.23.

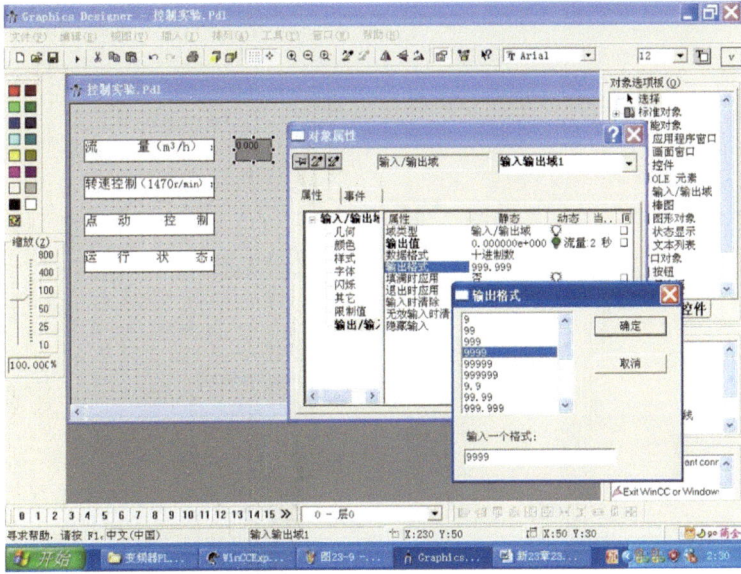

Fig. 10.21 Define the traffic display format

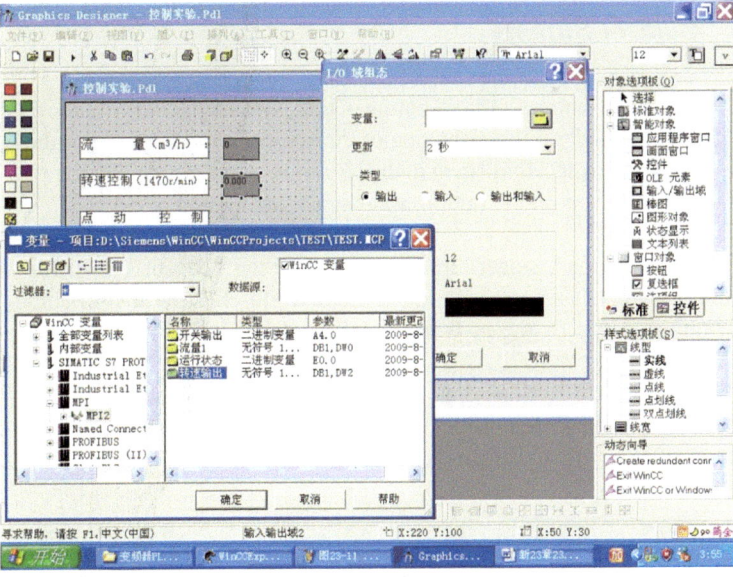

Fig. 10.22 Speed control

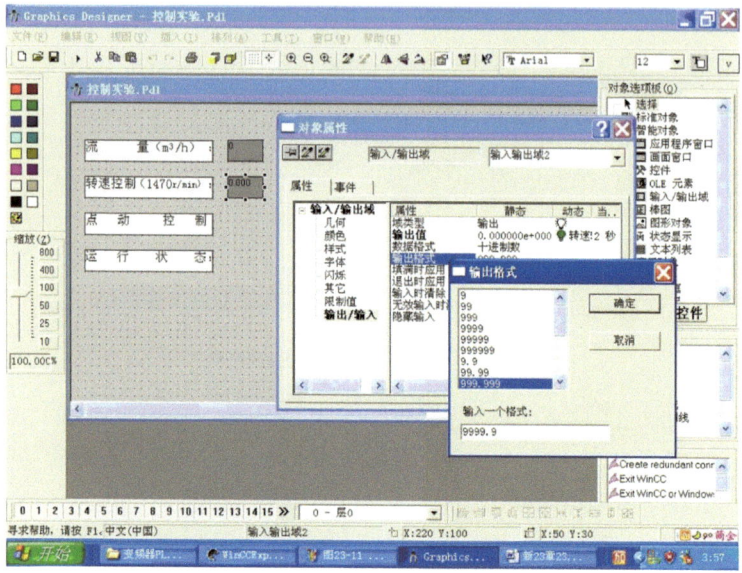

Fig. 10.23 Define the speed display format

10.3.9 Control Button

Insert the variable "Switch Control" button. Click "Button" under "Window Object" in the "Object Palette" on the right side of the screen, put the button at the desired position, and write in the "Text" item in the "Button Configuration" dialog box. Enter the text to be displayed on the button, write it as "Start", and click "OK", as shown in Fig. 10.24.

Right-click "Button", select "Properties", open the "Object Properties" dialog box, in the "Event" tab, select the "Mouse" item. Right-click the arrow behind "Press the left button" in the dialog box on the left, and select "C Action", as shown in Fig. 10.25.

In the pop-up "Edit Action" dialog box. Open "Internal Function"—"tag"—"set" on the left side of the screen, select the set function "SetTagBit", open the "Assign Parameters" dialog box of this function, select the bit to be set "Variable", as shown in Fig. 10.26.

In the "Assign Parameters" dialog box, click "Value" of the "Tag Name" item. Right-click "Variable Selection", select "Switch Output", and enter 1 in the "value" item, indicating that the variable value of "Switch Output" is set to 1. Click "OK". In the pop-up dialog box, confirm the choice to recompile the source code of the action, as shown in Fig. 10.27.

In the "Object Properties" dialog box of "Button", on the "Event" tab, select "Mouse". Right-click the arrow behind "Release Left Button" on the left, select "C Action". In the pop-up "Edit Action" dialog box, open the "internal

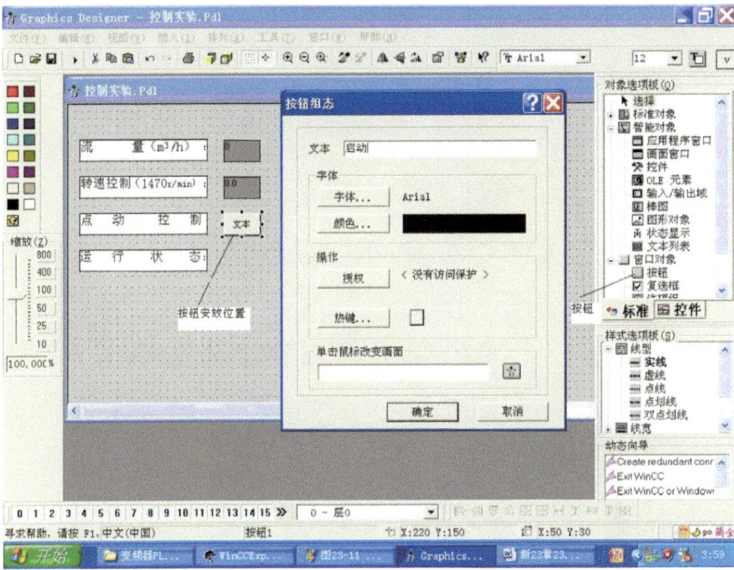

Fig. 10.24   Add button

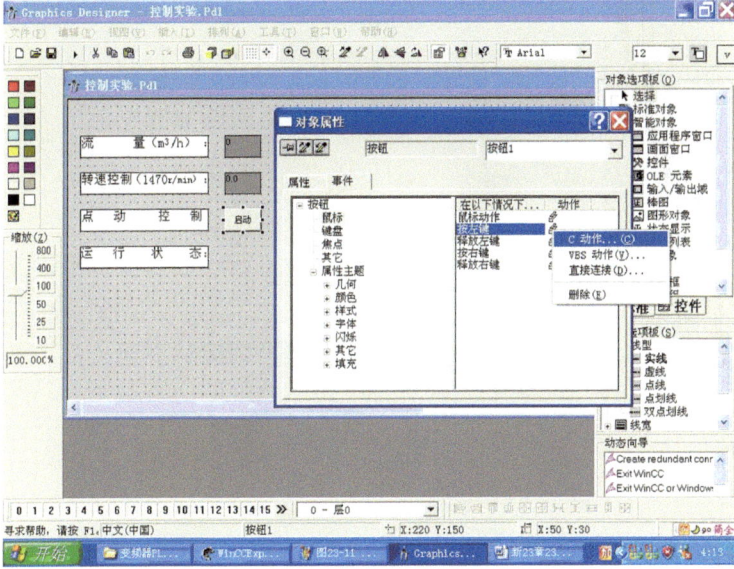

Fig. 10.25   Defining the action of the button

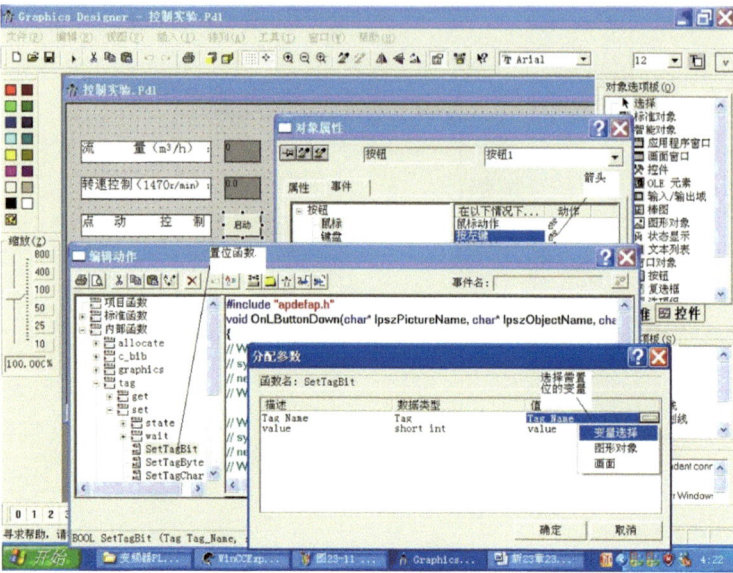

Fig. 10.26 Select the set function

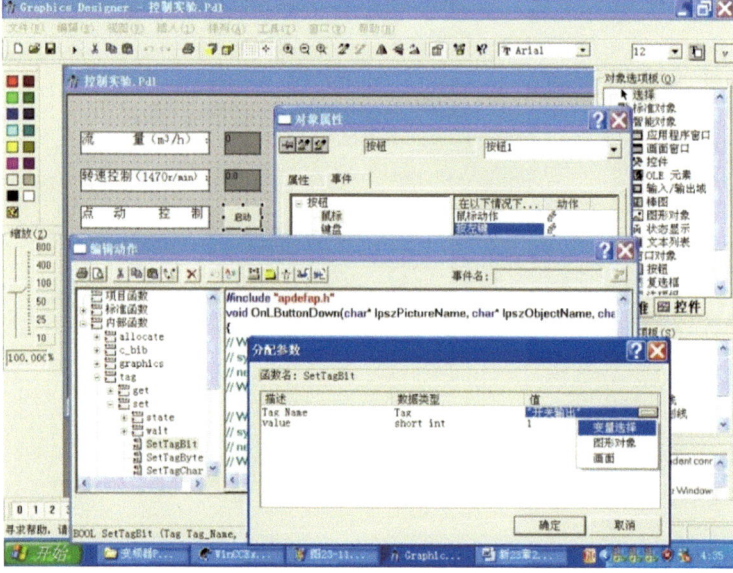

Fig. 10.27 Define allocation parameters

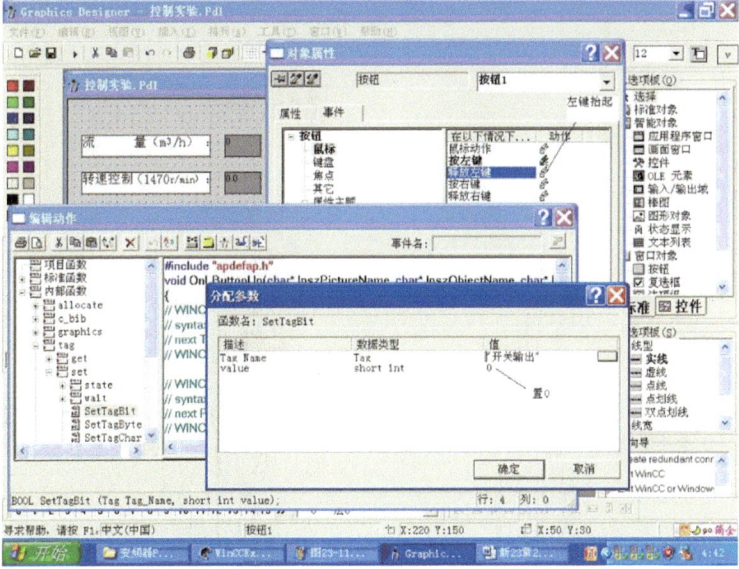

Fig. 10.28 Defining mouse actions

function"—"tag"—"set" on the left side of the screen. Select and click the set function "SetTagBit", open the "Assign Parameters" dialog box of this function, in the "Tag Name" item, select "Variable". Click "Variable Selection", select the variable "Switch Output", enter 0 in the "value" item, which means that the "Switch Output" variable is set to 0, click "OK", and in the pop-up dialog box, confirm the selection to recompile the action source code, click "OK", the action of the button is defined. After Wincc runs, press the button with the mouse, "switch output" $Q4.0 = 1$, when the mouse is lifted, "switch output" $Q4.0 = 0$, as shown in Fig. 10.28.

10.3.10 Device Run/stop Display

Insert the "running status" indicator light, click the "circle" under the "standard object" in the "object palette" on the right side of the screen, place the circle at the desired placement position. Right click the "circle" and select "properties", in the "object in the "Properties" dialog box, select the "Color"—"Background Color" item. Right-click the arrow behind "Change" on the left, and select "C Action". In the pop-up "Edit Action" dialog box, open the "Internal Function" "—"tag"—"get", select and click the read bit function "GetTagBit". Open the "Assign Parameters" dialog box of the function, select the "variable" to be read, click "Variable Selection", select the variable "Run State", the address corresponding to the PLC is I0.0. Click "OK", and in the pop-up dialog box, confirm the choice to recompile the source code of the

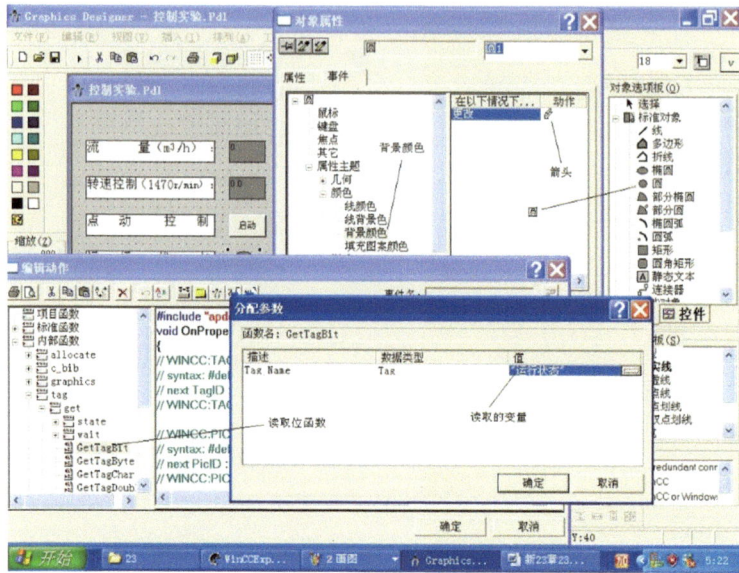

Fig. 10.29 Add and define the "running status" indicator light

action, click "OK". The indicator light is defined. After Wincc runs, I0.0 is input, the color of the indicator light changes at the same time, as shown in Fig. 10.29.

In this way, the configuration screen of this case is basically completed. Next, define the size of the screen when WINCC is running. Right-click the screen and select "Properties". ", determine the size of the screen display, and then click "Save" for the file, as shown in Fig. 10.30.

10.3.11 PC Communication Address and Running Start Screen

Define the first start screen after startup, right-click the "Computer" item, in the opened "Computer Properties"–"Graphics Runtime" dialog box, click the "Browse" button on the right side of the "Start Screen". In the pop-up "Screen" dialog box, select "Control Experiment.Pdl". In this way, the "Start Screen" displayed after Winccc starts up is the "Control Experiment.Pdl" screen we compiled above, as shown in Fig. 10.31. Press the run button on the top of the software screen to start the program.

In the computer installed with the "Wincc" running software, it is also necessary to configure the communication hardware and software of the MPI bus in the computer "Control Panel", open the "Set PG/PC Interface" item, and install the "PC Adapter" adapter, as shown in Fig. 10.32. Select the MPI address. Generally, the MPI address

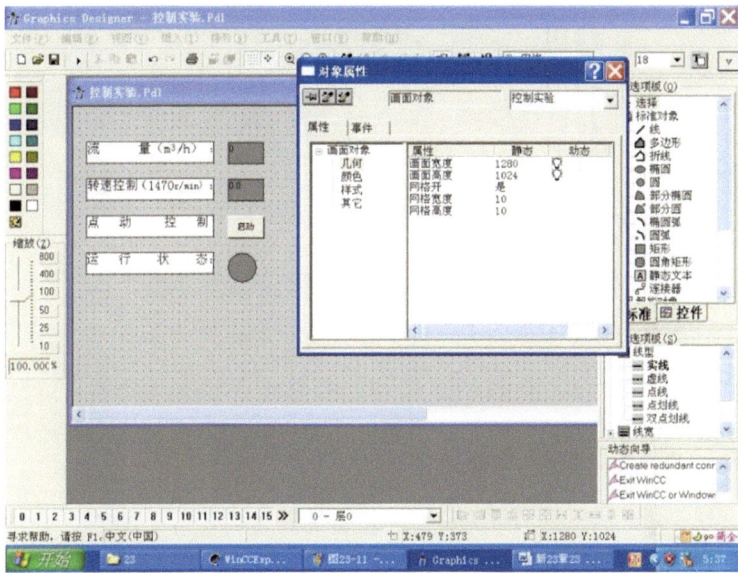

Fig. 10.30 Define the size of the screen

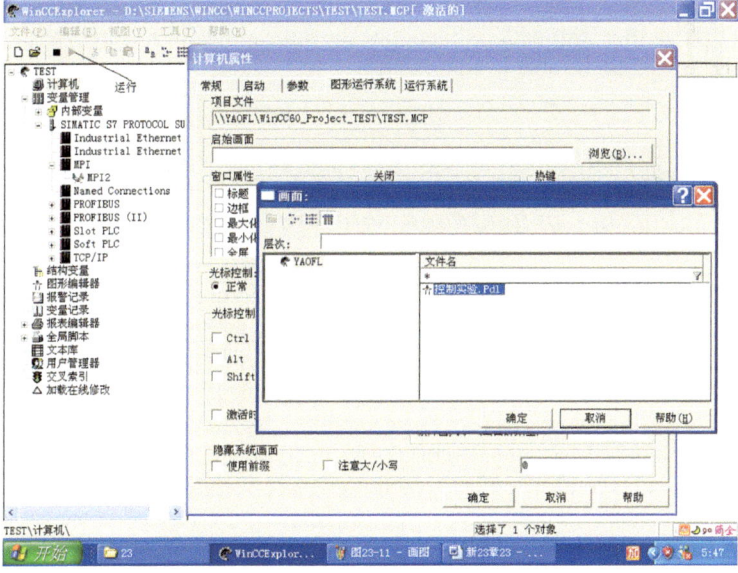

Fig. 10.31 Define the start screen

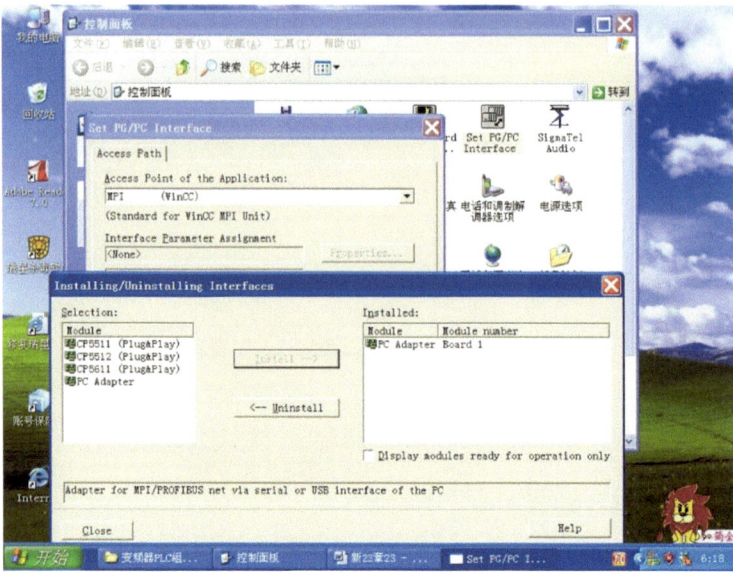

Fig. 10.32 Configure the MPI communication hardware and address of the computer

of the PC is set to 0. This address cannot be the same as the MPI address of the PLC, otherwise it will not be able to communicate.

10.3.12 Techniques for Reducing the Number of Variable Tags in Configuration Software

The sales price of the configuration software is directly related to the number of variables allowed by the software. In order to reduce the total number of variables, we can reduce the total number of variables by combining multiple switch variables into one variable, such as defining a digital input module IW4 is a variable. When the "status display" on the screen uses these variables, use the "GetTagBit" function to extract these bits and use them. Define the digital output module QW8 as a variable. When the "switch control" on the screen uses these variables, use the "SetTagBit" function to operate on a certain bit of the variable.

Chapter 11
Calculation and Selection of Motor Parameters in Automation System

Motors are the most important executive components in electrical automation equipment, including AC motors, DC motors, AC servo motors, DC servo motors, stepping motors, etc. In the design of the automation system, the selection of the motor is a frequently encountered problem. You need to calculate the parameters of the required motor. Novices often do not know where to start. Below we give a simple calculation and selection method.

11.1 Determination of the Rated Torque N_e of the Motor

The selection of the motor is mainly based on the calculation of the rated torque, and the rated torque Ne of the motor should meet:

$$N_e > N_w + N_f \tag{11.1}$$

Among them, N_W is the load torque, and N_f is the resistance torque of the equipment.

The load torque N_w includes the constant speed load torque N_c and the variable speed load torque N_{ad}. the variable speed load torque N_{ad} includes the acceleration load torque N_a and the deceleration load torque N_d. The variable speed load torque N_{ad} takes the maximum value of N_a and N_d.

$$N_w = N_c + N_{ad}$$
$$N_{ad} = \max(N_a, N_d) \tag{11.2}$$

The equipment resistance torque N_f includes the equipment static resistance torque N_{fs} and the equipment motion resistance torque N_{fw}. Generally, the equipment static resistance torque N_{fs} is greater than the equipment movement resistance torque

F. Yao and Y. Yao, *Efficient Energy-Saving Control and Optimization for Multi-Unit Systems*, https://doi.org/10.1007/978-981-97-4492-3_11

N_{fw}, and the resistance moment of the equipment is considered based on the static resistance moment of the equipment.

$$N_f = N_{fs.} \tag{11.3}$$

1. Calculation and determination of resistance torque

For the improvement of existing equipment, it is obtained by experimental methods. If it is new equipment, according to past experience or references, if no relevant data can be found, the static resistance torque of the equipment must be obtained by calculation methods. It is recommended that novices do not continue because the calculation process is complicated and the calculated results may not be correct. We suggest that if readers are inexperienced, they should not consider the resistance moment first, only the load moment, and then increase margin coefficient, which will make it easier to get started.

When obtaining the static resistance torque of the equipment by experiment, first remove the coupling between the motor and the connecting shaft. Three simple measurement methods are as follows:

(1) Use a torque wrench (or torque electric wrench) to rotate the connecting shaft, and the torque value that makes the connecting shaft just rotate is the resistance torque value of the equipment. Write down the torque reading. The torque wrench is shown in Fig. 11.1.
(2) Tie one end of a rod (such as a piece of wood or iron rod) to the connecting shaft with a nylon cable tie (or wire), or clamp a large wrench on the connecting shaft, as shown in Fig. 11.2.

The rod is placed horizontally and perpendicular to the connecting shaft. The weight of the rod is W_1 (kg, kilogram force). An object (such as several nuts, or small pieces) is hung on the other end of the rod. The length of the connecting shaft to the object is L (m, m), increase the weight of the object to W_2 (kg), until the connecting shaft starts to rotate. If there is a tension scale, it can also be replaced by a tension scale. Read the kilogram force W_2 of the tension scale when the connection

Fig. 11.1 Torque wrench   Torque display

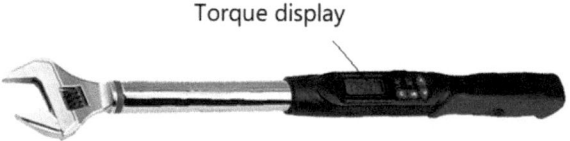

Fig. 11.2 Test resistance
torque

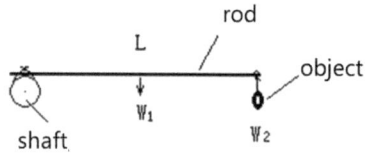

Fig. 11.3 Test resistance
torque

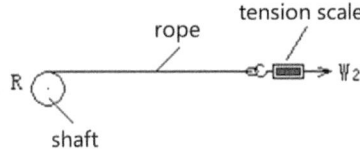

shaft just starts to rotate the conversion of kg into N needs to be multiplied by the
acceleration of gravity g, then the resistance moment N_f (N · m) is equal to

$$N_f = W_2 * 9.81 * L + W_1 * 9.81 * L/2 \tag{11.4}$$

(3) If the method of fixing a rod on the connecting shaft is limited, a binding rope
(or material tape) can also be wound on the connecting shaft. The radius of the
connecting shaft is R (m), and the binding rope (or material tape) The other end
is connected to a tension scale, as shown in Fig. 11.3.

Pull the binding rope (or material belt) hard, and record the kilogram force reading
W_2 (kg) of the tension scale when the connecting shaft starts to rotate, then the
resistance torque N_f (N · m) is equal to

$$N_f = W_2 * 9.81 * R \tag{11.5}$$

If it is inconvenient to remove the coupling between the motor and the connecting
shaft, or it is inconvenient to fix the rod (or wrap the packing rope) on the connecting
shaft. It is necessary to consider other shafts (connected with the connecting shaft) to
measure. The transmission mode and the transmission ratio will affect the calculation
of resistance torque, which will be explained later when we talk about load torque
measurement.

2. **Calculation and determination of constant speed load torque**

(1) For rotating loads directly driven by the connecting shaft;
  (1) If it is cutting head of cutting machine tool or digging head of mining
  excavation equipment, etc., as shown in Fig. 11.4.
The constant speed load torque N_c is equal to the cutting force F multiplied by
the cutting head radius R_1.

$$N_c = F * R_1 \tag{11.6}$$

Fig. 11.4 Cutting torque

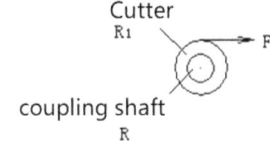

Fig. 11.5 Uniform speed
load torque

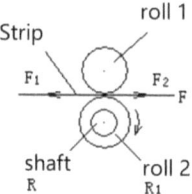

(2) If it is a printing machine, rotary die-cutting machine or paper machine, etc., as
shown in Fig. 11.5.

The total pulling force F_1 minus the total pulling force F_2, the constant speed
load torque N_c is

$$N_c = (F_1 - F_2) * R_1 \tag{11.7}$$

where F_1 is the sum of the tensions of the various strips passing through the work
roll 1, and the force opposite to the rotation direction of the coupling shaft. F_2 is the
total amount of tension of strips and scraps after processing.

The force relationship in gear transmission or rubber roller transmission is shown
in Fig. 11.6. The force at the edge of each gear is constant, but the torque of each
gear is different.

(2) For linear motion loads, such as CNC machine tools, machining centers,
assembly machines, etc., the connecting shaft drives the ball screw, and the
screw nut drives the worktable to perform linear motion. The pitch of the ball
screw is h, as shown in Fig. 11.7.

Fig. 11.6 The relationship
between gear transmission
and force

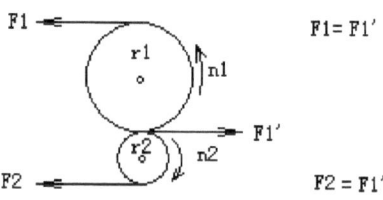

Fig. 11.7 Linear motion
load

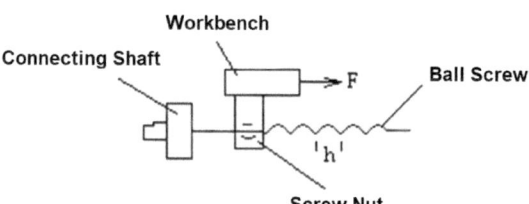

The connecting shaft rotates once, and the movement angle is 2π radians. At the same time, the ball screw rotates once, and the screw nut drives the worktable to move linearly with a pitch h (m), and the thrust of the worktable is F (N). According to the principle of energy conservation, the uniform speed load moment N_c (N · m) is

$$N_c = \frac{F * h_1}{2\pi} = F\frac{h_1}{2\pi}.$$ (11.8)

3. Calculation and determination of variable speed load torque

The calculation of acceleration torque and deceleration torque is the same, but the torque direction is opposite, so we only discuss the calculation of acceleration torque.

(1) For rotating loads.

An object with a mass of m (kg) rotates around a radius r (m), its linear velocity is V (m/s), its angular velocity is ε, and its kinetic energy W is

$$W = \frac{1}{2}mV^2 = \frac{1}{2}m(\varepsilon r)^2 = \frac{1}{2}(mr^2)\varepsilon^2 = \frac{1}{2}J\varepsilon^2$$ (11.9)

The moment of inertia $J(kg \cdot m^2)$ is defined as

$$J = mr^2$$ (11.10)

For a uniform disk with a diameter of R (m) and a total mass of m (kg) or a uniform circular shaft with a length of L, its moment of inertia J (kg · m^2) is

$$J = \frac{1}{2}mR^2$$ (11.11)

An object with a moment of inertia J and an angular velocity ε passes through a reducer (or gear, synchronous belt), the reduction ratio is n, the angular velocity is n_ε, and the equivalent moment of inertia is J'. Regardless of friction, the kinetic energy on both sides is equal. Out of the equivalent moment of inertia J' on the other side of the reducer,

$$\frac{1}{2}J\varepsilon^2 = \frac{1}{2}J'(n\varepsilon)^2$$
$$J' = \frac{J}{n^2}$$ (11.12)

All the k work rolls that are driven by the connecting shaft, the transmission ratio with the connecting shaft is $n_1, n_2, \dots n_k$. Their moments of inertia are $J_1, J_2, \dots J_k$. The moment of inertia converted to the connecting shaft on the motor side is $J_{1d}, J_{2d}, \dots J_{kd}$. The moment of inertia of the motor is J_m, then the total moment of inertia J converted to the motor side is

$$J = J_m + J_{1d} + J_{2d} + + J_{kd} = J_m + \frac{J_1}{n_1^2} + \frac{J_2}{n_2^2} + \cdots + \frac{J_k}{n_k^2} \tag{11.13}$$

According to the speed requirements of the equipment, converted to the connecting shaft, the time for the angular velocity of the connecting shaft to accelerate from 0 to ε is t (s), and the torque required on the connecting shaft is N_{ad} (N · m), which is the speed change torque that the motor needs to provide.

$$N_{ad} = F * r = (ma)r = m\frac{v}{t}r = m\frac{\varepsilon r}{t}r = mr^2\frac{\varepsilon}{t} = J\frac{\varepsilon}{t} \tag{11.14}$$

If there is a reducer between the connecting shaft and the motor, the connecting shaft torque N_{ad} (N · m) speed n_1 (rpm). After passing through the reducer with a reduction ratio of n, the torque converted to the motor side is $N_{ad'}$ (N · m) and the speed is n_2 (rpm). Regardless of friction, the power on both sides of the reducer is equal, and the speed change torque $N_{ad'}$ (N · m) on the motor side is obtained

$$\frac{N_{ad}n_1}{9550} = \frac{N_{ad'}n_2}{9550}$$
$$n_2 = n_1 n \tag{11.15}$$
$$N_{ad'} = \frac{N_{ad}}{n}.$$

(2)   For linear motion loads.

The connecting shaft drives the ball screw, and the screw nut drives the worktable. The mass of the screw nut and the worktable is m1 (kg), and the worktable drags an object with a mass of m_2 (kg) to make a linear motion, as shown in Fig. 11.8.

The time for the object to accelerate from 0 to V_{max} (m/s) is t (s), and the force F (N) that the workbench needs to apply is

$$F = (m_1 + m_2)a = (m_1 + m_2)\frac{v_{v_{max}}}{t} \tag{11.16}$$

Regardless of the friction, the work done on the side of the connecting shaft and the side of the table is equal, and the acceleration torque N_{ad1} converted to the side of the connecting shaft is given by

Fig. 11.8 Linear motion load

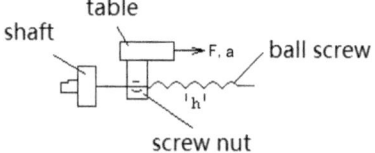

$$N_{ad1} = F\frac{h}{2\pi} \tag{11.17}$$

The moment of inertia of the connecting shaft, lead screw and motor is J. When the speed of the object accelerates from 0 to V_{max} (m/s) within time t (s). The angular velocity of the connecting shaft and lead screw accelerates from 0 to ε_{max} (rad/s). The acceleration torque N_{ad2} required on the connecting shaft

$$\varepsilon_{max} = v_{max}\frac{2\pi}{h}$$
$$N_{ad2} = J\frac{\varepsilon_{max}}{t} \tag{11.18}$$

The total acceleration torque N_{ad} required by the motor

$$N_{ad} = N_{ad1} + N_{ad2}. \tag{11.19}$$

11.2 Determination of Motor Speed

The motor speed is determined according to the highest working speed required by the automation equipment.

For the load shown in Fig. 11.9, the maximum speed V (m/min) of the material belt corresponds to the maximum speed n_{max} (rpm) of the work roll 1.

The maximum speed of the connecting shaft and work roll 1 is the same, equal to

$$v_{max} = n_{max}2\pi R_1$$
$$n_{max} = \frac{v_{max}}{2\pi R_1} \tag{11.20}$$

For the load shown in Fig. 11.8, if the maximum linear motion velocity of the known object is V_{max}, the maximum angular velocity converted to the motor side of the connecting shaft is ε_{max} (rad/s). The maximum rotational speed is n_{max} (rpm).

Fig. 11.9 Tape load

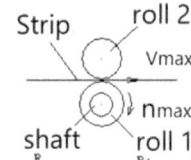

$$\varepsilon_{max} = v_{max}\frac{2\pi}{h}$$

$$n_{max} = \varepsilon_{max}\frac{60}{2\pi}$$

(11.21)

Select the rated speed of the motor n_e

$$n_e \geq n_{max}.$$

(11.22)

11.3   Determination of the Maximum Acceleration of the Servo Motor

According to the process conditions, general motion loads have an acceleration requirement. For example, the time required to increase the speed of the material belt (or paper) from 0 to the maximum speed V_{max} (m/s) does not exceed t (s). The acceleration of the material belt a (m/s^2) for

$$a = \frac{V_{max}}{t}$$

(11.23)

Referring to the method in Sect. 11.2, calculate the maximum angular velocity ε_{max} (rad/s) of the motor connection shaft according to the maximum vehicle speed V_{max} (m/s). Then the angular acceleration ω (rad/s^2) of the connection shaft is

$$\omega = \frac{\varepsilon_{max}}{t}$$

(11.24)

The maximum angular acceleration ω_{max} of the servo motor must satisfy

$$\omega_{max} > \frac{\varepsilon_{max}}{t}$$

(11.25)

If a reducer is installed on the front of the servo motor, and the reduction ratio is n, the maximum angular acceleration ω_{max} of the servo motor must satisfy

$$\omega_{max} > n\frac{\varepsilon_{max}}{t}$$

(11.26)

Otherwise, the driver of the servo motor will give an alarm during acceleration.

11.4 Determination of Motor Power

The power $P(kw)$ of the motor is equal to

$$P(\text{kw}) = \frac{N_e(n \cdot m)n_e(rpm)}{9550}. \tag{11.27}$$

11.5 Determination of Encoder Resolution

The selection of encoder resolution is determined according to the processing accuracy required by the automation equipment. For example, in Fig. 11.7, the positioning accuracy A of the material strip is required to be 0.1 mm. The diameter R1 of the work roll 1 is 50 mm, the circumference L of the work roll 1 is $2\pi R1 = 314.16$ mm, and the connecting shaft of the work roll 1 is installed for the motor. The minimum resolution Re of the encoder should be $L/A = 3140$. In order to leave room for adjustment of the PID and feed-forward parameters of the positioning link, the resolution of the encoder should be increased by an order of magnitude, that is, greater than or equal to 31,400.

$$R_e \geq 10\frac{L}{A}. \tag{11.28}$$

11.6 Servo Motor Inertia Ratio

Servo motor inertia ratio λ: refers to the ratio of the load's moment J of inertia to the motor's moment J_m of inertia.

$$\lambda = \frac{J}{J_m} \tag{11.29}$$

For a fixed load, when λ increases, the response speed of the system will slow down and the control accuracy will decrease. λ is generally between 5:1 and 10:1. For some low-speed motion situations, λ can be selected between 3:1 and 5:1.

It is best to select this parameter according to the technical specifications provided by the servo motor manufacturer, otherwise it will seriously affect the control accuracy and dynamic response speed of the system.

Chapter 12
Parameter Design in High-Speed High-Precision Motion Control

The first major task of automation is to replace labor to achieve faster and more precise production. In systems such as CNC machine tools, machining centers, CNC engraving, printing, die-cutting, papermaking, steel rolling, and high-speed rail, it is necessary to control the speed of multiple motors. Or position control, at this time need to use motion control (or follow-up control), in motion control, how to design the structure of the control system? How to determine the PID parameters in the control system? How to determine the feedforward parameters? How to set the parameters of the frequency converter to synchronize the speed between the frequency converters in the speed chain? Only by rationally designing and debugging the parameters of motion control can the system run in a faster and more precise working state.

12.1 Determination of Feedforward Parameters—"Yao's Trial and Error Method"

The accuracy and speed of many CNC equipment produced by some manufacturers are better than those of other manufacturers. This is not accidental. The design and debugging of the electronic control system are often very critical.

Many people know the PID controller. In fact, in motion control, if only PID is used for adjustment, the control speed and accuracy are not optimal. The reason is that PID control is feedback control, and only the feedback value is measured before passing. Calculate the error between the given value and the feedback value to obtain the control output after PID calculation, so the output always lags behind the generation of the error. For high-precision motion control, you also need to configure feed-forward parameters to improve control accuracy and speed. In this way, the integration depth of PID parameters can be reduced, making the motion control more precise and faster. The block diagram of the CNC system with feedforward unit is shown in Fig. 12.1.

© The Author(s) 2024
F. Yao and Y. Yao, *Efficient Energy-Saving Control and Optimization for Multi-Unit Systems*, https://doi.org/10.1007/978-981-97-4492-3_12

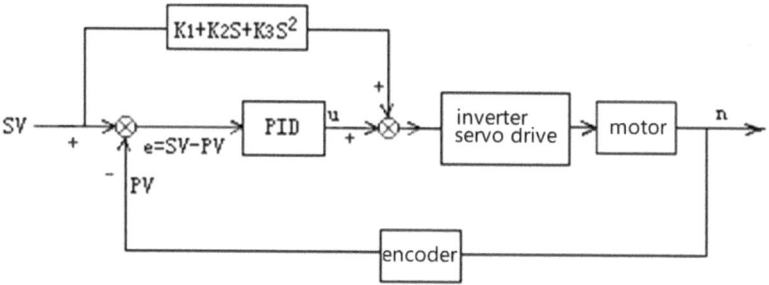

Fig. 12.1 Feedforward + feedback control block diagram

Feedforward parameters include feedforward proportional parameter K_1, feedforward speed parameter K_2 and feedforward acceleration parameter K_3. In general, take $K_2 = 0$, $K_3 = 0$, and only use feedforward proportional parameter K_1 to achieve high control accuracy.

Feedforward proportional parameter K_1 can be obtained by calculation. K_1 and encoder resolution R_{e1} (10000P/r), rated speed n_e (such as 1000 rpm), feedforward unit (such as D/A output unit of motion controller) output full scale voltage (such as ± 10 V) corresponds to the resolution R_{e2}. As shown in the following formula, where k_0 is a coefficient, which varies with different motion controllers.

$$K_1 = k_0 \frac{R_{e2}}{R_{e1} n_e} \tag{12.1}$$

The feed-forward proportional parameter K_1 can also be measured by the "trial and error method", and the method is as follows.

Simply set the three parameters of PID, as long as there is no vibration, set a given speed SV_1, record the speed n_1 of the motor, then set the three parameters of PID to zero, keep the given speed SV_1. The value of K_1 starts from 0 increase until the output speed of the motor is equal to n_1, record the corresponding K_1 value K_{10}, then

$$K_1 = K_{10} \tag{12.2}$$

If several motors run synchronously, set the three parameters of a PID to zero, change the value of K_1 to make it basically equal to the speed of other motors, then the value of K_{10} is the result.

"Porcelain trial and error method" can also be used to determine the feed-forward parameters of process control.

12.2 A Simple Adjustment Method of PID Parameters—"Two-Four Rule"

In the motion control system, if the motion controller provides the PID parameter self-tuning algorithm, it can perform self-tuning. If there is no self-tuning function in the motion controller, novices need to adjust the PID parameters by themselves. There are many books that explain this. Finally, let the front and rear attenuation ratio of the overshoot part of the response curve be 4:1. Since most methods require professional equipment for observation, it is sometimes inconvenient to apply in the field, and many novices will be at a loss.

The following introduces a simple method PID parameter determination method—"two-four rule" can be used as a reference for novices, the method is as follows:

1. The three parameters of PID are zero, and P gradually increases from 0 until the slight vibration of the motor is heard. At this time, it is P_0, and the ratio P is taken as

$$P = \frac{P_0}{2} \qquad (12.3)$$

2. $P = 0.5 * P_0, D = 0$, I gradually increases from 0 until the slight vibration of the motor is heard, at this time it is I_0, then the integral I is taken as

$$I = \frac{I_0}{2} \qquad (12.4)$$

3. $P = 0.5 * P_0, I = 0.5 * I_0$, D gradually increases from 0 until the slight vibration of the motor is heard, at this time it is D_0, and the differential D is taken as

$$D = \frac{D_0}{4}. \qquad (12.5)$$

12.3 "Yao's Speed up and Down Rules" of the Frequency Converter in the Speed Chain

In steel rolling, papermaking, textile and other fields, many motion control systems use frequency converters to drive motors, which often fail to maintain synchronization during the speed up and down process, and novices are even at a loss for this. A very important reason here is that the speed-up and down-speed time of each inverter in the speed chain is not adjusted properly. The speed-up and down-speed time of each inverter must be determined by the "Yao's speed-up and down-speed rule".

Speed-up and down-speed rule: For multiple frequency converters operating according to the speed chain, the output frequency of each frequency converter is inversely proportional to its speed-up and down-speed time. That is, the speed-up and speed-down time is short for high frequency and long for low frequency.

Since the same paper belt or steel belt is dragged, the linear speed of each link in the speed chain is the same. When the system is debugged, it is not necessary to pull the material first, and use the speedometer to measure the linear speed to adjust the linear speed of each link to be consistent, and close to the normal operating speed. Record the frequency of n frequency converters at this time as $f_1, f_2, ..., f_k, ... f_n$.

Select a motor with the largest load inertia, and the output frequency of the frequency converter corresponding to the motor is f_k, and set the acceleration time T_{ka} and deceleration time T_{kd} of the frequency converter corresponding to the motor according to the process requirements.

The acceleration time T_{1a} and deceleration time T_{1d} of the first inverter are determined as follows.

$$T_{1a} = \frac{f_k}{f_1} T_{ka}$$
$$T_{1d} = \frac{f_k}{f_1} T_{kd} \tag{12.6}$$

The acceleration time T_{2a} and deceleration time T_{2d} of the second inverter are adjusted as follows.

$$T_{2a} = \frac{f_k}{f_2} T_{ka}$$
$$T_{2d} = \frac{f_k}{f_2} T_{kd} \tag{12.7}$$

The acceleration time T_{na} and deceleration time T_{nd} of the nth frequency converter are adjusted by the following method.

$$T_{na} = \frac{f_k}{f_n} T_{ka}$$
$$T_{nd} = \frac{f_k}{f_n} T_{kd}. \tag{12.8}$$

12.4 The Wonderful Effect of "Virtual Axis" in Speed Synchronous Control

In the fields of printing, die-cutting, steel rolling, papermaking, and textile, it is necessary to operate at a certain speed ratio between motors of n stations. The synchronization performance of many synchronous motion control systems is very poor. The phenomenon of asynchronous operation may be caused by your poor synchronous control strategy. Many beginners often use an intuitive control mode to let the rear motor follow the movement of the front motor. The disadvantage of this method is that it is easy to cause the motor at the back are always in a state of micro-oscillation operation. Any motor vibration affects the movement of all motors behind it, a bit like the "look right" command in military training. The latter one looks at the previous one, and after a while everyone stands well, if there is a problem in the front, there will be a problem later, and it will take a long time to adjust.

In order to avoid this problem, you can use the running mode of the virtual axis to solve it.

The specific method is:

Define an internal imaginary axis in the motion control unit. The set speed of the production line controls this imaginary axis. All motors follow this imaginary axis. Because the imaginary axis has no vibration problems, all motors follow an internal imaginary axis that runs stably. Therefore, there will be no situation where everyone suffers after a problem, and the adjustment speed of the whole line is also fast.

12.5 Approximate Feedforward Parameter K_1

In some control systems, it is impossible or not allowed to measure the feedforward parameter K_1 through field experiments. We can also approximately calculate the feedforward parameter K_1 based on the characteristic parameters of the device.

For example, in a large variable frequency constant pressure water supply pumping station, the pumps are all centrifugal pumps and the rated frequency of the motors is f_e. The controller realizes constant pressure control of the pumping station by controlling the start and stop of the pump and the operating frequency of the frequency converter. Assume that the total water supply head of the pumping station is H_{sv}. According to the water pump characteristic curve Q-H provided by the pump manufacturer, as shown in Fig. 12.2.

Fig. 12.2 The characteristic curve Q-H of the pump

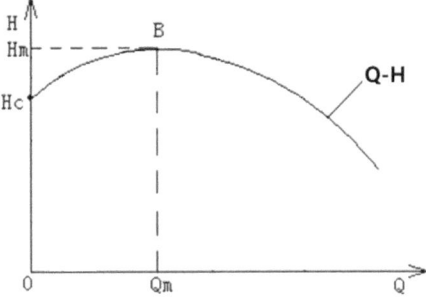

where H_c is the closed head and H_m is the maximum head. The feedforward parameter K_1 (converted to frequency) is approximately calculated as follows:

$$K_1 \approx f_e \sqrt{\frac{H_{SV}}{H_m}} \tag{12.9}$$

If $H_m = H_c$, then

$$K_1 \approx f_e \sqrt{\frac{H_{SV}}{H_C}} \tag{12.10}$$

This approximate K_1 can greatly reduce the adjustment amplitude of PID, improve the anti-saturation performance of PID, and increase the speed and stability of adjustment.

This method was proposed by Dr. Yao.

Chapter 13
Anti-interference and Fault Analysis of Control System

13.1 Anti-interference Measures

There are various reasons for interference. The most important principle is that the wiring of strong current and weak point should be separated. There are many other requirements. It is impossible for us to list them here. Here are some common anti-interference methods.

13.1.1 Common Mode Interference

In the practical application of automation engineering, it often occurs that the sensor signal needs to be sent to several places. For example, the local display instrument needs to display important pressure signals on site, and the PLC also needs to collect the pressure signal, so that there will be sharing question of sensor signals. When the 4–20 mA, 0–10 mA, 1–5 V measurement signals of more than one channel in the field need to be sent to multiple controllers at the same time. If no appropriate measures are taken, the controller may not be able to detect the correct signal due to common mode interference. Input signal, signal overflow or abnormal phenomenon. In Fig. 13.1, two 4–20 mA sensor signals are sent to PLC and RTU at the same time.

For the current signal output by the sensor, as long as the sum of the input impedance of PLC1 and RTU1 is not greater than the output load impedance required by the pressure sensor. The signal of the pressure sensor P1 can be input to PLC1 and RTU1 in series at the same time. If there is only P1 signal, PLC1 and RTU1 It can also receive the signal of P1 normally. When the 4–20 mA signal of the liquid level sensor is also sent to PLC1 and RTU1 at the same time, the problem may come, because the (−) end of the signal received by PLC1 is at the (+) end of RTU1 terminal. When the signal currents of P1 and L1 are different, it is likely to cause a large difference in the potential of the two (−) terminals of PLC1, which forms a

F. Yao and Y. Yao, *Efficient Energy-Saving Control and Optimization for Multi-Unit Systems*, https://doi.org/10.1007/978-981-97-4492-3_13

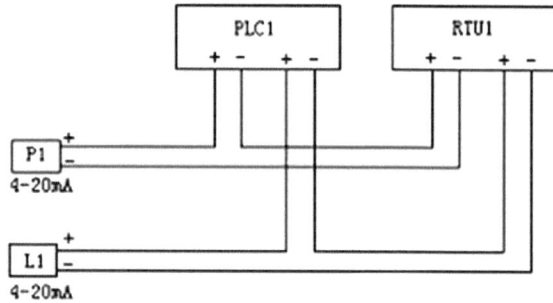

common-mode interference voltage. If the input terminals of PLC1 are not isolated
from each other, it will cause the analog signal input terminal of PLC1 to receive
abnormally. The two receiving signals of RTU1 are behind PLC1, and the potential
of its (−) terminal can be kept at the state set by the internal circuit of RTU1, which
is not a big problem.

The example shown in Fig. 13.1 can be solved by installing an isolation transmission module on the PLC1 side, and the isolation transmission module is connected
as shown in Fig. 13.2. The potential between the input and output of the isolated
transmission module is isolated, and the input, output, and power supply of some
isolated modules are all isolated.

After the isolation transmission module is connected, one of the two (−) terminals
of PLC1 is in an isolated state, so there will be no two (−) terminals with one high
and one low. Which will cause the reference potential of the PLC internal circuit to
be disordered. The power supply voltage of the isolation module is mostly DC24V,
and some are AC220V.

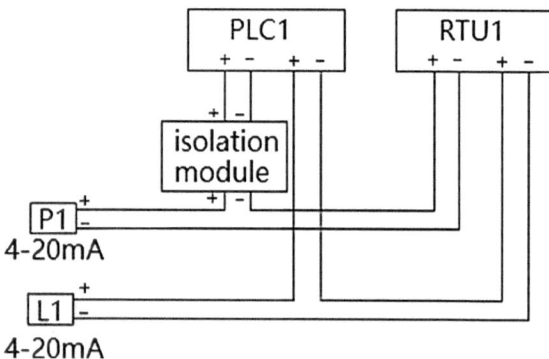

13.1.2   Signal Transmission Interference by Other Means

Since the input and output of the isolation module are isolated, the isolation methods include photoelectric methods, and isolation transformers are also used to isolate the metal signal lines from the field. Due to electromagnetic induction or coupling between lines, the lines carry electromagnetic waves. For interference, we can also use isolation methods to keep these interference signals away from the control system. However, the photoelectric isolation method has a limited effect on limiting the interference that has already generated current reflection on the line. This is because the current generated by the interference on the input side has become part of the signal, and the isolation module cannot distinguish whether it is an interference signal or a useful signal. High-frequency interference can be solved by filtering, but for low-frequency interference, the isolation module is not easy to deal with.

If funds permit, an isolated transmission module can be connected to each signal input terminal. The isolation module has various forms such as 1–5 V input 4–20 mA output, 4–20 mA input 1–5 V output, 1–5 V input 1–5 V output, 4–20 mA input 4–20 mA output, etc.

There are also many signal transmission modules themselves with photoelectric isolation function, it is no longer necessary to add isolation modules for such lines. The connection method of the single-channel isolation module is shown in Fig. 13.3. The power supply of the isolation module has specifications such as DC12V, DC24V and AC220V. In actual use, generally for a set of control systems, it is more convenient to use only one isolation module in the form of power supply. The output signal of the isolation module is sent to the PLC and generally converted into the same signal.

Fig. 13.3 Multiple use of isolation modules

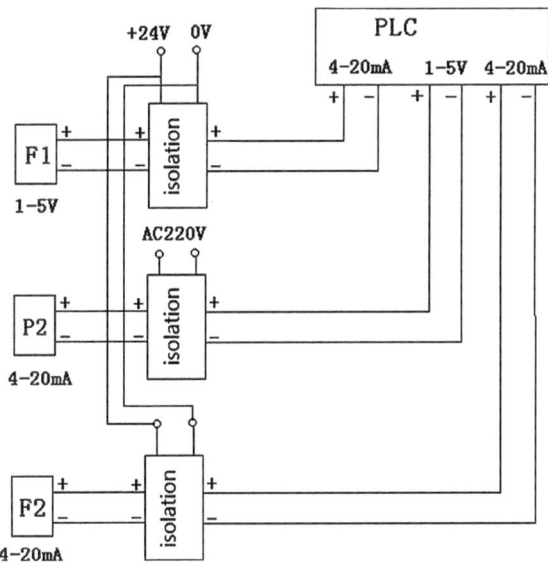

13.1.3   Communication Interference

1. Communication ports RS232, RS485, etc. between controllers such as PLCs and touch screens, etc., are isolated and then connected by photoelectric isolation modules to avoid system signals not normally transmitted due to ground potential differences between multiple devices.
2. Ground the 0 V of each device directly to the nearest ground, which can solve the problem that the 0 V of each device has a large voltage to the ground caused by induction. However, the application of this method requires that all analog signal types of the system allow 0 V to be grounded, otherwise it needs to be grounded. After signal isolation, 0 V is grounded, and the 0 V of each power supply can also be connected to the ground terminal through a 0.47 μF and a 100 μF in parallel to reduce the excessively high voltage floating between 0 V.
3. When multiple PLCs, touch screens, computers, synchronizers and other devices communicate with each other, it is often easy to cause unreliable communication. Using products from the same manufacturer can improve reliability. Due to different 0 V, poor shielding or Due to the interference caused by the communication design of the manufacturer's product itself. When the transmitted data is disturbed and defective, resulting in unreliable data and random data, you can put a number in two addresses, and the numbers in the two addresses are equal. The method used removes unreliable data.
4. In intermittent communication such as wireless communication, the same number is transmitted twice, and each time the number is placed in two addresses. In this way, only when the four numbers are equal can the control command be used to execute the action, which can avoid misoperation.
5. The same data can only be used within a credible range. When it is greater than the possible maximum value allowed by the process, or less than the possible minimum value allowed by the process, use the value passed last time.
6. Strictly restrict the execution of the control function. Only when the conditions are accurately met, the program will be executed. Some irrelevant conditions may affect the output control. When the conditions are not satisfied, the control will not be executed, even if there is interference. There will be no false action to reduce the probability of false action.
7. Many buses with RS485 hardware protocol as the bus structure, such as MPI, Profibus, CAN, etc., cannot fork too long. For analysis and solutions, see the previous bus part.
8. For the bus structure, if there are repeaters, the bifurcated lines of two adjacent repeaters are not allowed to be too long, otherwise there will be abnormal communication problems. For the analysis and solution, see the previous bus part.

13.1.4 Signal Connection and Conversion Between 4-Wire Sensor and 2-Wire Sensor

The four-wire sensor means that the 2 power lines and the 2 signal lines of the sensor are separated. If the (−) end of the power line and the (−) end of the signal line share (such as 0 V), the four-wire sensor can also be used. There are 3 wires, as shown in Fig. 13.4.

The two-wire sensor means that the two wires on the sensor provide power to the sensor, and at the same time, the two wires provide a 4–20 mA signal output. The wiring method of the two-wire sensor is shown in Fig. 13.5.

If the 2-wire pressure sensor is connected to the analog input module of the PLC, the signal input selection block on the analog input module must be pried up and turned to the 2-wire position. If the 4-wire pressure sensor is connected to the analog input module of S7-300, the signal input selection block on the analog input module must be turned to the 4-wire position.

For two-wire sensors, the (−) terminal is also the signal output terminal, and the signal current 4–20 mA is output through this terminal. The (+) terminal of the two-wire sensor is equivalent to the (+) terminal of the power supply, so where are the (−) terminal of the power supply terminal and the (−) terminal of the signal? In fact, the (−) end of the power supply and the (−) end of the signal are moved to the PLC on the signal receiving side.

Take the connection between the two-wire sensor and the S7-300 analog input block as an example, as shown in Fig. 13.6.

In Fig. 13.6, the (+) terminal of the two-wire sensor is connected to the (+) of the SM331 block on the S7-300, the voltage of the (+) terminal is close to the internal power supply +24 V. The (−) terminal of the SM331 passes through a resistor R is

Fig. 13.4 Four-wire sensor

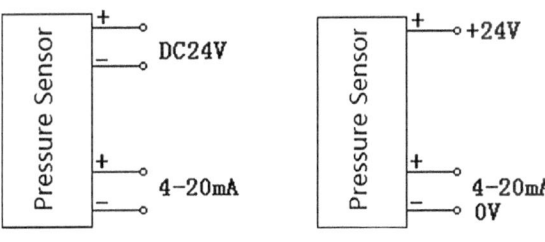

Fig. 13.5 Two-wire sensor

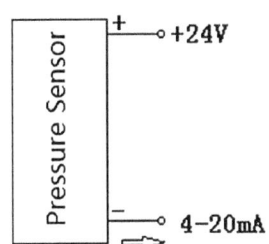

Fig. 13.6 Connection of
two-wire sensor signals

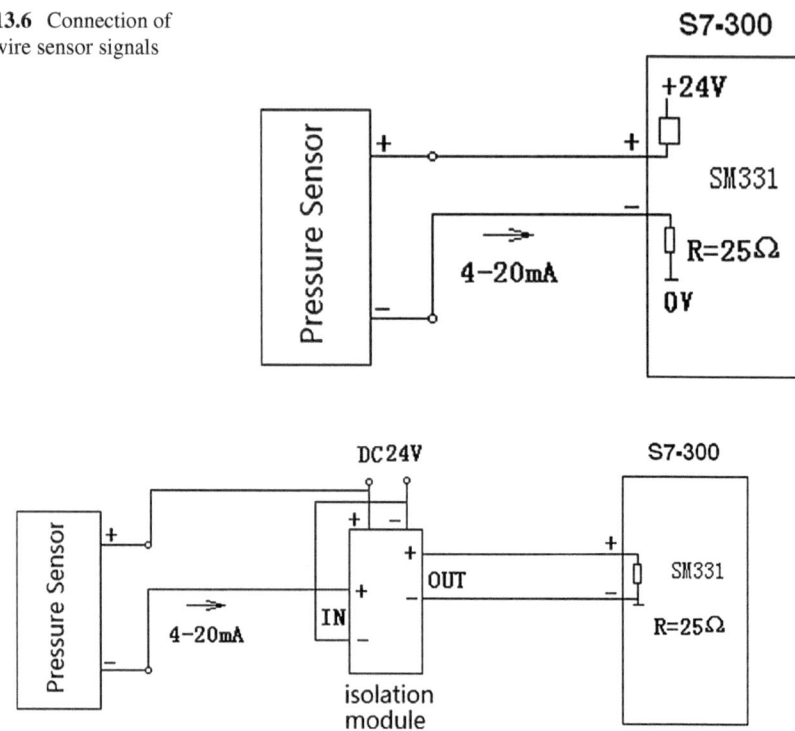

Fig. 13.7 Connection of two-wire sensor and isolation module

connected to the internal power supply 0 V. The current signal output by the pressure
sensor flows through the resistor R, the signal is collected and measured on the
resistor R, and the 0 V of the power supply and the (−) terminal of the signal are
moved to the PLC.

If the isolation module is used to receive 2-wire signals, the sensor and the isolation
module share the same power supply, and the SM331 module of the S7-300 is set to
the 4-wire receiving position. The wiring diagram is shown in Fig. 13.7.

If the isolation module is used to receive 2-wire signals, the sensor and the isolation
module do not share the same power supply, and the SM331 module of the S7-300
is set to the 4-wire receiving position. The wiring diagram is shown in Fig. 13.8.

13.1.5 Power Isolation and Sharing of Isolation Modules

After adding the isolation module, the power supply on the sensor can no longer
be connected casually, otherwise, the isolation module will not work at all, why?
Because the signal input and signal output of the isolation module must be isolated,
but there are three possibilities for the power supply on the isolation module.

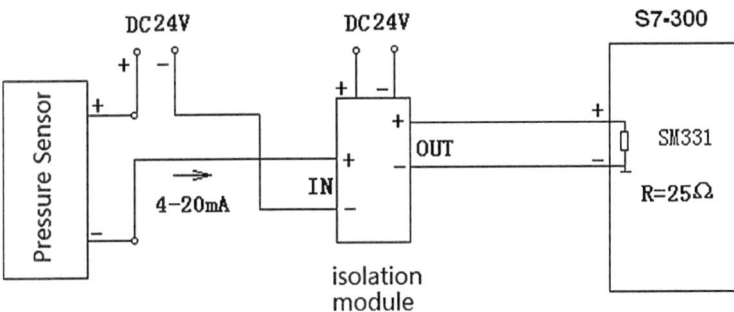

Fig. 13.8 Connection of two-wire sensor and isolation module (dual power supply)

Depending on the products of different manufacturers, the power supply is isolated from the input signal, the power supply is isolated from the output signal, and the power supply is isolated from the input signal and the output signals.

1. If the power supply of the sensor and the power supply on the isolation module share the same power supply, after the signal is input to the isolation module. If the signal input of the isolation module is connected to the public power supply, the input signal of the isolation module is no longer isolated from its own power supply.

    (1) If the isolation module is selected as the one that does not isolate the output signal and power supply, the use of such an isolation module will have no effect, but only adds a possible failure point;

    (2) If the isolation module is selected If the signal input and power supply are not isolated, then it is no problem to use such an isolation module;

    (3) If the isolation module is the one that isolates the input, output and power supply, then such an isolation module is even more problematic to use, It's just that the input and power supply are no longer isolated.

2. If the power supply of the sensor and the power supply on the PLC share the same power supply, the isolation module has no isolation meaning, because the signal of the field sensor has been connected to the analog input module of the PLC on the signal receiving side through the power supply of the PLC. This is the most important thing to be noticed. Some enterprise technicians have encountered such problems and found that adding an isolation module has no effect. In places where the interference is not strong and isolation is not required, the isolation module may not be used.

3. If an isolation module is used, if possible, the power supplies of all isolation modules can share one power supply. All on-site sensors can share one or more power supplies (depending on the distance of the sensor, the strength of on-site interference, etc.). One or more power supplies are used on the PLC side (depending on the safety and current load conditions), and the three-party power supplies are best not to be electrically connected.

13.1.6 Inverter Interference

The sine wave output by the frequency converter is superimposed by many high-frequency square waves, which contains a large number of harmonic components. The equipment cannot work normally, and sometimes the frequency converter itself often has ground faults, resulting in failure to work normally. The high-frequency harmonic interference of the frequency converter has become a public nuisance.

The line from the frequency converter to the controlled motor is long, the insulation of the cable is good. Since the metal wire in the cable and the ground at the cable form a distributed capacitance, the high-frequency harmonics of the frequency converter will still form a displacement current and enter the ground. No matter how well insulated the cable is, it will not help, because this is not a problem caused by the insulation strength of the cable, which will cause the sum of the three-phase current vectors entering the inverter to be non-zero, as shown in Fig. 13.9, which may cause the leakage protection switch to trip. Malfunction or cause the inverter to stop due to a motor ground fault, and other electrical equipment connected to the line will also introduce interference through the ground terminal.

To reduce the frequency converter interference, the measures that can be taken are:

1. The inverter must be grounded correctly and reliably according to the instructions.
2. Set the carrier frequency of the inverter as low as possible to reduce the harmonic radiation intensity, reduce the displacement current formed between the inverter output and the earth, avoid tripping of the leakage switch. Reduce the potential formed by the inverter output to other ground terminals through radiation Interference and the interference introduced into the system, and also reduce the frequency converter's interference to video signals (such as closed-circuit television).
3. The output side of the inverter is connected to an output reactor to reduce the electromagnetic radiation of the cable, reduce the displacement current formed between the inverter output and the earth, avoid tripping of the leakage switch, and reduce the potential formed by the inverter output to other ground terminals through radiation Interference and the interference introduced into the system, and also reduce the interference of the frequency converter to the video signal, see the chapter of the frequency converter.

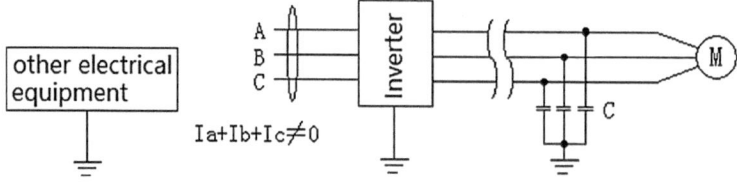

Fig. 13.9 Inverter output interference

4. The analog input and output signals connected to the frequency converter are isolated with an isolation module, and the switch signals are isolated with an intermediate relay.

5. Add an input reactor and a radio interference suppressor to the power input side of the frequency converter, see the chapter on the frequency converter, to reduce the harmonic pollution of the frequency converter to the grid side. When multiple frequency converters share a power grid, the input reactor can also reduce mutual influence and malfunction.

6. The power cable on the input side of the inverter is armored cable, and the metal armored inverter side is grounded to reduce electromagnetic radiation and also reduce the interference of the inverter to video signals.

7. The power cable on the output side of the inverter is armored cable, and the metal armored inverter side is grounded to reduce electromagnetic radiation and also reduce the interference of the inverter to the video signal.

8. If the frequency converter is used, the interference to the closed-circuit television monitoring signal is very serious. The technical personnel responsible for the closed-circuit television project should consider whether to add an anti-interference frequency conversion device, and first modulate the on-site video signal to the interference band of the frequency converter. In addition, make it much higher than the frequency of the interference signal, and then add a demodulator at the receiving end to restore the high-frequency signal back to the normal video signal.

13.1.7 Power Interference

Many interference signals are propagated through power lines. For control lines and control devices, the following measures can be used to reduce the impact of interference on the system.

1. Its power supply can be powered by a 1:1 isolation transformer and the shield of the isolation transformer is grounded reliably. Note that the AC output after isolation has no ground terminal, so there is no fire or zero, so touching any one of the wires will not cause electric shock, as shown in Fig. 13.10.

2. For a more complex system, if there are several controllers such as PLC or many signals from different sites inside, in order to increase the anti-interference ability, it is best to use its own independent power supply for each part. Try to avoid each part make an electrical connection, because any electrical connection, whether it is 0 V or +24 V, will cause interference signals to enter each other! If you need to use five 1A power supplies to separate the electrical connections of each part,

Fig. 13.10 Isolation transformer

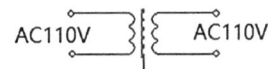

Fig. 13.11 RC filter

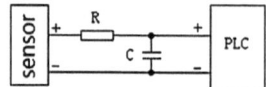

you cannot use one 5A power supply to make the controllers of each part public to replace.

3. Use a single-phase or three-phase power filter to block the interference from the power system.

13.1.8   Anti-interference of Sensor Output Signal

The weak current signal from the sensor to the PLC (or other controllers) can be filtered by resistance–capacitance to reduce the influence of interference (signal noise), as shown in Fig. 13.11.

The output signal of the sensor, whether it is voltage or current, is filtered by R and C resistors, the high-frequency interference signal in the signal is filtered out, and the output signal is smooth. If the sensor is a voltage signal, the resistance R can be larger, ranging from 1 K to hundreds of K, and the capacitance C is from 0.1 to 10 μF. If the sensor outputs a current signal, the sum of the resistance R and the input resistance of the PLC side cannot be greater than the sensor the maximum load resistance value. In most cases $R \leq 500\Omega$, the value of capacitor C is between 0.10 and 10 Mf. After adding resistance–capacitance filtering, the signal reflection speed will slow down, for occasions with high reflection speed requirements (such as fast precision transmission), cannot be handled in this way.

13.1.9   Digital Input of the Controller

Sometimes the switching value input of PLC or other controllers is affected by external interference, and the input error occurs instantaneously, causing the PLC to malfunction. Eliminate interference such as electrostatic induction or electric field coupling, as shown in Fig. 13.12. For an effective input connected to +24 V, the capacitor and resistor are grounded, and for an effective input connected to 0 V, connect the resistor and capacitor in parallel and then connect to +24 V.

Fig. 13.12 Eliminate the
interference of digital input

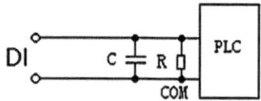

Fig. 13.13 Remote AC
contactor control failure

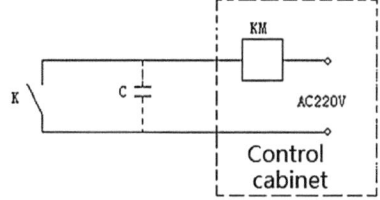

13.1.10 Electrical Circuit Control Failure

When using buttons and switches to start and stop an AC motor (or equipment) far away, sometimes you want to turn it off but you can't. The circuit is shown in Fig. 13.13.

Since the switch K is far away from the AC contactor KM, the two wires are very long and the distributed capacitance C will become larger. In the AC circuit, there will be a displacement current flowing in this distributed capacitance, even if the switch K is turned off. It may be that the KM cannot be released because the energy maintained by the KM does not need to be too large, and the equipment cannot stop. In this case, the following solutions can be adopted:

1. It can be remotely controlled by DC signal.
2. Add a resistor to the coil of the AC contactor KM, so that the current flowing through the distributed capacitance is not enough to maintain the pull-in of the coil.

13.2 Selection of Signal Lines and Shielding Grounding Issues

For the transmission of weak signals, if a pair of loops with equal currents and opposite directions can be formed, it is best to use twisted-pair wires, so that the wire itself has a certain anti-interference ability. Because the induced voltage formed by two similar paired wires is just opposite, itself offsets the interference signal coupled in from the outside world, and if coupled with good shielding, its anti-interference ability will be stronger.

The shielding layer of most weak current signal lines can be grounded at one point on the receiving signal side (such as the PLC side), or not grounded on both sides, depending on the anti-interference effect of the actual site. Most books and materials introducing anti-interference emphasize shielding A little grounding principle, but we have also found cases where the grounding effect is better on both sides of the shielded wire in practice. If one end cannot be grounded, the other end of the shield can also be grounded with a 0.1 μ capacitor.

13.3   Failure Analysis

The sequence of failure analysis is generally as follows:

1. The first is to analyze the power supply part: measure the power supply to see if there is electricity or lack of phase. If the power supply is abnormal, check whether the circuit breaker of the power supply is tripped, whether the fuse or fuse of the secondary control line is blown, and whether the power switch Whether the contacts are good. In actual work, many people often ignore this step. The equipment that has not been repaired for several days may be that a fuse is broken or the contacts of the power switch are in poor contact, or there is no power at all. If the power supply of the device is normal, this step can be skipped.

2. Check the input part of the equipment: In the closed-loop automatic control system, if there is no input signal or is abnormal, the system cannot work normally, just like a normal person walking. If his eyes have problems, he will not be able to walk naturally. Normally, check whether the input sensor is faulty or disconnected. In electrical control, if the contact of the function input button is abnormal or the self-protection contact of the relay is not in good contact, the electrical control system cannot work normally. This step can be skipped if all the input signals are normal or the function control buttons and self-protection contacts are normal.

3. Check the output part of the equipment: If the output signal of the controller is available, but the actuator (such as a frequency converter) does not act. It means that there is a problem with the execution part or the connection to the actuator. In electrical control, check whether the power supply to the motor and other electrical equipment is normal, and whether there is a phase loss problem. If it is normal, it means that there is a problem with the electrical equipment (or motor). If the output signal of the controller is normal, skip this step.

4. Check the intermediate circuit and the main controller: start from the power supply and start from top to bottom, check the intermediate circuit to see which component has the power failure or phase loss, and then solve it. For the main controller (such as PLC), first check whether the action of the output port and the output signal are normal, and if it is normal, then focus on checking the program to see where there is a problem.

5. After a fault occurs, unless there is danger, do not rush to reset, carefully check whether the signals of each link are normal or not, and use the detection function of the programming software on the computer to diagnose the cause of the fault.

13.4   Lightning Protection Measures

Lightning sometimes causes devastating damage to the automatic control system. Many lightning strikes are transmitted to the system through outdoor sensors, especially cables that transmit signals through high overhead lines are more likely to be

Fig. 13.14 Lightning protection measures

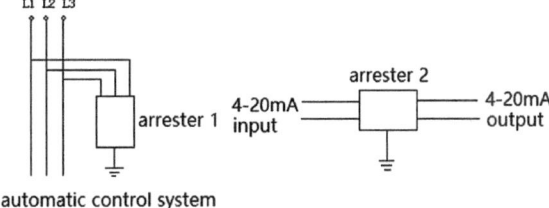

automatic control system

struck by lightning. The sensors are located near the liquid surface in a large area. The sensor is easy to damage the sensor because the charge induced by the liquid level does not have a good discharge channel, but more lightning strikes are transmitted through the power system.

The measures for power and signal lightning protection of the automatic control system are shown in Fig. 13.14. The ground terminal of the arrester must have a good ground connection. In the figure, arrester 1 is a three-phase power arrester, and arrester 2 is a signal lightning protection module.

13.5 Communication Port Crash Problem

When using a custom communication port, sometimes the communication port crash problem occurs, and resetting the communication port does not work, it can only be restored by disconnecting the power supply, such as the RS232/485 communication module of Siemens SMART and the CP340 equipped with S-300PLC series, this problem has occurred. The author finally found a solution, that is, once the communication port is found to be down, modify the parameters of the communication port to other parameters, and then modify the parameters of the communication port back to solve this problem.

Chapter 14
Energy Efficiency Optimization of Multi-Unit System

14.1 What is a Multi-Unit System?

Multi-unit systems refer to those systems that are composed of more than one device to complete the same task.

For example, a power plant composed of multiple generators, a pumping station composed of multiple water pumps, a high-speed train driven by multiple motors, a hydrogen production station composed of multiple hydrogen generators, an LNG station composed of multiple vaporizers, and multiple transformers composed of substations, and so on.

Only a system with multiple energy-consuming devices that work together to complete a task is called a multi-unit system. Multi-unit systems include numerous systems: power stations, transmission and distribution stations, energy-consuming systems, etc.

14.2 The Essence of Multi-Unit System Optimization

If you have on-site operating data and equipment data, can you complete the following work?

(1) With the same input of water volume and water pressure, can you still increase the power generation of the largest hydropower station in the world? How much? By what method? What is the basis?
(2) When transporting the same amount of water and the same water pressure, can you save more electricity for the world's largest water diversion pump station? How much can you save? By what method? What is the basis?
(3) Under the same wind volume and wind speed, can the world's largest wind power hydrogen production station still increase the hydrogen production capacity? How much? By what method? What is the basis?

F. Yao and Y. Yao, *Efficient Energy-Saving Control and Optimization for Multi-Unit Systems*, https://doi.org/10.1007/978-981-97-4492-3_14

There are many such examples. It can be said that as long as there is more than one power generation or power consumption equipment in a system, there will be such a problem. This is the problem of energy efficiency optimization.

The so-called energy efficiency optimization is to complete the same task with the least energy. It can also be said: input the same primary energy and produce the most secondary energy.

Obviously, this is not a simple job.

Can you do this confidently?

Can you accurately predict optimal overall energy efficiency?

How much is it?

This chapter is intended to address these issues.

Energy efficiency optimization of multi-unit system mainly solves three problems:

1. How many devices are optimal?
2. What is the optimal output of each device?
3. What is the maximum operating energy efficiency of the system?

The first two are methods, the last one is the result.

14.3 Energy Efficiency Optimization of Multi-Unit System

A system C that converts energy A into energy B has m devices in total, and n devices are running. The total amount of energy A consumed by system C is P_t, and the total amount of energy B produced is W_t. The i-th device produces energy B is W_i, the energy A consumed by the i-th device is P_i, and the energy efficiency optimization of system C has two expressions.

For a fixed total amount of energy-A input P_t, the maximization of the total amount of output energy B W_t is expressed as:

$$W_t = \max \sum_{i=1}^{n} P_i \eta_{Pi}(P_i)$$
$$s.t. \sum_{i=1}^{n} P_i = P_t \qquad (14.1)$$
$$P_{im} \geq P_i \geq 0$$

For a fixed total energy B output W_t, the minimization of the input energy A total P_t can be expressed as:

$$P_t = \min \sum_{i=1}^{n} \frac{W_i}{\eta_{wi}(W_i)}$$
$$s.t. \sum_{i=1}^{n} W_i = W_t \qquad (14.2)$$
$$W_{im} \geq W_i \geq 0$$

Energy efficiency functions $\eta_{Pi}(P_i)$ and $\eta_{wi}(W_i)$ are not the same function, their variables are different, and their function values are also different.

The Eq. (14.1) and Eq. (14.2) are two different manifestations of the same optimization problem, and both are running energy efficiency optimization problems. Only one of them is solved, the problem is solved. We choose to solve the optimal result of Eq. (14.1).

It can be seen from Eq. (14.1) that since the energy efficiency function is nonlinear and has a limit on the maximum input value P_{im}, this is a nonlinear optimization problem with constraints. Since the number n of optimized operating units is an integer, the optimal load distribution value P_i is a real number, which is another integer-real-number mixed optimization problem.

We call the method to solve this optimization problem the quantum optimization method.

14.4 Energy Efficiency Function

The energy efficiency function has a similar shape and some of the same characteristics. Let's take $\eta_{Pi}(P)$ as an example, for the convenience of writing, we use $\eta_i(P)$ instead of $\eta_{Pi}(P)$. The energy efficiency curve of $\eta_i(P)$ is shown in Fig. 14.1.

In Fig. 14.1, P_{im} is the maximum A energy input of the i-th device, η_{ie} is the highest operating energy efficiency of the i-th device, P_{ie} is the A energy input when the i-th device has the highest operating energy efficiency, and $\eta_i(P)$ has the following characteristics:

$$0 \leq P \leq P_{im}$$
$$\eta_{ie} = \eta_i(P_{ie})$$
$$0 \leq \eta_i(P) \leq \eta_{ie} \qquad (14.3)$$
$$\eta_i(0) = 0$$
$$\eta_i''(P) < 0$$

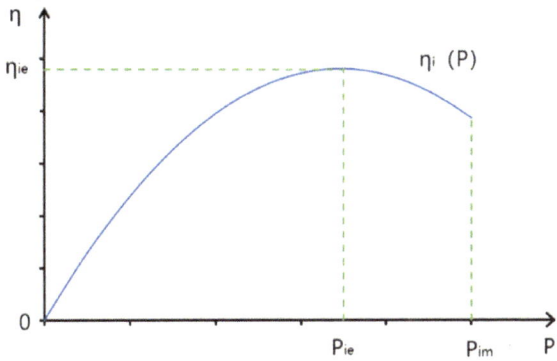

Fig. 14.1 Energy efficiency curve $\eta_i(P)$

Fig. 14.2 Energy efficiency curves $\eta_1(P)$ and $\eta_i(P)$

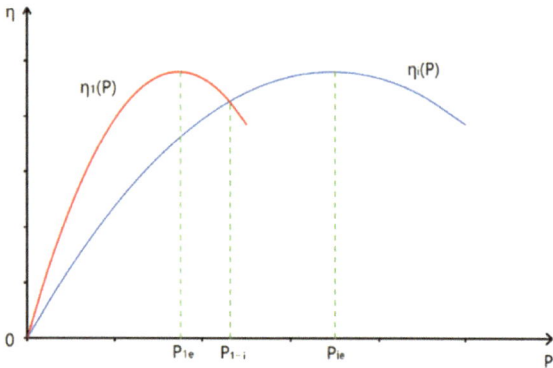

14.5 Similar Energy Efficiency Device

Assume the energy efficiency curves $\eta_1(P)$ and $\eta_i(P)$ of the first and i-th device, as shown in Fig. 14.2.

If the following equation holds:

$$\eta_i(P_i) = \eta_1\left(\frac{P_1}{\beta_i}\right) \tag{14.4}$$

where β_i is a constant, we say that the i-th device and the first device are devices with similar energy efficiency, referred to as "similar energy efficiency device".

$\beta_i < 1$, the i-th device is a device with a smaller output capability than the first device;

$\beta_i = 1$, then the i-th device and the first device are the same device, recorded as $\beta_1 = 1$;

$\beta_i > 1$, the i-th device has a larger output capability than the first device.

Device with similar energy efficiency has the following characteristics:

$$\begin{aligned} \eta_i'(P_i) &= \frac{\eta_1'(P_1)}{\beta_i} \\ P_{ie} &= \beta_i P_{1e} \\ \beta_i &= \frac{P_{ie}}{P_{1e}} \end{aligned} \tag{14.5}$$

14.6 Optimal Load Distribution Theorem of Multi-Unit System, Yao Theorem 1

Assume that Pi is greater than 0, that is, every operating device does work. The Eq. (14.1) is simplified as

$$W_t = \max \sum_{i=1}^{n} P_i \eta_i(P_i)$$
$$s.t. \sum_{i=1}^{n} P_i = P_t > 0 \tag{14.6}$$
$$P_{imax} \geq P_i > 0$$

Assuming that n is optimal, consider the following three situations:

(1) $n = 2$

System C has two variables P_1 and P_2

$$P_1 + P_2 = P_t$$
$$P_1 > 0 \tag{14.7}$$
$$P_2 > 0$$

The objective function can be expressed as

$$W_t = P_1 \eta_1(P_1) + P_2 \eta_2(P_2) \tag{14.8}$$

The optimization condition is

$$W_t'(P_1) = 0 \tag{14.9}$$

According to known conditions

$$P_2 = P_t - P_1 \tag{14.10}$$

We have

$$\eta_1(P_1) + P_1 \eta_1'(P_1) - \eta_2(P_2) - P_2 \eta_2'(P_2) = 0 \tag{14.11}$$

If two devices are devices with the identical energy efficiency, then

$$\eta_2(P) = \eta_1(P) \tag{14.12}$$

easy to see

$$P_1 = \frac{P_t}{2}$$
$$P_2 = P_1 = \frac{P_t}{2} \tag{14.13}$$

is an optimization point.

The optimal control method is to keep

$$P_1 = P_2 = \frac{P_t}{n} \tag{14.14}$$

The maximum total output energy W_t of system C is

$$\max W_t = P_t \eta_1 \left(\frac{P_t}{n} \right) \tag{14.15}$$

The maximum value of the overall operating energy efficiency $\eta_{t2}(P_t)$ of system C is

$$\max \eta_{t2}(P_t) = \eta_1 \left(\frac{P_t}{2} \right) \tag{14.16}$$

Since the overall energy efficiency of the system is the same as that of a single device, the second derivative is also less than zero.

$$W_t''(P_1) < 0 \tag{14.17}$$

W_t has the only maximum value, and the overall energy efficiency has the only maximum value also.

If two devices are similar energy efficiency devices, then we have

$$\eta_2(P_2) = \eta_1 \left(\frac{P_1}{\beta_2} \right) \tag{14.18}$$

and it is easy to see that

$$\begin{aligned} P_1 &= \frac{P_t}{1+\beta_2} \\ P_2 &= \beta_2 P_1 \end{aligned} \tag{14.19}$$

is an optimization point.

The optimal control method is to keep

$$\begin{aligned} P_1 &= \frac{P_t}{1+\beta_2} \\ P_2 &= \beta_2 P_1 \end{aligned} \tag{14.20}$$

The maximum total output energy W_t of system C is

$$\max W_t = P_t \eta_1 \left(\frac{P_t}{1+\beta_2} \right) \tag{14.21}$$

The maximum value of the overall operating energy efficiency $\eta_{t2}(P_t)$ of system C is

$$\max \eta_{t2}(P_t) = \eta_1 \left(\frac{P_t}{1+\beta_2} \right) \tag{14.22}$$

Since the shape of the overall efficiency curve of the system is the same as that of a single device, the second derivative is also less than zero. W_t has a maximum value, and the overall energy efficiency is a maximum value also.

(2) $n = 3$

The system has three variables P_1, P_2 and P_3

$$P_1 + P_2 + P_3 = P_t$$
$$P_1 > 0$$
$$P_2 > 0 \tag{14.23}$$
$$P_3 > 0$$

W_t expression is

$$W_t = P_1\eta_1(P_1) + P_2\eta_2(P_2) + P_3\eta_3(P_3) \tag{14.24}$$

If the three devices have the identical energy efficiency, using the commutative law of addition, assuming that P_3 is the optimal point and fixed, P_1 and P_2 are variables, there is

$$P_1 + P_2 = P_t - P_3 \tag{14.25}$$

Based on the conclusion of $n = 2$, there are

$$P_2 = P_1 \tag{14.26}$$

Similarly, assuming that P_2 is the optimal point and has been fixed, there is

$$P_3 = P_1 \tag{14.27}$$

Assuming that P_1 is the optimal point and it has been fixed, we have

$$P_3 = P_2 \tag{14.28}$$

where

$$P_1 = P_2 = P_3 = \frac{P_t}{3} \tag{14.29}$$

is an optimization point.
    The optimal control method is to keep

$$P_1 = P_2 = P_3 = \frac{P_t}{n} \tag{14.30}$$

The maximum total output energy W_t of system C is

$$\max W_t = P_t\eta_1\left(\frac{P_t}{n}\right) \tag{14.31}$$

The overall maximum operating energy efficiency $\eta_{t3}(P_t)$ of system C is

$$\max \eta_{t3}(P_t) = \eta_1\left(\frac{P_t}{n}\right) \tag{14.32}$$

If the three devices have similar energy efficiency, using the commutative law of addition, assuming that P_3 is the optimal point and fixed, based on the conclusion of $n = 2$, we have

$$P_2 = \beta_2 P_1 \tag{14.33}$$

Similarly, assuming that P_2 is the optimal point and has been fixed, there is

$$P_3 = \beta_3 P_1 \tag{14.34}$$

where

$$\begin{aligned} P_1 &= \frac{P_t}{1+\beta_2+\beta_3} \\ P_2 &= \beta_2 P_1 \\ P_3 &= \beta_3 P_1 \end{aligned} \tag{14.35}$$

is an optimization point.

The optimal control method is to keep

$$\begin{aligned} P_1 &= \frac{P_t}{\sum_{i=1}^{3}\beta_i} \\ P_2 &= \beta_2 P_1 \\ P_3 &= \beta_3 P_1 \end{aligned} \tag{14.36}$$

The maximum total output energy W_t of the system is

$$\max W_t = P_t \eta_1\left(\frac{P_t}{\sum_{i=1}^{3}\beta_i}\right) \tag{14.37}$$

The maximum value of the overall operating energy efficiency $\eta_{t3}(P_t)$ of system C is

$$\max \eta_{t3}(P_t) = \eta_1\left(\frac{P_t}{\sum_{i=1}^{3}\beta_i}\right) \tag{14.38}$$

(3) $n = k$

If k devices are devices with the identical energy efficiency, the above conclusion is extended to the case of $n = k$, and the optimal point is

$$P_1 = P_2 = \cdots = P_k = \frac{P_t}{k} \tag{14.39}$$

The optimal control method is to keep

$$P_1 = P_2 = \cdots = P_k = \frac{P_t}{k} \tag{14.40}$$

The maximum total output energy W_t of system C is

$$\max W_t = P_t \eta_1 \left(\frac{P_t}{k}\right) \tag{14.41}$$

The maximum value of the overall operating energy efficiency $\eta_{tk}(P_t)$ of system C is

$$\max \eta_{tk}(P_t) = \eta_1 \left(\frac{P_t}{k}\right) \tag{14.42}$$

If k devices are devices with similar energy efficiency, the above conclusion is extended to the case of $n = k$, and the optimal point is

$$
\begin{aligned}
P_1 &= \frac{P_t}{\sum_{i=1}^{k} \beta_i} \\
P_2 &= \beta_2 P_1 \\
&\cdots \\
P_k &= \beta_k P_1
\end{aligned}
\tag{14.43}
$$

is an optimization point.

The optimal control method is to keep

$$
\begin{aligned}
P_1 &= \frac{P_t}{\sum_{i=1}^{k} \beta_i} \\
P_2 &= \beta_2 P_1 \\
&\cdots \\
P_k &= \beta_k P_1
\end{aligned}
\tag{14.44}
$$

The maximum total output energy W_t of system C is

$$\max W_t = P_t \eta_1 \left(\frac{P_t}{\sum_{i=1}^{k} \beta_i}\right) \tag{14.45}$$

The maximum value of the overall operating energy efficiency $\eta_{tk}(P_t)$ of system C is

$$\max \eta_{tk}(P_t) = \eta_1 \left(\frac{P_t}{\sum_{i=1}^{k} \beta_i}\right) \tag{14.46}$$

Whether it is device with the identical energy efficiency or device with similar energy efficiency, their optimal load distribution methods have one thing in common, that is, the energy efficiency of all operating device is the same.

$$\eta_1(P_1) = \eta_2(P_2) = \cdots = \eta_n(P_n) \tag{14.47}$$

Optimal Load Distribution Theorem (Yao Theorem 1): The optimal load distribution method is to keep the operating energy efficiency of each operating device equal.

$$\eta_1(P_1) = \eta_2(P_2) = \cdots = \eta_n(P_n) \tag{14.48}$$

If all devices have the identical energy efficiency, the optimal load distribution method is to keep the load of every operating device equal.

$$P_1 = P_2 = \cdots = P_n \tag{14.49}$$

Using the inductive method, we can still draw the above conclusion

(1) $n = 2$

Same as above. (omitted).

(2) If $n = k$ holds, prove that $n = k + 1$ still holds.

For system C composed of n devices with the identical energy efficiency function, if the optimization conclusion holds,

$$\begin{aligned} W_t &= \max \sum_{i=1}^{k} P_i \eta(P_i) \\ s.t.\ &\sum_{i=1}^{k} P_i = P_t > 0 \\ &P_{imax} \geq P_i > 0 \end{aligned} \tag{14.50}$$

The optimal point is

$$P_1 = P_2 = \cdots = P_k = \frac{P_t}{k} \tag{14.51}$$

The maximum total output energy W_t of system C is

$$\max W_t = P_t \eta(\frac{P_t}{k}) \tag{14.52}$$

The maximum value of the overall operating energy efficiency $\eta_{tk}(P_t)$ of system C is

$$\max \eta_{tk}(P_t) = \eta(\frac{P_t}{k}) \tag{14.53}$$

For $n = k + 1$, we have

$$
\begin{aligned}
W_t &= \sum_{i=1}^{k+1} P_i \eta(P_i) = \sum_{i=1}^{k} P_i \eta(P_i) + P_{k+1} \eta(P_{k+1}) \\
&= (P_t - P_{k+1}) \eta\left(\frac{P_t - P_{k+1}}{k}\right) + P_{k+1} \eta(P_{k+1})
\end{aligned}
\tag{14.54}
$$

The optimization condition is

$$
W_t'(P_{k+1}) = 0 \tag{14.55}
$$

We have

$$
\begin{aligned}
& -\eta\left(\frac{P_t - P_{k+1}}{k}\right) - \frac{P_t - P_{k+1}}{k} \eta'\left(\frac{P_t - P_{k+1}}{k}\right) \\
& + \eta(P_{k+1}) + P_{k+1} \eta'(P_{k+1}) = 0
\end{aligned}
\tag{14.56}
$$

It is easy to see that, the optimal point is

$$
\begin{aligned}
P_{k+1} &= \frac{P_t - P_{k+1}}{k} \\
P_{k+1} &= \frac{P_t}{k+1} \\
P_1 = P_2 &= \dots P_k = \frac{P_t - P_{k+1}}{k} = \frac{P_t}{k+1}
\end{aligned}
\tag{14.57}
$$

The maximum total output energy W_t of system C is

$$
\max W_t = P_t \eta\left(\frac{P_t}{k+1}\right) \tag{14.58}
$$

The maximum value of the overall operating energy efficiency $\eta_{tk}(P_t)$ of system C is

$$
\max \eta_{t(k+1)}(P_t) = \eta\left(\frac{P_t}{k+1}\right) \tag{14.59}
$$

The above conclusion still works.

For system C composed of n devices with the similar energy efficiency function, if the optimization conclusion holds,

$$
\begin{aligned}
W_t &= \max \sum_{i=1}^{k} P_i \eta_i(P_i) \\
s.t. \sum_{i=1}^{k} P_i &= P_t > 0 \\
P_{imax} &\geq P_i > 0
\end{aligned}
\tag{14.60}
$$

The optimal point is

$$P_1 = \frac{P_t}{\sum_{i=1}^{k} \beta_i}$$
$$P_2 = \beta_2 P_1$$
$$\cdots$$
$$P_k = \beta_k P_1$$

(14.61)

The maximum total output energy W_t of system C is

$$\max W_t = P_t \eta_1 \left(\frac{P_t}{\sum_{i=1}^{k} \beta_i} \right)$$

(14.62)

The maximum value of the overall operating energy efficiency $\eta_{tk}(P_t)$ of system C is

$$\max \eta_{tk}(P_t) = \eta_1 \left(\frac{P_t}{\sum_{i=1}^{k} \beta_i} \right)$$

(14.63)

For $n = k + 1$, we have

$$W_t = \sum_{i=1}^{k+1} P_i \eta_i(P_i) = (P_t - P_{k+1}) \eta_1 \left(\frac{P_t - P_{k+1}}{\sum_{i=1}^{k} \beta_i} \right) + P_{k+1} \eta_{k+1}(P_{k+1})$$

$$= (P_t - P_{k+1}) \eta_1 \left(\frac{P_t - P_{k+1}}{\sum_{i=1}^{k} \beta_i} \right) + P_{k+1} \eta_1 \left(\frac{P_{k+1}}{\beta_{k+1}} \right)$$

(14.64)

The optimization condition is

$$W_t'(P_{k+1}) = 0$$

(14.65)

We have

$$- \eta_1 \left(\frac{P_t - P_{k+1}}{\sum_{i=1}^{k} \beta_i} \right) - \frac{P_t - P_{k+1}}{\sum_{i=1}^{k} \beta_i} \eta_1{}' \left(\frac{P_t - P_{k+1}}{\sum_{i=1}^{k} \beta_i} \right)$$
$$+ \eta_1 \left(\frac{P_{k+1}}{\beta_{k+1}} \right) + \frac{P_{k+1}}{\beta_{k+1}} \eta' \left(\frac{P_{k+1}}{\beta_{k+1}} \right) = 0$$

(14.66)

It is easy to see that, the optimal point is

$$\frac{P_{k+1}}{\beta_{k+1}} = \frac{P_t - P_{k+1}}{\sum_{i=1}^{k} \beta_i}$$
$$P_{k+1} = \frac{\beta_{k+1} P_t}{\sum_{i=1}^{k+1} \beta_i}$$
$$P_1 = \frac{P_t}{\sum_{i=1}^{k+1} \beta_i}$$

(14.67)

The maximum total output energy W_t of system C is

$$\max W_t = P_t \eta_1 \left(\frac{P_t}{\sum_{i=1}^{k+1} \beta_i} \right) \tag{14.68}$$

The maximum value of the overall operating energy efficiency $\eta_{tk}(P_t)$ of system C is

$$\max \eta_{t(k+1)}(P_t) = \eta_1 \left(\frac{P_t}{\sum_{i=1}^{k+1} \beta_i} \right) \tag{14.69}$$

The above conclusion works very well.

14.7 Optimal Switching Theorem for Multi-Unit System, Yao Theorem 2

The optimal methods of load distribution obtained above are all obtained under the assumption that n is already optimal, but is n optimal? We analyze two cases.

The total value of input energy A of system C is P_t, and there are n running devices with the identical energy efficiency, the highest overall energy efficiency of system C is

$$\max \eta_{tn}(P_t) = \eta_1 \left(\frac{P_t}{n} \right) \tag{14.70}$$

For the same P_t, there are $(n-1)$ running devices with the identical energy efficiency, and the highest overall energy efficiency of system C is

$$\max \eta_{t(n-1)}(P_t) = \eta_1 \left(\frac{P_t}{n-1} \right) \tag{14.71}$$

For the same P_t, there are $(n+1)$ running devices with the identical energy efficiency, and the highest overall energy efficiency of system C is

$$\max \eta_{t(n+1)}(P_t) = \eta_1 \left(\frac{P_t}{n+1} \right) \tag{14.72}$$

apparently

$$\frac{P_t}{n-1} > \frac{P_t}{n} > \frac{P_t}{n+1} \tag{14.73}$$

On the $\eta_1(P_1)$ energy efficiency curve, the $P_t/(n-1)$ point is on the right side of the P_t/n point, and the $P_t/(n+1)$ point is on the left side of the P_t/n point.

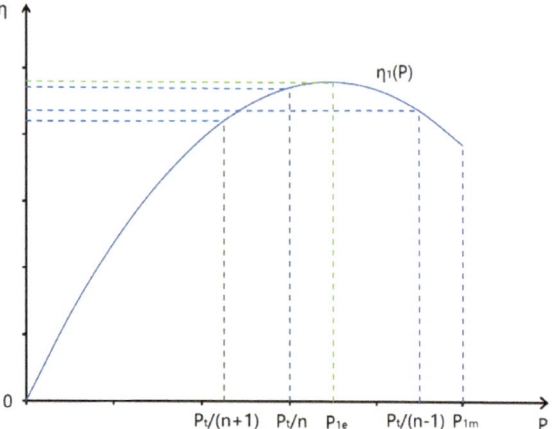

Fig. 14.3 Energy efficiency comparison curve of the multi-unit system with the identical devices

$$\eta_1\left(\frac{P_t}{n}\right) \geq \eta_1\left(\frac{P_t}{n-1}\right)$$
$$\eta_1\left(\frac{P_t}{n}\right) \geq \eta_1\left(\frac{P_t}{n+1}\right) \tag{14.74}$$

That is

$$\eta_1\left(\frac{P_t}{n}\right) = max\left\{\eta_1\left(\frac{P_t}{n-1}\right), \eta_1\left(\frac{P_t}{n}\right), \eta_1\left(\frac{P_t}{n+1}\right)\right\} \tag{14.75}$$

Then the running number n is truly optimal.

As shown in Fig. 14.3, $\eta_1(P_t/n)$ has the highest operating energy efficiency, and the P_t/n point is closer to the P_{1e} point than $P_t/(n-1)$ and $P_t/(n+1)$,

In the above discussion, we have always regarded P_t as an invariable constant. In practical applications, P_t is a quantity that changes with the process requirements.

When P_t increases, $\eta_1(P_t/(n+1))$ also increases, when the following conditions are met

$$\eta_1\left(\frac{P_t}{n+1}\right) = \eta_1\left(\frac{P_t}{n}\right) \tag{14.76}$$

The switching point is reached, if P_t continues to increase, then

$$\eta_1\left(\frac{P_t}{n+1}\right) > \eta_1\left(\frac{P_t}{n}\right) \tag{14.77}$$

It should be switched from running on n devices to running on $(n+1)$ devices.

Similarly, when P_t decreases, $\eta_1(P_t/(n-1))$ increases, when the following conditions are met

$$\eta_1\left(\frac{P_t}{n-1}\right) = \eta_1\left(\frac{P_t}{n}\right) \tag{14.78}$$

The switching point is reached, if P_t continues to decrease, then

$$\eta_1\left(\frac{P_t}{n-1}\right) > \eta_1\left(\frac{P_t}{n}\right) \tag{14.79}$$

we should switch from running on n devices to running on n−1 devices.

Due to the limitation of P_{1m}, when P_t increases, P_t/n also increases until $P_t/n = P_{1m}$, which is still not satisfied

$$\eta_1\left(\frac{P_t}{n+1}\right) = \eta_1\left(\frac{P_t}{n}\right) \tag{14.80}$$

If P_t/n continues to increase, the device will be overloaded, and it is necessary to switch to $(n + 1)$ devices forcibly at the point $P_t/n = P_{1m}$.

Due to the limitation of P_{1m}, when P_t decreases, P_t/n also increases until $P_t/(n-1) = P_{1m}$, which is still not satisfied

$$\eta_1\left(\frac{P_t}{n-1}\right) = \eta_1\left(\frac{P_t}{n}\right) \tag{14.81}$$

It is necessary to switch to (n–1) devices at the point $P_t/(n-1) = P_{1m}$.

The total value of the input energy A of system C is P_t, and there are n devices with similar energy efficiency running, the maximum value of the overall operating energy efficiency $\eta_{tn}(P_t)$ of system C is

$$\max\eta_{tn}(P_t) = \eta_1\left(\frac{P_t}{\sum_{i=1}^{n}\beta_i}\right) \tag{14.82}$$

For the same P_t, there are k devices with similar energy efficiency operating, k is any feasible combination except n, and the highest overall operating energy efficiency of system C is

$$\max\eta_{tk}(P_t) = \eta_1\left(\frac{P_t}{\sum_{i=1}^{k}\beta_i}\right) \tag{14.83}$$

If the number of running units n is optimal, it must satisfy

$$\eta_1\left(\frac{P_t}{n\sum_{i=1}^{n}\beta_i}\right) = max\left\{\eta_1\left(\frac{P_t}{\sum_{i=1}^{n}\beta_i}\right), \eta_1\left(\frac{P_t}{\sum_{i=1}^{k1}\beta_i}\right), \eta_1\left(\frac{P_t}{\sum_{i=1}^{k2}\beta_i}\right), \cdots\right\} \tag{14.84}$$

k1 and k2 are any combination other than the optimal combination of n units this time, and also include other combinations of n units.

Fig. 14.4 Energy efficiency comparison curve of the multi-unit system with the similar efficiency devices

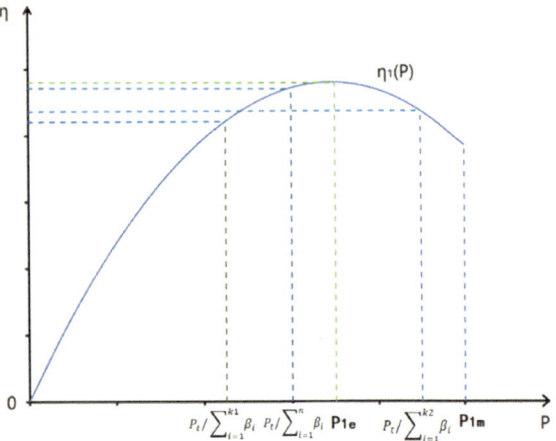

Point $P_t/\sum_{i=1}^{k1}\beta_i$ is the point closest to $P_t/\sum_{i=1}^{n}\beta_i$ to the left of point $P_t/\sum_{i=1}^{n}\beta_i$. Point $P_t/\sum_{i=1}^{k2}\beta_i$ is the point closest to $P_t/\sum_{i=1}^{n}\beta_i$ to the right of point $P_t/\sum_{i=1}^{n}\beta_i$, as shown in Fig. 14.4.

Point A is the point closest to B to the left of point B.

When P_t increases, the condition is satisfied

$$\eta_1\left(\frac{P_t}{\sum_{i=1}^{n}\beta_i}\right) = \eta_1\left(\frac{P_t}{\sum_{i=1}^{k_1}\beta_i}\right) \qquad (14.85)$$

The switching point has been reached. If P_t continues to increase, it should switch from n operating devices to k_1 operating devices.

If there is no point $P_t/\sum_{i=1}^{k1}\beta_i$ that meets the conditions, the switching point is $P_t/\sum_{i=1}^{n}\beta_i = P_{1m}$.

When P_t decreases, the condition is satisfied

$$\eta_1\left(\frac{P_t}{\sum_{i=1}^{n}\beta_i}\right) = \eta_1\left(\frac{P_t}{\sum_{i=1}^{k_2}\beta_i}\right) \qquad (14.86)$$

The switching point has been reached. If P_t continues to decrease, it should switch from n operating devices to k_2 operating devices.

If there is no point $P_t/\sum_{i=1}^{k2}\beta_i$ that meets the conditions, the switching point is $P_t/\sum_{i=1}^{k2}\beta_i = P_{1m}$.

According to the system energy efficiency curve, the optimal switching method is given by observing the changing trend of the curve.

For the control process completed by such optimized switching, when the total load changes, the overall energy efficiency curve between two adjacent load switching points is the amplification of the energy efficiency curve of a single device, and the second derivative is less than zero, so such the solution is the only global optimal

Fig. 14.5 The overall energy efficiency curve of multi-unit system

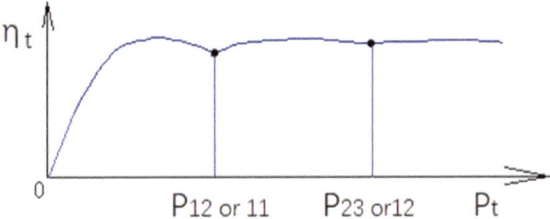

solution, and the overall energy efficiency value is also the global maximum efficiency value. The overall energy efficiency curve as shown in Fig. 14.5.

If there is no point of equal energy efficiency, it is at the point of maximum load of the devices.

Optimal Switching Theorem (Yao Theorem 2): The optimal switching point for the number of operating units is at the point of equal efficiency or at the maximum output point of the devices.

$$\eta_1\left(\frac{P_t}{\sum_{i=1}^{n}\beta_i}\right) = \eta_1\left(\frac{P_t}{\sum_{i=1}^{k}\beta_i}\right) or \frac{P_t}{\sum_{i=1}^{n}\beta_i} = P_{1m} or \frac{P_t}{\sum_{i=1}^{k}\beta_i} = P_{1m} \qquad (14.87)$$

$P_t/\sum_{i=1}^{k}\beta_i$ is the closest point to $P_t/\sum_{i=1}^{n}\beta_i$. If all devices have the identical energy efficiency, then $k = n + 1$ or $k = n-1$.

In practice, many devices need to be warmed up before they can be used, so preparations should be made in advance for starting up the device. Preheating also requires some energy.

14.8 Simulation Results

1. Assume system C has 3 devices in total, and all devices are devices with the same efficiency, there are

$$\eta_1(P) = \eta_2(P) = \eta_3(P) = 2.6P - 2P^2$$
$$\eta_{1e} = \eta_{2e} = \eta_{3e} = 0.845$$
$$P_{1e} = P_{2e} = P_{3e} = 0.65 \qquad (14.88)$$
$$P_{1m} = P_{2m} = P_{3m} = 1.1$$

The total output W_t maximization expression of system C is

$$W_t = \max \sum_{i=1}^{3} P_i\eta_i(P_i)$$
$$s.t. \sum_{i=1}^{3} P_i = P_t > 0 \qquad (14.89)$$
$$P_{im} \geq P_i > 0$$

Based on the above conclusions, the optimal control method when the two devices are running is to keep

$$P_1 = P_2 = \frac{P_t}{2} \tag{14.90}$$

The maximum value of W_t is

$$P_t \eta_1 \left(\frac{P_t}{2} \right) \tag{14.91}$$

The maximum value of the overall operating energy efficiency $\eta_{t2}(P_t)$ is

$$\eta_1 \left(\frac{P_t}{2} \right) \tag{14.92}$$

Based on the above conclusions, the optimal control method for the operation of the three devices is to maintain

$$P_1 = P_2 = P_3 = \frac{P_t}{3} \tag{14.93}$$

The maximum value of W_t is

$$P_t \eta_1 \left(\frac{P_t}{3} \right) \tag{14.94}$$

The maximum value of the overall operating energy efficiency $\eta_{t2}(P_t)$ is

$$\eta_1 \left(\frac{P_t}{3} \right) \tag{14.95}$$

Based on the above conclusions, the switching point P_{12} of one operating device and two operating devices satisfies

$$\eta_1(P_{12}) = \eta_1 \left(\frac{P_{12}}{2} \right) \tag{14.96}$$

inferred

$$P_{12} = 0.8667 \tag{14.97}$$

Based on the above conclusions, the switching point P_{23} of 2 operating devices and 3 operating devices satisfies

$$\eta_1 \left(\frac{P_{23}}{2} \right) = \eta_1 \left(\frac{P_{23}}{3} \right) \tag{14.98}$$

Fig. 14.6 The overall energy efficiency curve of system C with three devices

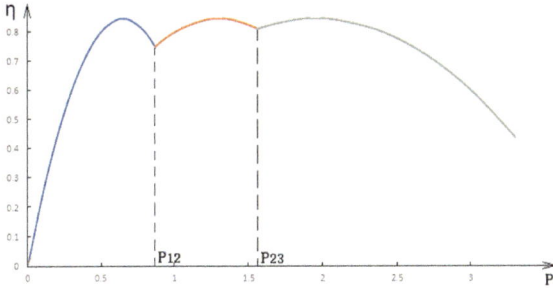

inferred

$$P_{23} = 1.56 \tag{14.99}$$

That is, when Pt changes:
$0 < P_t < \ = 0.8667$, run with 1 device;
$0.8667 < P_t < \ = 1.56$, run with 2 devices;
$1.56 < P_t < \ = 3.3$, run with 3 devices.
When $0 < P_t < \ = 0.8667$, $0.8667 < P_t < \ = 1.56$ and $1.56 < P_t < \ = 3.3$, $W_t^{''} <$ 0, so the obtained W_t is the maximum value.

In this example, if $P_{1m} = 0.8$, there is no equivalent switching point between the operation of one device and the operation of two devices, and only the maximum output point P_{im} of one device can be used as the switching point:
When $0 < P_t < \ = 0.8$, run with 1 device;
When $0.8 < P_t < \ = 1.56$, run with 2 devices;
When $1.56 < P_t < \ = 2.4$, run with 3 devices.
The overall energy efficiency curve of the system C is as shown in Fig. 14.6.

2. Assume that system C has 2 devices in total, and the two devices are devices with similar efficiency

Efficiency characteristics of device No. 1:

$$\eta_1(P_1) = 2.6P_1 - 2P_1^2$$
$$\eta_{1e} = 0.845$$
$$P_{1e} = 0.65 \tag{14.100}$$
$$P_{1m} = 1.1$$

Defined as $\beta_1 = 1$.
Efficiency characteristics of device No. 2:

$$\eta_2(P_2) = 1.3P_2 - 0.5P_2^2$$
$$\eta_{2e} = 0.845$$
$$P_{2e} = 1.3 \tag{14.101}$$
$$P_{2m} = 2.2$$

$\beta_2 = 2$, the output capacity of device No. 2 is larger than that of device No. 1. The total output W_t maximization expression of system C is

$$W_t = \max \sum_{i=1}^{2} P_i \eta_i(P_i)$$
$$s.t. \sum_{i=1}^{2} P_i = P_t > 0 \tag{14.102}$$
$$P_{im} \geq P_i > 0$$

In this system, n = 1 has two combinations, and n = 2 has one combination.

When n = 1, there is a switching point between the first small device No. 1 and the second large device No. 2. According to the equivalent switching theorem, we have

$$\eta_1(P_t) = \eta_2(P_t) \tag{14.103}$$

Obtain the switching point P_{11} of a No. 1 small device and No. 2 large device

$$P_{11} = 0.8667 \tag{14.104}$$

When n = 2, the optimal control method is to keep

$$P_1 = \frac{P_t}{\sum_{i=1}^{2} \beta_i} = \frac{P_t}{3}$$
$$P_2 = \beta_2 P_1 = \frac{2P_t}{3} \tag{14.105}$$

The maximum total output energy W_t of system C is

$$\max W_t = P_t \eta_1 \left(\frac{P_t}{\sum_{i=1}^{2} \beta_i} \right) = P_t \eta_1 \left(\frac{P_t}{3} \right) \tag{14.106}$$

The maximum value of the overall operating energy efficiency $\eta_{t2}(P_t)$ of system C is

$$\eta_{t2}(P_t) = \eta_1 \left(\frac{P_t}{\sum_{i=1}^{2} \beta_i} \right) = \eta_1 \left(\frac{P_t}{3} \right) \tag{14.107}$$

Switching point P_{12} of No. 1 equipment of 1 large device and 2 devices satisfies

$$\eta_2(P_t) = \eta_1 \left(\frac{P_t}{3} \right) \tag{14.108}$$

Fig. 14.7 The overall energy efficiency curve of system C with two devices

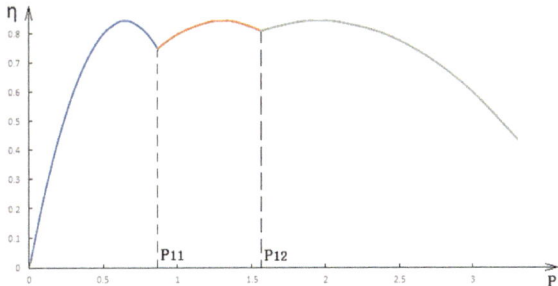

from which we have

$$P_{12} = 1.56 \tag{14.109}$$

When $P_{11} < P_t < \ = P_{12}$ and $P_{12} < P_t < \ = 3.3$, $W_t{}'' < 0$, so the obtained W_t is the maximum value.

The overall energy efficiency curve of the system C as shown in Fig. 14.7.

14.9   Quantum Optimization Method and Energy Efficiency Predictive Theory

There is no need to establish an accurate mathematical model of the system, based on the characteristics of the energy efficiency function, this chapter presents a constrained, nonlinear, integer-real-number hybrid energy efficiency optimization method for multi-unit systems.

This optimization method includes two theorems: optimal load distribution theorem and optimal switching theorem.

Optimal load distribution theorem: The optimal load distribution method of a multi-unit system is to keep the operating energy efficiency of each operating device equal, **Yao Theorem 1**.

$$\eta_1(P_1) = \eta_2(P_2) = \cdots = \eta_n(P_n) \tag{14.110}$$

Optimal switching theorem: The optimal switching point for the number of operating units is at the point of equal efficiency or at the maximum output point of the devices, **Yao Theorem 2**.

$$\eta_1\left(\frac{P_t}{\sum_{i=1}^{n}\beta_i}\right) = \eta_1\left(\frac{P_t}{\sum_{i=1}^{k}\beta_i}\right) or \eta_1\left(\frac{P_t}{\sum_{i=1}^{n}\beta_i}\right)$$

$$= \eta_{1m} or \eta_1 \left(\frac{P_t}{\sum_{i=1}^{k} \beta_i} \right) = \eta_{1m} \qquad (14.111)$$

We call this optimization method as **quantum optimization method of multi-unit system**.

We call this theory to solve energy efficiency optimization as **energy efficiency predictive theory of multi-unit system**.

These methods have the following advantages:

(1) Easy to use, no needing to establish an accurate mathematical model of the system;
(2) Strong versatility, including linear systems, nonlinear systems, multivariable systems, time invariant systems, time varying systems;
(3) Integer and real optimizations are solved together.

14.10   The Second Definition of Similar Energy Efficiency Devices

We define the load rate γ_i of the i-th device as

$$\gamma_i = \frac{P_i}{P_{ie}} \qquad (14.112)$$

We call $\eta_{Ni}(\gamma_i)$ as the normalization efficiency function of the i-th device. The normalization efficiency function $\eta_{Ni}(\gamma)$ has a shape shown in Fig. 14.8.

In Fig. 14.8, γ is a variable, η_{Ni} is the efficiency. η_{Ni} and η_i have the relation as following.

$$\eta_i(P_i) = \eta_i(\gamma_i P_{ie}) = \eta_{Ni}(\gamma_i) \qquad (14.113)$$

Fig. 14.8 The normalization efficiency function $\eta_{Ni}(\gamma)$

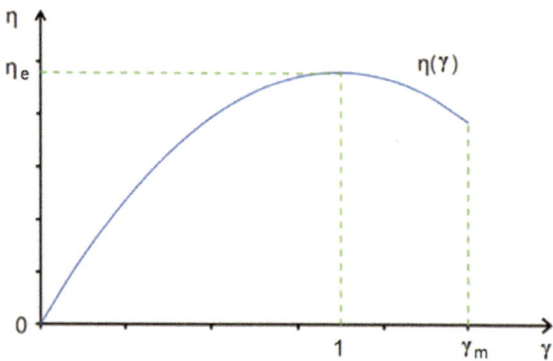

If the normalization efficiency functions of the first device and the i-th device are the same, we have

$$\eta_{N1}(\gamma_1) = \eta_{Ni}(\gamma_i) \qquad (14.114)$$

We call them the similar energy efficiency devices.

14.11 Another Way to Prove Yao's Theorem

Assume that P_i is greater than 0, that is, every operating device does work. The formula (14.1) is simplified as

$$\begin{aligned} W_t &= \max \sum_{i=1}^{n} P_i \eta_i(P_i) \\ s.t. \; \sum_{i=1}^{n} P_i &= P_t > 0 \\ P_{imax} &\geq P_i > 0 \end{aligned} \qquad (14.115)$$

Assuming that n is optimal, consider the following two situations:

(1) n = 2

System C has two variables P_1 and P_2

$$\begin{aligned} P_1 + P_2 &= P_t \\ P_1 &> 0 \\ P_2 &> 0 \end{aligned} \qquad (14.116)$$

The objective function can be expressed as

$$W_t = P_1 \eta_1(P_1) + P_2 \eta_2(P_2) \qquad (14.117)$$

The optimization condition is

$$W_t'(P_1) = 0 \qquad (14.118)$$

According to known conditions

$$P_2 = \gamma_2 P_{1e} = P_t - \gamma_1 P_{1e} \qquad (14.119)$$

We have

$$\eta_1(P_1) + P_1 \eta_1{}'(P_1) - \eta_2(P_2) - P_2 \eta_2{}'(P_2) = 0 \qquad (14.120)$$

If two devices are similar energy efficiency devices, then we have

$$\eta_1(P_1) = \eta_1(\gamma_1 P_{1e}) = \eta_{N1}(\gamma_1)$$
$$\eta_2(P_2) = \eta_2(\gamma_2 P_{2e}) = \eta_{N2}(\gamma_2) \qquad (14.121)$$
$$\eta_{N1}(\gamma) = \eta_{N2}(\gamma)$$

The objective function can be expressed as

$$W_t = P_1\eta_1(P_1) + P_2\eta_2(P_2) = \gamma_1 P_{1e}\eta_{N1}(\gamma_1) + \gamma_2 P_{2e}\eta_{N2}(\gamma_2) \qquad (14.122)$$

The optimization condition is

$$W_t'(\gamma_1) = 0 \qquad (14.123)$$

According to known conditions, we have

$$\gamma_2 = \frac{P_t - \gamma_1 P_{1e}}{P_{2e}} \qquad (14.124)$$

from which we know

$$\gamma_1 = \gamma_2 = \frac{P_t}{P_{1e} + P_{2e}} \qquad (14.125)$$

is an optimization point.

The optimal control method is to keep

$$P_1 = \gamma_1 P_{1e} = \frac{P_{1e}}{P_{1e}+P_{2e}} P_t$$
$$P_2 = \gamma_2 P_{2e} = \frac{P_{2e}}{P_{1e}+P_{2e}} P_t \qquad (14.126)$$

The maximum total output energy W_t of system C is

$$\max W_t = P_t \eta_{N1}\left(\frac{P_t}{P_{1e} + P_{2e}}\right) \qquad (14.127)$$

The maximum value of the overall operating energy efficiency $\eta_{t2}(P_t)$ of system C is

$$\max \eta_{t2}(P_t) = \eta_{N1}\left(\frac{P_t}{P_{1e} + P_{2e}}\right) \qquad (14.128)$$

Since the shape of the overall efficiency curve of the system is the same as that of a single device, the second derivative is also less than zero. W_t has a maximum value, and the overall energy efficiency is a maximum value as well.

(2)   If $n = k$ holds, prove that $n = k + 1$ still holds.

For system C composed of n devices with the similar energy efficiency function, if the optimization conclusion holds,

$$W_t = \max \sum_{i=1}^{k} P_i \eta_i(P_i)$$
$$s.t. \sum_{i=1}^{k} P_i = P_t > 0 \qquad (14.129)$$
$$P_{imax} \geq P_i > 0$$

The optimal point is

$$\gamma_1 = \gamma_2 = \cdots = \gamma_k = \frac{P_t}{\sum_{i=1}^{k} P_{ie}} \qquad (14.130)$$

The optimal control method is to keep

$$P_1 = \frac{P_{1e}}{\sum_{i=1}^{k} P_{ie}} P_t$$
$$P_2 = \frac{P_{2e}}{\sum_{i=1}^{k} P_{ie}} P_t \qquad (14.131)$$
$$\cdots$$
$$P_k = \frac{P_{ke}}{\sum_{i=1}^{k} P_{ie}} P_t$$

The maximum total output energy W_t of system C is

$$\max W_t = P_t \eta_{N1} \left(\frac{P_t}{\sum_{i=1}^{k} P_{ie}} \right) = P_t \eta_1 \left(\frac{P_{1e}}{\sum_{i=1}^{k} P_{ie}} P_t \right) \qquad (14.132)$$

The maximum value of the overall operating energy efficiency $\eta_{tk}(P_t)$ of system C is

$$\max \eta_{tk}(P_t) = \eta_{N1} \left(\frac{P_t}{\sum_{i=1}^{k} P_{ie}} \right) = \eta_1 \left(\frac{P_{1e}}{\sum_{i=1}^{k} P_{ie}} P_t \right) \qquad (14.133)$$

For n = k + 1, we have

$$W_t = \sum_{i=1}^{k+1} P_i \eta_i(P_i) = \left(P_t - \gamma_{k+1} P_{(k+1)e} \right) \eta_{N1} \left(\frac{P_t - \gamma_{k+1} P_{(k+1)e}}{\sum_{i=1}^{k} P_{ie}} \right)$$
$$+ \gamma_{k+1} P_{(k+1)e} \eta_{N(k+1)}(\gamma_{k+1}) \qquad (14.134)$$

The optimization condition is

$$W_t'(\gamma_{k+1}) = 0 \qquad (14.135)$$

It is easy to see that, the optimal point is

$$\gamma_{k+1} = \frac{P_t}{\sum_{i=1}^{k+1} P_{ie}}$$
$$\gamma_1 = \gamma_2 = \cdots = \gamma_k = \frac{P_t}{\sum_{i=1}^{k+1} P_{ie}} \qquad (14.136)$$

The optimal control method is to keep

$$P_1 = \frac{P_{1e}}{\sum_{i=1}^{k+1} P_{ie}} P_t$$
$$P_2 = \frac{P_{2e}}{\sum_{i=1}^{k+1} P_{ie}} P_t$$
$$\cdots$$
$$P_{k+1} = \frac{P_{(k+1)e}}{\sum_{i=1}^{k+1} P_{ie}} P_t \tag{14.137}$$

The maximum total output energy W_t of system C is

$$\max W_t = P_t \eta_{N1} \left(\frac{P_t}{\sum_{i=1}^{k+1} P_{ie}} \right) = P_t \eta_1 \left(\frac{P_{1e}}{\sum_{i=1}^{k+1} P_{ie}} P_t \right) \tag{14.138}$$

The maximum value of the overall operating energy efficiency $\eta_{tk}(P_t)$ of system C is

$$\max \eta_{t(k+1)}(P_t) = \eta_{N1} \left(\frac{P_t}{\sum_{i=1}^{k+1} P_{ie}} \right) \eta_1 \left(\frac{P_{1e}}{\sum_{i=1}^{k+1} P_{ie}} P_t \right) \tag{14.139}$$

Acknowledgements In order to solve the energy efficiency optimization problem of multi-unit system, the author has conducted long-term research. Thanks to Yanfang Zhang from Hebei Automation Company and Bosheng Yao from Beijing IAO Technology Development Company for their valuable help and advice during the development and experiment of this theory. I would like to thank my doctoral supervisor, Professor Hexu Sun, my postdoctoral supervisor, Dr. Chengyu Cao, and my supervisor, Dr. Qingbin Gao and Dr. Robert X. Gao, for their valuable help and suggestions.

Chapter 15
Optimal Control of Hydropower Stations Composed of Identical Generators

There are many hydropower stations (HPS) around the world, which consist of hydropower units that convert the potential energy of water in rivers or reservoirs into electrical energy (Fig. 15.1). Global hydropower generation in 2022 totaled 4,334.190 billion kilowatt-hours, a year-on-year increase of 1.1%. The proportion of global hydropower generation in total global power generation was 16.0% in 2022.

In a hydropower station, under a certain water head and capacity, there is a maximum power supply. To achieve the maximum energy output, we should decide how many generators should be used in the system and how much capacity should be allocated to each generator. This issue is becoming increasingly important with the increasing awareness of environmental protection and the need for energy-saving systems.

At present, various optimization methods have been widely studied, such as linear programming [1], recursive quadratic programming [2], stochastic dynamic programming [3], Lagrangian relaxation [4], genetic algorithm [5], simulated annealing [6] etc.], particle swarm optimization [7], ant colony optimization [8], quasi-Newton [9], neural network [10], evolutionary strategy [11], equal increment principle [12], etc. [13−15].

Almost all of these optimization methods require the establishment of mathematical models of the system. However, since accurate models of real systems are difficult to establish, these optimization methods quickly become difficult to apply in real systems, considering the complexity of the models and algorithms, the curse of dimensionality, and the computational time.

Optimization solutions that avoid these problems would be very useful for business and industry.

© The Author(s) 2024
F. Yao and Y. Yao, *Efficient Energy-Saving Control and Optimization for Multi-Unit Systems*, https://doi.org/10.1007/978-981-97-4492-3_15

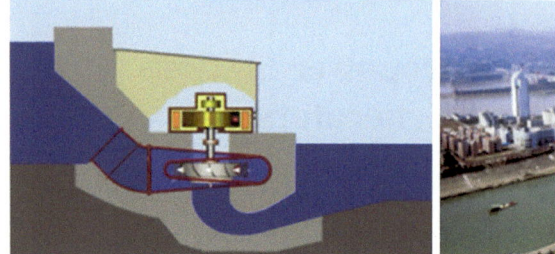

Fig. 15.1 Hydropower stations

15.1   Energy Efficiency of a Hydroelectric Generating Set

A hydroelectric generating set must be accelerated to the rated speed n_0 before it can be connected to the grid. The energy efficiency of a hydroelectric generating set is shown in Fig. 15.2.

In Fig. 15.2, Q is the flow rate through the hydroelectric generating set, η is the efficiency function of the hydroelectric generating set at rated speed n_0 and given water head H_0, Q_0 is the zero-load flow rate of the hydroelectric generating set, and η_e is the maximum efficiency value, Q_e is the flow rate corresponding to the maximum efficiency value η_e, Q_m is the maximum flow rate, and

$$Q_m \geq Q \geq Q_0 \tag{15.1}$$

$$\eta_e = \max(\eta) = \eta(Q_e)$$

$$\eta(Q) \geq 0$$

$$\eta''(Q) < 0$$

Fig. 15.2 Energy efficiency of a hydroelectric generating set

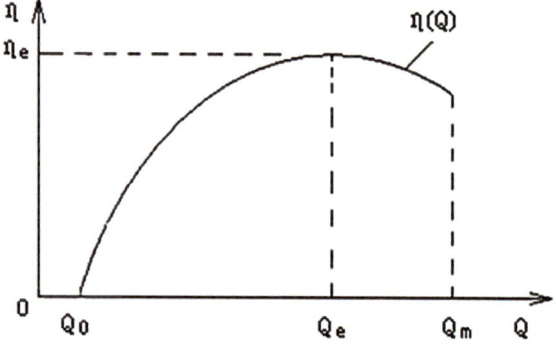

The energy efficiency function η can be expressed as

$$\eta(Q) = \sum_{i=0}^{\infty} a_i Q^i \tag{15.2}$$

15.2 HPS Power Generation

The water heads of each hydroelectric generating set in a hydropower station are identical. The electrical energy generated by the hydropower station is expressed as

$$
\begin{aligned}
W_t &= \sum_{i=1}^{n}(9.81 H_0 Q_i \eta_i) = 9.81 Q_t H_0 \sum_{i=1}^{n} \frac{Q_i}{Q_t} \eta_i \\
&= 9.81 Q_t H_0 \sum_{i=1}^{n} \theta_i \eta_i = 9.81 Q_t H_0 \eta_0 = W_0 \eta_o
\end{aligned}
\tag{15.3}
$$

$$W_0 = 9.81 Q_t H_0$$

where n is the total number of hydroelectric generating sets in the hydropower station, W_0 is the potential energy of water flowing through the hydropower station, kW, W_t is the total electric energy output by the hydropower station, kW, and Q_t is the total flow rate of water flowing through the hydroelectric generating sets, m^3/s, H_0 is the water head, m, η_0 is the overall energy efficiency of the hydropower station, η_i is the operating efficiency of the i-th hydroelectric generating set, Q_i represents the flow through the i-th hydroelectric generating set, Q_{i0} denotes the zero-load flow of ith hydroelectric generating set, θ_i represents the load rate of the i-th hydroelectric generating set, θ_{i0} denotes the load rate of the i-th hydroelectric generating set at Q_{i0}, expressed as

$$Q_t = \sum_{i=1}^{n} Q_i > \sum_{i=1}^{n} Q_{i0} \tag{15.4}$$

$$\theta_i = \frac{Q_i}{Q_t}$$

$$\theta_{i0} = \frac{Q_{i0}}{Q_t}$$

$$\sum_{i=1}^{n} \theta_i = 1$$

$$\eta_0 = \sum_{i=1}^{n} \theta_i \eta_i$$

15.3 Optimal Control in a Hydropower Station

Assuming that all hydroelectric generating sets are the identical model, that is

$$Q_{10} = Q_{20} = \cdots = Q_{n0} = Q_0 \tag{15.5}$$

$$\eta_1(Q) = \eta_2(Q) = \cdots = \eta_n(Q) = \eta(Q)$$

Theorem *For the optimization problem of W_t at the fixed head H_0 and Q_t, the maximization of the total electricity energy output W_t of the power station*

$$\begin{array}{c} \max \quad W_t \\ s.t.Q_i > Q_0 \\ \sum_{i=1}^{n} Q_i = Q_t \end{array} \tag{15.6}$$

is given by

$$\max W_t = W_0 \eta\left(\frac{Q_t}{n}\right) \tag{15.7}$$

The maximization of the overall efficiency η_0 of the hydropower station

$$\begin{array}{c} \max \quad \eta_0 \\ s.t.Q_i > Q_{i0} \\ \sum_{i=1}^{n} Q_i = Q_t \end{array} \tag{15.8}$$

is given by

$$\max \eta_0 = \eta\left(\frac{Q_t}{n}\right) \tag{15.9}$$

Proof We begin our inductive proof by considering the case where n $= 2$.

The constraint condition then becomes

$$Q_1 + Q_2 = Q_t \tag{15.10}$$

where

$$Q_1 > Q_0 \tag{15.11}$$

$$Q_2 > Q_0$$

The objective function W_t is expressed as

$$W_t = 9.81H_0(Q_1\eta(Q_1) + Q_2\eta(Q_2)) \tag{15.12}$$

The optimal condition is given for

$$W_t'(Q_1) = 0 \tag{15.13}$$

We have

$$\eta(Q_1) - \eta(Q_t - Q_1) + Q_1\eta'(Q_1) - (Q_t - Q_1)\eta'(Q_2) = 0 \tag{15.14}$$

It is then easily verified that

$$Q_1 = Q_2 = \frac{Q_t}{2} \tag{15.15}$$

is an optimal point.

We then check the second derivative,

$$W_t'' < 0 \tag{15.16}$$

So, the optimal point is only maximum.

The maximal value of the total electricity energy output W_t of the power station is

$$\max W_t = W_0\eta(\frac{Q_t}{2}) \tag{15.17}$$

The maximal value of the overall efficiency η_0 of the power station is

$$\max \eta_0 = \eta(\frac{Q_t}{2}) \tag{15.18}$$

We then assume that this holds for $n = k$. The above conclusion is readily extended to the case of $n = k$, and the optimal point is then

$$Q_1 = Q_2 = \cdots = Q_k = \frac{Q_t}{k} \tag{15.19}$$

The maximal value of W_t is

$$\max W_t = W_0\eta(\frac{Q_t}{k}) \tag{15.20}$$

The maximal value of the overall efficiency η_0 of the power station is

$$\max \eta_0 = \eta(\frac{Q_t}{k}) \tag{15.21}$$

Our inductive case is then given by $n = k + 1$. For the total electricity energy output W_t of the power station we have

$$W_t = 9.81H_0\left(\sum_{i=1}^{k} Q_i\eta(Q_i) + Q_{k+1}\eta(Q_{k+1})\right) \tag{15.22}$$

and the maximum of the first item is

$$\max 9.81H_0\sum_{i=1}^{k} Q_i\eta(Q_i) = 9.81H_0(Q_t - Q_{k+1})\eta\left(\frac{Q_t - Q_{k+1}}{k}\right) \tag{15.23}$$

where

$$Q_1 = Q_2 = \cdots = Q_k \tag{15.24}$$

The expression for W_t becomes

$$W_t = 9.81H_0\left((Q_t - Q_{k+1})\eta\left(\frac{Q_t - Q_{k+1}}{k}\right) + Q_{k+1}\eta(Q_{k+1})\right) \tag{15.25}$$

Based on the above conclusion for $n = 2$, the optimal point is

$$Q_{k+1} = \frac{Q_t - Q_{k+1}}{k} \tag{15.26}$$

The solution is

$$Q_{k+1} = \frac{Q_t}{k+1} \tag{15.27}$$

and

$$Q_1 = Q_2 = \cdots = Q_k = \frac{Q_t - Q_{k+1}}{k} = \frac{Q_t}{k+1} \tag{15.28}$$

The optimal point is then

$$Q_1 = Q_2 = \cdots = Q_k = Q_{k+1} = \frac{Q_t}{k+1} \tag{15.29}$$

and the maximal value of the total electricity energy output W_t of the power station is

$$\max W_t = W_0\eta\left(\frac{Q_t}{k+1}\right) \tag{15.30}$$

The maximal value of the overall energy efficiency η_0 of the power station is

Fig. 15.3 The n is the optimal

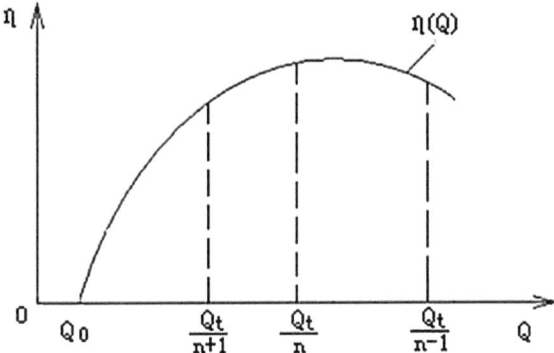

$$\max \eta_0 = \eta \left(\frac{Q_t}{k+1} \right) \tag{15.31}$$

Load distribution theorem: In a hydropower station which consists of n hydro-electric generating sets that are the identical model, the optimal control method is to keep each hydroelectric generating set to have the same load.

15.4 Optimization Discriminants

In a hydropower station with the identical model hydroelectric generating sets, if n is the optimal, as shown in Fig. 15.3, there must be

$$W_0 \eta \left(\frac{Q_t}{n} \right) = \max \left(W_0 \eta \left(\frac{Q_t}{n-1} \right), W_0 \eta \left(\frac{Q_t}{n} \right), W_0 \eta \left(\frac{Q_t}{n+1} \right) \right) \tag{15.32}$$

Namely

$$\eta \left(\frac{Q_t}{n} \right) = \max \left(\eta \left(\frac{Q_t}{n-1} \right), \eta \left(\frac{Q_t}{n} \right), \eta \left(\frac{Q_t}{n+1} \right) \right) \tag{15.33}$$

15.5 Optimal Switch in the Hydropower Station

Now we consider the optimal switch point for a hydropower station. Suppose the hydropower station has M-unit hydroelectric generating sets in total. Then we take n to be less than or equal to M and the optimum, the total electricity energy output W_t of the power station is the maximum.

$$\begin{array}{c} \max \\ s.t. Q_m \ge Q_i > Q_0 \end{array} \quad W_t = W_0 \eta\left(\frac{Q_t}{n}\right) \qquad (15.34)$$
$$\sum_{i=1}^{n} Q_i = Q_t$$
$$n \le M$$

The optimal overall efficiency is then

$$\eta\left(\frac{Q_t}{n}\right) = \max\left(\eta\left(\frac{Q_t}{n-1}\right), \eta\left(\frac{Q_t}{n}\right), \eta\left(\frac{Q_t}{n+1}\right)\right) \qquad (15.35)$$

If the total flow Q_t varies, the optimal n may change also. The optimal switch point is dependent upon whether Q_t is increasing or decreasing.

Theorem *In a hydropower station with the identical model hydroelectric generating sets, if n is the optimal, when Q_t is increasing, the optimal switch point is at*

$$\eta\left(\frac{Q_t}{n}\right) = \eta\left(\frac{Q_t}{n+1}\right) \qquad (15.36)$$

when Q_t is decreasing, the optimal switch point is at

$$\eta\left(\frac{Q_t}{n}\right) = \eta\left(\frac{Q_t}{n-1}\right) \qquad (15.37)$$

Proof In Figs. 15.4 and 15.5, the η_e is the maximum efficiency. Q_e is the flow rate at η_e. The $\eta\,(Q_t/(n-1))$ is less than $\eta(Q_t/n)$ on the right side of Q_e, the $\eta\,(Q_t/(n+1))$ is less than $\eta\,(Q_t/n)$ on the left side of Q_e.

From Figs. 15.4 and 15.5, we see that

Fig. 15.4 Q_t/n is equal or less than Q_e

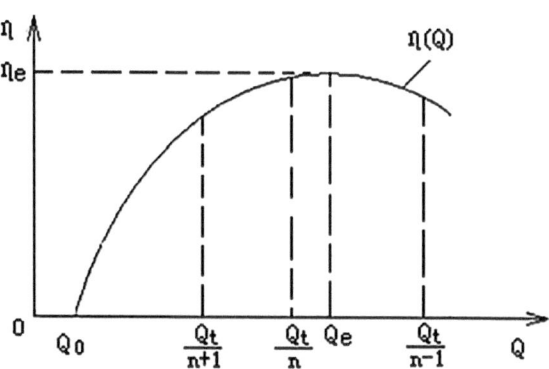

Fig. 15.5 Q_t/n is greater than the flow Q_e

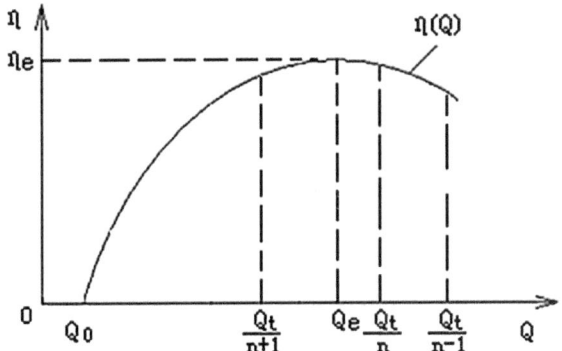

$$\eta\left(\frac{Q_t}{n}\right) = \max\left(\eta\left(\frac{Q_t}{n-1}\right), \eta\left(\frac{Q_t}{n}\right), \eta\left(\frac{Q_t}{n+1}\right)\right) \tag{15.38}$$

As seen in Fig. 15.4, if $Q_t/n < Q_e$ and Q_t increases, then η (Q_t/n) will continue to increase until Q_t/n reach Q_e, at which point immediately after it will begin to decrease. At the same time, η $(Q_t/(n+1))$ will increase until it will eventually become greater than $\eta(Q_t/n)$ and will be the new maximum.

$$\eta\left(\frac{Q_t}{n+1}\right) = \max\left(\eta\left(\frac{Q_t}{n-1}\right), \eta\left(\frac{Q_t}{n}\right), \eta\left(\frac{Q_t}{n+1}\right)\right) \tag{15.39}$$

This change of the maximum efficiency defines the optimal switch point at

$$\eta\left(\frac{Q_t}{n}\right) = \eta\left(\frac{Q_t}{n+1}\right) \tag{15.40}$$

If $Q_t/n > Q_e$, then η (Q_t/n) will decrease with the increase of Q_t while $\eta(Q_t/(n+1))$ will increase as can be inferred from Fig. 15.5. Ultimately, $\eta(Q_t/(n+1))$ will increase to such a point that it will become the new maximum

$$\eta\left(\frac{Q_t}{n+1}\right) = \max\left(\eta\left(\frac{Q_t}{n-1}\right), \eta\left(\frac{Q_t}{n}\right), \eta\left(\frac{Q_t}{n+1}\right)\right) \tag{15.41}$$

and the optimal switch point is at

$$\eta\left(\frac{Q_t}{n}\right) = \eta\left(\frac{Q_t}{n+1}\right) \tag{15.42}$$

In a manner akin to our arguments for an increasing Q_t, we begin by looking at $Q_t/n > Q_e$. If Q_t is decreasing, then Fig. 15.5 shows that η (Q_t/n) will increase until Q_t/n reaches the value of Q_e, where it will immediately begin to decrease. Simultaneously, $\eta(Q_t/(n-1))$ will increase. As before, there will be point in which

$\eta\left(Q_t/(n-1)\right)$ overtakes $\eta(Q_t/n)$ as the new maximum.

$$\eta\left(\frac{Q_t}{n-1}\right) = \max\left(\eta\left(\frac{Q_t}{n-1}\right), \eta\left(\frac{Q_t}{n}\right), \eta\left(\frac{Q_t}{n+1}\right)\right) \qquad (15.43)$$

This is our optimal switch point and is given by

$$\eta\left(\frac{Q_t}{n}\right) = \eta\left(\frac{Q_t}{n-1}\right) \qquad (15.44)$$

As seen in Fig. 15.4, for a given $Q_t/n < Q_e$, $\eta(Q_t/n)$ will decrease as Q_t decreases. We observe that $\eta\left(Q_t/(n-1)\right)$ will increase to the point where it will be the maximum efficiency.

$$\eta\left(\frac{Q_t}{n-1}\right) = \max\left(\eta\left(\frac{Q_t}{n-1}\right), \eta\left(\frac{Q_t}{n}\right), \eta\left(\frac{Q_t}{n+1}\right)\right) \qquad (15.45)$$

The optimal switch point is then

$$\eta\left(\frac{Q_t}{n}\right) = \eta\left(\frac{Q_t}{n-1}\right) \qquad (15.46)$$

15.6 Operation at the Maximum Efficiency

For the total flow rate $Q_t(\text{m}^3/\text{s})$ of the river, the optimal run-unit of a hydropower station is n as described above. The maximal value of the overall efficiency η_0 of the power station is

$$\max\eta_0 = \eta\left(\frac{Q_t}{n}\right) \qquad (15.47)$$

and

$$\eta\left(\frac{Q_t}{n}\right) \leq \eta_e \qquad (15.48)$$

If a hydropower station is located at a reservoir which has a certain storage capacity, then the hydropower station can operate at the rate efficiency η_e. As shown in Fig. 3.1, η_e is the maximum, and greater than any efficiency value.

The total flow of each day in the river is $24*3600*Q_t$, and can be resolved into t_1*Q_{t1} and t_2*Q_{t2} as following

$$24 \times 3600 \times Q_t = t_1 \times Q_{t1} + t_2 \times Q_{t2} \tag{15.49}$$

$$24 \times 3600 = t_1 + t_2$$

where t_1 denotes running time (s) under run-unit n_1, Q_{t1} denotes the flow rate (m³/s) through all hydroelectric generating sets during t_1 under run-unit n_1, t_2 denotes time (s) under run-unit n_2, Q_{t2} denotes the flow rate (m³/s) through all hydroelectric generating sets during t_2 under run-unit n_2. n_1 and n_2 satisfy the following relationship

$$\eta\left(\frac{Q_t}{n_1}\right) = \eta_e \tag{15.50}$$

$$\eta\left(\frac{Q_t}{n_2}\right) = \eta_e$$

Thus, the hydropower station can work under the maximum efficiency η_e.

15.7 Conclusion

By supposing the total flow Q_t fixed value, we propose an optimal control method, this method does not depend on the exact model of a hydropower station. By changing the total flow Q_t, we also propose an optimal switch method, this method only depends on the efficiency function of the generator at constant head.

The proof of the optimal control and switch method given by this chapter are mainly based on the efficiency function characteristics which can be approximately considered a concave, non-negative function. Thusly, this optimal method has the following features:

(1) Both liner and non-liner systems are included,
(2) The system's mathematical model is not needed,
(3) The method is high universal.

References

1. Shern CM, Laughton MA (1970) Power system load scheduling with security constraints using dual linear programming [J]. Proc IEEE 117(1):2714–2127
2. Bartholomew-Biggs MC (1987) Recursive quadratic programming method based on the augmented Lagrangian [J]. Math Program Study 31:21–41
3. Feng L (1993) A parametric iteration method of stochastic dynamic programming for optimal dispatch of hydroelectric plants [C]. In: IEEE 2nd international conference on advances in power system control, operation and management. Hong Kong, vol 12, pp 1304–1324

4. Balci HH, Valenzuela JF (2004) Scheduling electric power generators using particle swarm optimization combined with the Lagrangain relaxation method [J]. Int J Appl Math Comput Sci 14:411–421
5. Orero SO, Irving MR (1998) A genetic algorithm modeling framework and solution technique for short term optimal hydrothermal scheduling [J]. IEEE Trans Power Syst 5(2):1254–1265
6. Rajan A, Christober C (2011) Hy dro-thermal unit commitment problem using simulated annealing embedded evolutionary programming approah [J]. Int J Electr Power & Energy Syst 33(4):939–946
7. Nagesh Kumar D, Janga RM (2007) Multipurpose reservoir operation using particle swarm optimization [J]. J Water Resour Plan Manag 133(3):192–201
8. Lopezibanez M, Prasad TD, Paechter B (2008) Ant colony optimization for optimal control of pumps in water distribution networks. J Water Resour Plan Manage 134(4):337–346
9. Giras TC, Talukdar SN (1981) Quasi-Newton method for optimal power flows [J]. Int J Electr Power Energy Syst 3(2):59–64
10. Basu M (2003) Hopfield neural networks for optimal scheduling of fixed head hydrothermal power station [J]. Electr Power Syst Res 64(1):11–15
11. Yuan X, Zhang Y, Wang L et al (2008) An enhance differential evolution algorithm for daily optimal hydro generation scheduling [J]. Comput Math Appl 55(11):2458–2468
12. Huang H, Peng D, Zhang Y, Liang Y (2013) Research on load optimal distribution based on equal incremental principle [J]. J Comput Inf Syst 9(18):7477–7484
13. Simopoulos DN, Kavatza SD, Vournas CD (2006) Unit commitment by an enhanced simulated annealing algorithm [J]. Power Syst, IEEE Trans on 21(1):68–76
14. Yao F, Sun H (2012) Efficiency optimal control and dispatching method for general equipment. China Machine Press, China
15. Yao FL, Zhang YF (2009) Electrical energy saving control method and practice. China Electric Power Press, Beijing

Chapter 16
Efficiency Optimization of Power Stations with Different Generating Units

There are many power stations in the world with different generators. These include thermal power stations, hydropower stations, etc (Fig. 16.1). They convert the energy of raw materials into electrical energy. In these power stations, with the same input of raw materials, there is maximum power generation. To achieve the maximum, we should decide how many generators to use and how many tasks should be assigned to each generator.

16.1 Power Generation in ESPS

A generalized generator must reach the rated speed n_0 before it can be connected to the grid. Taking the input amount of raw materials as a variable, such as the amount of coal input in a thermal power station or the water flow in a hydropower station, the efficiency of a generalized generator is shown in Fig. 16.2.

In Fig. 16.2, L is the adjustable input of the generator, η is the efficiency function of the generator, L_0 is L at zero load of the generator, η_e is the maximum efficiency corresponding to L_e, L_e is the optimal input, and L_m is the maximum L.

We define the load factor γ as

$$\gamma = \frac{L}{L_e} \tag{16.1}$$

We define $\eta_N(\gamma)$ to be the normalized efficiency function. The normalized efficiency function $\eta_N(\gamma)$ has a shape shown in Fig. 16.3.

In Fig. 16.3, γ is the load rate and a variable, $\eta_N(\gamma)$ is the normalized efficiency. $\eta_N(\gamma)$ and $\eta(L)$ have the following relationship

$$\eta_N(\gamma) = \eta(\gamma L_e) \tag{16.2}$$

F. Yao and Y. Yao, *Efficient Energy-Saving Control and Optimization for Multi-Unit Systems*, https://doi.org/10.1007/978-981-97-4492-3_16

Fig. 16.1 Power stations

Fig. 16.2 The efficiency of a generalized power generator

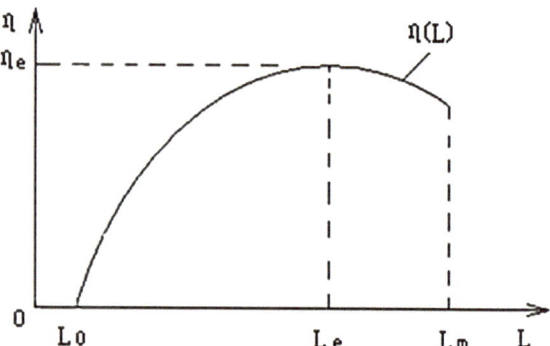

Fig. 16.3 The normalized efficiency function $\eta_N(\gamma)$

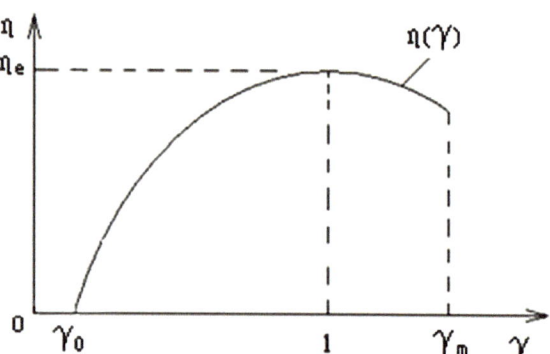

If the normalized efficiency functions of two generators are identical, then

$$\eta_{N1}(\gamma) = \eta_{N2}(\gamma) \tag{16.3}$$

We define them as efficiency similarity generators. The power station which includes the efficiency similarity generators is called the efficiency similarity power station (ESPS).

The electricity energy generated by an ESPS is expressed as

$$W_t = \sum_{i=1}^{n} W_i = \frac{W_0}{L_t} \sum_{i=1}^{n} L_i \eta_i(L_i) = \frac{W_0}{L_t} \sum_{i=1}^{n} \gamma_i L_{ie} \eta_i(\gamma_i L_{ie}) = \frac{W_0}{L_t} \sum_{i=1}^{n} \gamma_i L_{ie} \eta_N(\gamma_i)$$
$$= W_0 \eta_0 \tag{16.4}$$

$$\gamma_i = \frac{L_i}{L_{ie}}$$

$$\sum_{i=1}^{n} \gamma_i L_{ie} = L_t$$

$$\eta_0 = \frac{1}{L_t} \sum_{i=1}^{n} \gamma_i L_{ie} \eta(\gamma_i L_{ie}) = \frac{1}{L_t} \sum_{i=1}^{n} \gamma_i L_{ie} \eta_N(\gamma_i)$$

where n is the total number of generators, W_0 is the ideal energy input from raw materials, W_t is the total power generation of the power station, W_i is the power generation of the i-th generator, L_t is the total input of the power station, η_0 represents the total power generation efficiency, $\eta_i(L_i)$ represents the operating efficiency of the i-th generator, $\eta_N(\gamma_i)$ represents the normalized efficiency of the i-th generator, L_i represents the adjustable input of the i-th generator, and L_{ie} represents the maximum input of the i-th generator. Optimal input, L_{i0} represents L when the i-th generator is zero load, γ_i represents the load rate of the i-th generator, there is

$$L_t = \sum_{i=1}^{n} L_i > \sum_{i=1}^{n} L_{i0} \tag{16.5}$$

16.2 Optimal Control in an ESPS

Load distribution theorem: For the optimization problem of W_t at the fixed L_t, the maximization of the total electricity energy output W_t of the ESPS

$$\max_{\substack{s.t. L_m \geq L_i > L_0 \\ \sum_{i=1}^{n} L_i = L_t}} W_t \tag{16.6}$$

is given by

$$\max W_t = W_0 \eta_N \left(\frac{L_t}{\sum_{i=1}^{n} L_{ie}} \right) \tag{16.7}$$

That is the maximization of the overall efficiency η_0 of the power station

$$\max_{\substack{s.t.L_m \geq L_i > L_0 \\ \sum_{i=1}^{n} L_i = L_t}} \eta_0 \tag{16.8}$$

is given by

$$\max \eta_0 = \eta_N \left(\frac{L_t}{\sum_{i=1}^{n} L_{ie}} \right) \tag{16.9}$$

Proof We begin our inductive proof by considering the case where n = 2.

The constraint condition then becomes

$$L_1 + L_2 = L_t \tag{16.10}$$

where

$$L_m \geq L_1 > L_0 \tag{16.11}$$

$$L_m \geq L_2 > L_0$$

That is

$$\gamma_1 L_{1e} + \gamma_2 L_{2e} = L_t \tag{16.12}$$

The objective function W_t is expressed as

$$W_t = \frac{W_0}{L_t}(\gamma_1 L_{1e} \eta_N(\gamma_1) + \gamma_2 L_{2e} \eta_N(\gamma_2)) \tag{16.13}$$

The optimal condition is given for

$$W_t'(\gamma_1) = 0 \tag{16.14}$$

We have

$$L_{1e}\eta_N(\gamma_1) + \gamma_1 L_{1e}\eta_N'(\gamma_1) - L_{1e}\eta_N(\gamma_2) - (\gamma_2 L_{1e})n_N'(\gamma_2) = 0 \tag{16.15}$$

It is then easily verified that

$$\gamma_1 = \gamma_2 = \frac{L_t}{\sum_{i=1}^{2} L_{ie}} \tag{16.16}$$

is an optimal point.

We then check the second derivative,

$$W_t'' < 0 \tag{16.17}$$

So, the optimal point is only maximum.

The maximal value of the total electricity energy output W_t of the power station is

$$\max W_t = W_0 \eta_N \left(\frac{L_t}{\sum_{i=1}^{2} L_{ie}} \right) \tag{16.18}$$

The maximal value of the overall efficiency η_0 of the power station is

$$\max \eta_0 = \eta_N \left(\frac{L_t}{\sum_{i=1}^{2} L_{ie}} \right) \tag{16.19}$$

We then assume that this holds for $n = k$. The above conclusion is readily extended to the case of $n = k$, and the optimal point is then

$$\gamma_1 = \gamma_2 = \ldots = \gamma_k = \frac{L_t}{\sum_{i=1}^{k} L_{ie}} \tag{16.20}$$

The maximal value of W_t is

$$\max W_t = W_0 \eta_N \left(\frac{L_t}{\sum_{i=1}^{k} L_{ie}} \right) \tag{16.21}$$

The maximal value of the overall efficiency η_0 of the power station is

$$\max \eta_0 = \eta_N \left(\frac{L_t}{\sum_{i=1}^{2} L_{ie}} \right) \tag{16.22}$$

Our inductive case is then given by $n = k + 1$. For the total electricity energy output W_t we have

$$W_t = \frac{W_0}{L_t} \left(\sum_{i=1}^{k} \gamma_i L_{ie} \eta_N (\gamma_i) + \gamma_{k+1} L_{(k+1)e} \eta_N (\gamma_{k+1}) \right) \tag{16.23}$$

and the maximum of the first item is

$$\max \frac{W_0}{L_t} \sum_{i=1}^{k} \gamma_i L_{ie} \eta_N(\gamma_i) = \frac{W_0}{L_t}\left(L_t - \gamma_{k+1}L_{(k+1)e}\right)\eta_N\left(\frac{L_t - \gamma_{k+1}L_{(k+1)e}}{\sum_{i=1}^{k} L_{ie}}\right) \quad (16.24)$$

where

$$\gamma_1 = \gamma_2 = \ldots = \gamma_k \quad (16.25)$$

The expression for W_t becomes

$$W_t = \frac{W_0}{L_t}\left(\left(L_t - \gamma_{k+1}L_{(k+1)e}\right)\eta_N\left(\frac{L_t - \gamma_{k+1}L_{(k+1)e}}{\sum_{i=1}^{k} L_{ie}}\right) + \gamma_{k+1}L_{(k+1)e}\eta_N(\gamma_{k+1})\right) \quad (16.26)$$

The optimal condition is given for

$$W_t'(\gamma_{k+1}) = 0 \quad (16.27)$$

We have

$$-L_{(k+1)e}\eta_N\left(\frac{L_t - \gamma_{k+1}L_{(k+1)e}}{\sum_{i=1}^{k} L_{ie}}\right) - \left(L_t - \gamma_{k+1}L_{(k+1)e}\right)\eta_N'\left(\frac{L_t - \gamma_{k+1}L_{(k+1)e}}{\sum_{i=1}^{k} L_{ie}}\right)\frac{L_{(k+1)e}}{\sum_{i=1}^{k} L_{ie}}$$
$$+L_{(k+1)e}\,\eta_N(\gamma_{k+1}) + \gamma_{k+1}L_{(k+1)e}\eta_N'(\gamma_{k+1}) = 0 \quad (16.28)$$

It is then easily verified that

$$\gamma_{k+1} = \frac{L_t - \gamma_{k+1}L_{(k+1)e}}{\sum_{i=1}^{k} L_{ie}} \quad (16.29)$$

is an optimal point.
That is

$$\gamma_{k+1} = \frac{L_t}{\sum_{i=1}^{k+1} L_{ie}} \quad (16.30)$$

and

$$\gamma_1 \sum_{i=1}^{k} L_{ie} = L_t - \gamma_{k+1}L_{(k+1)e} \quad (16.31)$$

$$\gamma_1 = \frac{L_t - \gamma_{k+1}L_{(k+1)e}}{\sum_{i=1}^{k} L_{ie}} = \frac{L_t}{\sum_{i=1}^{k+1} L_{ie}}$$

Therefore, the optimal point is then

$$\gamma_1 = \gamma_2 = \ldots = \gamma_k = \gamma_{k+1} = \frac{L_t}{\sum_{i=1}^{k+1} L_{ie}} \tag{16.32}$$

and the maximal value of the total electricity energy output W_t of the power station is

$$\max W_t = W_0 \eta_N \left(\frac{L_t}{\sum_{i=1}^{k+1} L_{ie}} \right) \tag{16.33}$$

The maximal value of the overall efficiency η_0 of the power station is

$$\max \eta_0 = \eta_N \left(\frac{L_t}{\sum_{i=1}^{k+1} L_{ie}} \right) \tag{16.34}$$

Load distribution theorem: In an ESPS which consists of n generators that do not have all the identical model, the optimal control method is to keep each generator to have the same load rate.

16.3 Optimization Discriminant of the ESPS

We define $\gamma(L_t, n)$ as

$$\gamma(L_t, n) = \frac{L_t}{\sum_{i=1}^{n} L_{ie}} \tag{16.35}$$

In an ESPS, suppose that it has M-unit generators in total, if n is the optimal, as shown in Fig. 16.4, there must be

$$W_0 \eta_N \left(\frac{L_t}{\sum_{i=1}^{n} L_{ie}} \right) = \max \left(W_0 \eta_N \left(\frac{L_t}{\sum_{i=1}^{1} L_{ie}} \right), W_0 \eta_N \left(\frac{L_t}{\sum_{i=1}^{2} L_{ie}} \right), \ldots, \right.$$

$$\left. W_0 \eta_N \left(\frac{L_t}{\sum_{i=1}^{M} L_{ie}} \right) \right) \tag{16.36}$$

Namely

$$\eta_N \left(\frac{L_t}{\sum_{i=1}^{n} L_{ie}} \right) = \max \left(\eta_N \left(\frac{L_t}{\sum_{i=1}^{1} L_{ie}} \right), \eta_N \left(\frac{L_t}{\sum_{i=1}^{2} L_{ie}} \right), \ldots, \eta_N \left(\frac{L_t}{\sum_{i=1}^{M} L_{ie}} \right) \right) \tag{16.37}$$

Fig. 16.4 The n is the optimal

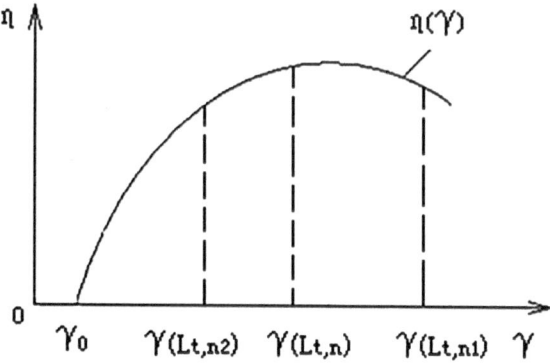

In Fig. 16.4, n1 and n2 are arbitrary number which corresponds to different device combinations. n1 is less than or equal to M, and n2 is less than or equal to M also.

Note: Even though the same n_1, there are many different generator combinations.

16.4   Optimal Switch in the ESPS

Now we consider the optimal switch point for an ESPS. Then we take n to be less than or equal to M and the optimum, the total electricity energy output W_t of the ESPS is the maximum.

$$\max_{\substack{s.t.L_m \geq L_i > L_0 \\ \sum_{i=1}^{n} L_i = L_t \\ n \leq M}} W_t = W_0 \eta_N \left(\frac{L_t}{\sum_{i=1}^{n} L_{ie}} \right) \tag{16.38}$$

The optimal overall efficiency is then

$$\max \eta_0 = \eta_N \left(\frac{L_t}{\sum_{i=1}^{n} L_{ie}} \right) \tag{16.39}$$

If the L_t varies, the optimal n may change as well. The optimal switching point depends on whether L_t is increasing or decreasing.

Theorem *The optimal switch point is at $\eta_N(L_t, n) = \eta_N(L_t, n_2)$, if L_t is increasing.*

Proof From Figs. 16.5 and 16.6, the $\eta_N(L_t, n_1)$ is less than $\eta_N(L_t, n)$ and greater than other efficiency values on the right side of $\gamma = 1$, the $\eta_N(L_t, n_2)$ is less than $\eta_N(L_t, n)$ and greater than other efficiency values on the left side of $\gamma = 1$, we see that

$$\eta_N(L_t, n) = \max(\eta_N(L_t, n1), \eta_N(L_t, n), \eta_N(L_t, n2)) \tag{16.40}$$

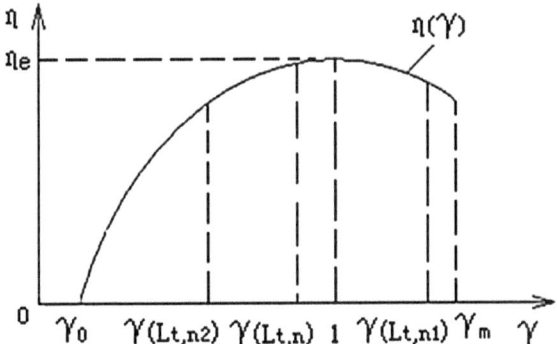

Fig. 16.5 $\eta_N(L_t, n)$ curve when $\gamma(L_t, n)$ is less than 1

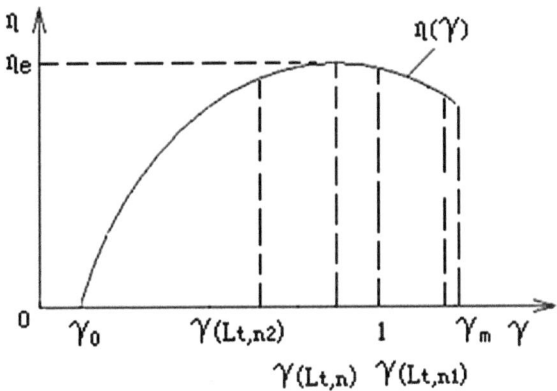

Fig. 16.6 $\eta_N(L_t, n)$ curve when $\gamma(L_t, n)$ is greater than 1

As seen in Fig. 16.5, if $\gamma(L_t, n) < 1$ and L_t increases, then $\eta_N(L_t, n)$ will continue to increase until $\gamma(L_t, n)$ reaches 1, immediately after which point it will begin to decrease. At the same time, $\eta_N(L_t, n_2)$ will increase until it will eventually become greater than $\eta_N(L_t, n)$ and will be the new maximum. This change of the maximum efficiency defines the optimal switch point at $\eta_N(L_t, n) = \eta_N(L_t, n_2)$.

If $\gamma(L_t, n) > 1$, then $\eta_N(L_t, n)$ will decrease with the increase of L_t while $\eta_N(L_t, n_2)$ will increase as can be inferred from Fig. 16.6. Ultimately, $\eta_N(L_t, n_2)$ will increase to such a point that it will become the new maximum and $\eta_N(L_t, n) = \eta_N(L_t, n_2)$ is the optimal switch point.

If $\gamma(L_t, n) > 1$, then $\eta_N(L_t, n)$ will decrease as L_t increases, while $\eta_N(L_t, n_2)$ will increase, as shown in Fig. 16.6. Eventually, $\eta_N(L_t, n_2)$ will increase to the point where it becomes the new maximum, and $\eta_N(L_t, n) = \eta_N(L_t, n_2)$ is the optimal switching point.

Theorem *The optimal switch point for an M-unit system is at $\eta_N(L_t, n) = \eta_N(L_t, n_1)$ if L_t is decreasing.*

Proof Similarly, if L_t decreases and $\gamma(L_t, n) > 1$, Fig. 16.6 shows that $\eta_N(L_t, n)$ will increase until $\gamma(L_t, n)$ reaches the value 1, at which point it will immediately start decreasing. At the same time, $\eta_N(L_t, n_1)$ will increase. As before, there is a point where $\eta_N(L_t, n_1)$ exceeds $\eta_N(L_t, n)$ as a new maximum. This is our optimal switching point, given by: $\eta_N(L_t, n) = \eta_N(L_t, n_1)$.

As shown in Fig. 16.5, for a given $\gamma(L_t, n) < 1$, $\eta_N(L_t, n)$ will decrease as L_t decreases. We also find that $\eta_N(L_t, n_1)$ will increase to the point of maximum efficiency. Then the optimal switching point is $\eta_N(L_t, n) = \eta_N(L_t, n_1)$.

16.5 ESPS Operates at Maximum Efficiency

For the total L_t, if the optimal run-unit of the ESPS is n as described above, the maximal value of the overall efficiency η_0 of the power station is

$$\max \eta_0 = \eta_N\left(\frac{L_t}{\sum_{i=1}^{n} L_{ie}}\right) \tag{16.41}$$

If the following equation holds

$$\frac{L_t}{\sum_{i=1}^{n} L_{ie}} = L_e \tag{16.42}$$

Then

$$\eta_N\left(\frac{L_t}{\sum_{i=1}^{n} L_{ie}}\right) = \eta_e \tag{16.43}$$

Thus, the ESPS can work at the maximum efficiency η_e.

16.6 Conclusion

By assuming a fixed value of the total load, we propose an optimal control method that does not rely on an accurate model of the ESPS. By varying the total load, we also propose an optimal switching method. This method depends only on the efficiency function of the generator.

The proofs of optimal control and switching methods given in this chapter are mainly based on the characteristics of the efficiency function, which can be approximately considered as a concave and non-negative function. Therefore, this optimal method has the following characteristics:

(1) High versatility,
(2) Includes linear and nonlinear systems,
(3) No mathematical model of the system is required.

Acknowledgements In order to solve the optimization problem of general equipment, the author has conducted long-term research. We would like to thank Zhang Yanfang from Hebei Automation Company and Yao Bosheng from Beijing IAO Technology Development Company for their valuable help and suggestions during the development and experiment of this theory.

Chapter 17
Optimal Control and Scheduling of Distribution Stations

The electricity generated by the power plant is sent to the distribution station, and then transmitted to factories, mines and households through the public power grid. An enterprise power distribution station may also have many transformers, and the number of transformers used and power distribution are adjusted to transmit power to each workshop (Fig. 17.1). The electricity consumed by power transmission and distribution is very considerable, and a large number of scientists, scholars, students and engineers are engaged in research in this field. This chapter will discuss this issue.

17.1 Energy Relations for Multiple Transformers in a Distribution Station

Assume that there are n transformers in a distribution station to supply power to all devices in a factory area, the total power output of the station is P_0, assuming that P_0 is a fixed value, the power output by the i-th transformer is P_i, and all operating transformers consume the total electric energy is P_t, P_t is the total input of the station, and the expression of P_t is

$$P_t = \sum_{i=1}^{n} \frac{P_i}{\eta_i(P_i)} \tag{17.1}$$

$$P_0 = \sum_{i=1}^{n} P_i$$

The unit of measurement for electric energy is watts.

Assume the energy efficiency curve of the i-th transformer is shown in Fig. 17.2.

F. Yao and Y. Yao, *Efficient Energy-Saving Control and Optimization for Multi-Unit Systems*, https://doi.org/10.1007/978-981-97-4492-3_17

Fig. 17.1 Power transmission and distribution

Fig. 17.2 The energy efficiency curve of the i-th transformer

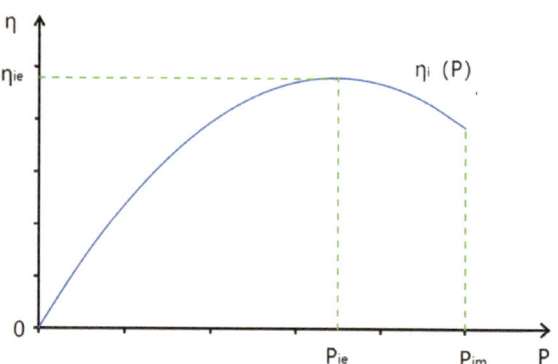

In Fig. 17.2, P_{im} is the maximum power output of the i-th transformer, η_{ie} is the maximum energy efficiency of the i-th transformer, Pie is the power output when the i-th transformer has the maximum energy efficiency, and $\eta_i(P)$ has the following characteristics:

$$0 \leq P \leq P_{im}$$
$$\eta_{ie} = \eta_i(P_{ie})$$
$$0 \leq \eta_i(P) \leq \eta_{ie} \qquad (17.2)$$
$$\eta_i(0) = 0$$
$$\eta_i''(P) < 0$$

Assume the energy efficiency functions $\eta_1(P)$ and $\eta_i(P)$ of the first and the i-th transformer as shown in Fig. 17.3.

If the following formula holds:

$$\eta_i(P_i) = \eta_1\left(\frac{P_1}{\beta_i}\right) \qquad (17.3)$$

where β_i is a constant, we say that the i-th transformer and the first transformer are the transformers with similar energy efficiency, referred to as "similar energy efficiency transformer", and $\beta_1 = 1$.

Fig. 17.3 The energy
efficiency curves of the first
and the i-th transformer

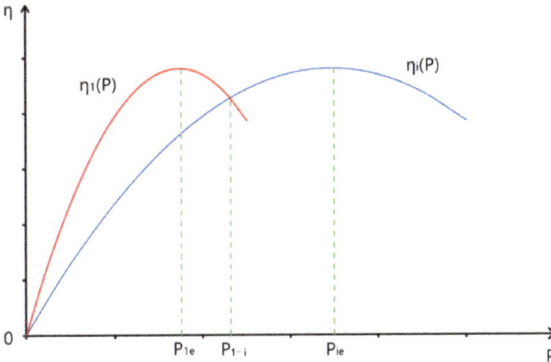

The transformer with similar energy efficiency has the following characteristics:

$$\eta_i'(P_i) = \frac{\eta_1'(P_1)}{\beta_i}$$
$$P_{ie} = \beta_i P_{1e} \tag{17.4}$$
$$\beta_i = \frac{P_{ie}}{P_{1e}}$$

17.2 Optimal Scheduling of Multiple Transformers in a Grid

1. If the n transformers are identical, the energy expression of the power supply
   becomes

$$P_t = P_0 \sum_{i=1}^{n} \frac{\theta_i}{\eta(\theta_i P_0)} \tag{17.5}$$

where

$$\theta_i = \frac{P_i}{P_0} \tag{17.6}$$

$$\sum_{i=1}^{n} \theta_i = 1$$

Consider the minimization problem of total power consumption

$$\min P_t$$

$$s.t.\ \theta_i > 0,\ i = 1, 2, ...n$$

$$\sum_{i=1}^{n} \theta_i = 1 \tag{17.7}$$

$$P_0 = cons\ tan\ t$$

This problem can also be written as

$$\min P_0 \sum_{i=1}^{n} \frac{\theta_i}{\eta(\theta_i P_0)}$$

$$s.t.\ \theta_i > 0,\ i = 1, 2, ...n$$

$$\sum_{i=1}^{n} \theta_i = 1 \tag{17.8}$$

$$P_0 = cons\ tan\ t$$

We consider three cases:

(1) n = 2

The system has two variables and has

$$\begin{aligned} \theta_1 + \theta_2 &= 1 \\ \theta_1 &> 0 \\ \theta_2 &> 0 \end{aligned} \tag{17.9}$$

The objective function P_t can be expressed as

$$P_t = P_0\left(\frac{\theta_1}{\eta(\theta_1 P_0)} + \frac{\theta_2}{\eta(\theta_2 P_0)}\right) = P_0\left(\frac{\theta_1}{\eta(\theta_1 P_0)} + \frac{1 - \theta_1}{\eta((1 - \theta_1)P_0)}\right) \tag{17.10}$$

The optimization condition is

$$P_t'(\theta_1) = 0 \tag{17.11}$$

and it is easy to see that

$$\theta_1 = \frac{1}{2} \tag{17.12}$$

is an optimization point. Then we have

$$\theta_2 = 1 - \theta_1 = \theta_1 = \frac{1}{2} = \frac{1}{n} \tag{17.13}$$

That is, the optimal control method is to keep

$$P_1 = P_2 = \frac{P_0}{2} = \frac{P_0}{n} \tag{17.14}$$

The total power dissipation is

$$P_t = \frac{P_0}{\eta\left(\frac{P_0}{2}\right)} = \frac{P_0}{\eta\left(\frac{P_0}{n}\right)} \tag{17.15}$$

Since the shape of the overall efficiency curve of the distribution station is the same as that of a single transformer, so the second derivative of the P_t is also greater than zero

$$P_t''(\theta_1) > 0 \tag{17.16}$$

P_t is the only minimum value.

$$\min P_t = \frac{P_0}{\eta\left(\frac{P_0}{n}\right)} \tag{17.17}$$

The overall energy efficiency η_t of the distribution station is the only maximum value.

$$\max \eta_t = \eta\left(\frac{P_0}{n}\right) \tag{17.18}$$

(2) n = 3

The system has three variables, based on known conditions, we have

$$\begin{aligned}
\theta_1 + \theta_2 + \theta_3 &= 1 \\
\theta_1 &> 0 \\
\theta_2 &> 0 \\
\theta_3 &> 0
\end{aligned} \tag{17.19}$$

The P_t expression becomes

$$P_t = P_0\left(\frac{\theta_1}{\eta(\theta_1 P_0)} + \frac{\theta_2}{\eta(\theta_2 P_0)} + \frac{\theta_3}{\eta((\theta_3 P_0)}\right) \tag{17.20}$$

Assuming that θ_3 is fixed and an optimization point, only θ_1 and θ_2 are variables, we have

$$\theta_1 + \theta_2 = 1 - \theta_3 = constant \tag{17.21}$$

Based on the conclusion of n = 2 above, we have

$$\theta_1 = \theta_2 \tag{17.22}$$

to be the optimal point.

Assuming that θ_2 is fixed and is an optimization point, only θ_1 and θ_3 are variables, there are

$$\theta_1 + \theta_3 = 1 - \theta_2 = constant \tag{17.23}$$

According to the conclusion of n = 2 above, we have

$$\theta_1 = \theta_3 \tag{17.24}$$

to be the optimal point.

Similarly, assuming that θ_1 is fixed and is an optimization point, only θ_2 and θ_3 are variables, we have

$$\theta_2 + \theta_3 = 1 - \theta_1 = constant \tag{17.25}$$

According to the conclusion of n = 2 above, we have

$$\theta_2 = \theta_3 \tag{17.26}$$

to be the optimal point.

So, we have the optimal point

$$\theta_1 = \theta_2 = \theta_3 = \frac{1}{3} = \frac{1}{n} \tag{17.27}$$

That is, the optimal control method is to keep

$$P_1 = P_2 = P_3 = \frac{P_0}{3} = \frac{P_0}{n} \tag{17.28}$$

The minimum total power dissipation is

$$\min P_t = \frac{P_0}{\eta\left(\frac{P_0}{3}\right)} = \frac{P_0}{\eta\left(\frac{P_0}{n}\right)} \tag{17.29}$$

The maximum overall energy efficiency η_t is

$$max\eta_t = \eta\left(\frac{P_0}{3}\right) = \eta\left(\frac{P_0}{n}\right) \tag{17.30}$$

(3) $n = k$

The system has n variables, and the above conclusion can be extended to the case of $n = k$, the optimal point is

$$\theta_1 = \theta_2 = \ldots = \theta_k = \frac{1}{k} \qquad (17.31)$$

That is, the optimal control method is to keep

$$P_1 = P_2 = \ldots = P_k = \frac{P_0}{k} \qquad (17.32)$$

The minimum total power dissipation is

$$\min P_t = P_0 \frac{1}{\eta\left(\frac{P_0}{n}\right)} \qquad (17.33)$$

The maximum overall energy efficiency η_t is

$$max\eta_t = \eta\left(\frac{P_0}{k}\right) = \eta\left(\frac{P_0}{n}\right) \qquad (17.34)$$

Conclusion: When there are n sets of identical transformers supplying power to a factory area on the same power grid, and the load allocated to each transformer is the same, it is the optimal dispatching scheme.

2. If the n transformers are the energy efficiency similarity transformers, the energy expression of the power supply becomes

$$P_t = P_0 \sum_{i=1}^{n} \frac{\theta_i}{\eta_i(\theta_i P_0)} \qquad (17.35)$$

where

$$\theta_i = \frac{P_i}{P_0} \qquad (17.36)$$

$$\sum_{i=1}^{n} \theta_i = 1$$

Consider the minimization problem of total power consumption

$$minP_t \qquad (17.37)$$

$$s.t. \theta_i > 0, i = 1, 2, \ldots n$$

$$\sum_{i=1}^{n} \theta_i = 1$$

$$P_0 = constant$$

This problem can also be written as

$$min P_0 \sum_{i=1}^{n} \frac{\theta_i}{\eta_i(\theta_i P_0)} \tag{17.38}$$

$$s.t. \theta_i > 0, i = 1, 2, \dots n$$

$$\sum_{i=1}^{n} \theta_i = 1$$

$$P_0 = constant$$

We consider three cases:

(1) $n = 2$

The system has two variables and has

$$\begin{aligned} \theta_1 + \theta_2 &= 1 \\ \theta_1 &> 0 \\ \theta_2 &> 0 \end{aligned} \tag{17.39}$$

The objective function P_t can be expressed as

$$P_t = P_0(\frac{\theta_1}{\eta_1(\theta_1 P_0)} + \frac{\theta_2}{\eta_2(\theta_2 P_0)}) = P_0(\frac{\theta_1}{\eta_1(\theta_1 P_0)} + \frac{1 - \theta_1}{\eta_2((1 - \theta_1)P_0)}) \tag{17.40}$$

The optimization condition is

$$P_t'(\theta_1) = 0 \tag{17.41}$$

and it is easy to see that

$$\theta_1 = \frac{1}{1 + \beta_2} = \frac{\beta_1}{\beta_1 + \beta_2} \tag{17.42}$$

for the optimization point. Then we have

$$\theta_2 = 1 - \theta_1 = \frac{\beta_2}{1 + \beta_2} = \frac{\beta_2}{\beta_1 + \beta_2} \tag{17.43}$$

That is, the optimal control method is to keep

$$P_1 = \frac{1}{1+\beta_2} P_0$$

$$\tag{17.44}$$

$$P_2 = \frac{\beta_2}{1+\beta_2} P_0$$

The total power dissipation is

$$P_t = \frac{P_0}{\eta_1(\frac{P_0}{1+\beta_2})} \tag{17.45}$$

Since the shape of the overall efficiency curve of the distribution station is the same as that of a single transformer, so the second derivative of the P_t is also greater than zero

$$P_t''(\theta_1) > 0 \tag{17.46}$$

P_t is the only minimum value.

$$\min P_t = \frac{P_0}{\eta_1(\frac{P_0}{1+\beta_2})} \tag{17.47}$$

The overall energy efficiency η_t of the distribution station is the only maximum value.

$$max\ \eta_t = \eta_1\left(\frac{P_0}{1 + \beta_2}\right) \tag{17.48}$$

(2) n = 3

The system has three variables, based on known conditions, we have

$$\theta_1 + \theta_2 + \theta_3 = 1 \tag{17.49}$$

$$\theta_1 > 0$$

$$\theta_2 > 0$$

$$\theta_3 > 0$$

The P_t expression becomes

$$P_t = P_0\left(\frac{\theta_1}{\eta_1(\theta_1 P_0)}\right) + \left(\frac{\theta_2}{\eta_2(\theta_2 P_0)}\right) + \left(\frac{\theta_3}{\eta_3(\theta_3 P_0)}\right) \qquad (17.50)$$

Assuming that θ_3 is fixed and is an optimization point, only θ_1 and θ_2 are variables, there are

$$\theta_1 + \theta_2 = 1 - \theta_3 = constant \qquad (17.51)$$

Based on the conclusion of n = 2 above, we have

$$\theta_1 = \frac{1}{1+\beta_2}$$
$$\qquad (17.52)$$
$$\theta_2 = \frac{\beta_2}{1+\beta_2}$$

to be is the optimal point.

Assuming that θ_2 is fixed and is an optimization point, only θ_1 and θ_3 are variables, there are

$$\theta_1 + \theta_3 = 1 - \theta_2 = constant \qquad (17.53)$$

According to the conclusion of n = 2 above, there are

$$\theta_1 = \frac{1}{1+\beta_3}$$
$$\qquad (17.54)$$
$$\theta_3 = \frac{\beta_3}{1+\beta_3}$$

is the optimal point.

Similarly, assuming that θ_1 is fixed and is an optimization point, only θ_2 and θ_3 are variables, we have

$$\theta_2 + \theta_3 = 1 - \theta_3 = constant \qquad (17.55)$$

According to the conclusion of n = 2 above, we have

$$\theta_2 = \frac{\beta_2}{\beta_2+\beta_3}$$
$$\qquad (17.56)$$
$$\theta_3 = \frac{\beta_3}{\beta_2+\beta_3}$$

to be the optimal point.

So, we have the optimal point, and the optimal control method is to keep

$$\theta_1 = \frac{\beta_1}{\beta_1+\beta_2+\beta_3}$$

$$\theta_2 = \frac{\beta_2}{\beta_1+\beta_2+\beta_3} \tag{17.57}$$

$$\theta_3 = \frac{\beta_3}{\beta_1+\beta_2+\beta_3}$$

The minimum total power consumption is

$$minP_t = \frac{P_0}{\eta_1\left(\frac{P_0}{1+\beta_2+\beta_3}\right)} \tag{17.58}$$

The maximum overall energy efficiency η_t is

$$max\eta_t = \eta_1\left(\frac{P_0}{1+\beta_2+\beta_3}\right) \tag{17.59}$$

(3) n = k

The system has n variables, the above conclusion can be extended to the case of n = k, the optimal point and the optimal control method is to keep

$$\theta_i = \frac{\beta_i}{\sum_{l=1}^{k}\beta_l} \tag{17.60}$$

The minimum total power dissipation is

$$minP_t = P_0\frac{1}{\eta_1\left(\frac{P_0}{\sum_{l=1}^{k}\beta_l}\right)} \tag{17.61}$$

The maximum overall energy efficiency η_t is

$$max\eta_t = \eta_1\left(\frac{P_0}{\sum_{l=1}^{k}\beta_l}\right) \tag{17.62}$$

17.3 Optimum Number of Transformers Operating in a Distribution Station

1. A distribution station has m identical transformers, if the following equations are satisfied

$$\eta\left(\frac{P_0}{n}\right) \geq \eta\left(\frac{P_0}{n-1}\right) \tag{17.63}$$

$$\eta\left(\frac{P_0}{n}\right) \geq \eta\left(\frac{P_0}{n+1}\right)$$

$$n \leq m$$

Then n is the number of transformers with optimal operation.

2. A distribution station has m energy efficiency similarity transformers, if the following equations are satisfied

$$\eta_1\left(\frac{P_0}{\sum_{l=1}^{n}\beta_l}\right) \geq \eta_1\left(\frac{P_0}{\sum_{l=1}^{n_1}\beta_l}\right) \tag{17.64}$$

$$n \leq m$$

$$n_1 \leq m$$

n_1 is any combination other than the optimal combination of n units this time, and also include other combinations of n units. The number n is the number of transformers with optimal operation.

17.4   Optimal Switching Rules for Multiple Transformers in Distribution Stations

1. A distribution station has m identical transformers, and the number n is the number of transformers currently in optimal operation. If P_0 increases to P_{01}, the following relation holds:

$$\eta\left(\frac{P_{01}}{n}\right) = \eta\left(\frac{P_{01}}{n+1}\right) \tag{17.65}$$

$$n \leq m$$

Then P_{01} is the optimal switching point between the operation of n transformers and the operation of n + 1 transformer. When the total required power P_0 is greater than P_{01}, the optimal number of transformers in operation is switched from n to n + 1. If P_0 increases until $P_0/n = P_{1m}$, there isn't the point P_{01}, then $P_0/n = P_{1m}$ is the switching point from n to n + 1.

If P_0 is reduced to P_{02}, the following relation is established

$$\eta\left(\frac{P_{02}}{n}\right) = \eta\left(\frac{P_{02}}{n-1}\right) \tag{17.66}$$

$$n \leq m$$

Then P_{02} is the optimal switching point between the operation of n transformers and the operation of n−1 transformers. When the total required power P_0 is less than P_{02}, the optimal number of transformers in operation is switched from n to n−1. If P_0 reduces until $P_0/(n-1) = P_{1m}$, there isn't the point P_{02}, then $P_0/(n-1) = P_{1m}$ is the switching point from n to n−1.

The analysis process is shown in Fig. 17.4.

2. A distribution station has m energy efficiency similarity transformers, and the number n is the number of transformers currently in optimal operation. If P_0 increases to P_{01}, the following relation holds:

$$\eta_1\left(\frac{P_0}{\sum_{l=1}^{n}\beta_l}\right) = \eta_1\left(\frac{P_0}{\sum_{l=1}^{k1}\beta_l}\right) \tag{17.67}$$

$$n \leq m$$

Then P_{01} is the optimal switching point between the operation of n transformers and the operation of k_1 transformers. When the total required power P_0 is greater than P_{01}, the optimal number of transformers in operation is switched from n to k_1. If P_0 increases until $P_0/\sum_{l=1}^{n}\beta_l=P_{1m}$, there isn't the point P_{01}, then $P_0/\sum_{l=1}^{n}\beta_l=P_{1m}$ is the switching point from n to k_1.

If P_0 is reduced to P_{02}, the following relation is established

$$\eta_1\left(\frac{P_0}{\sum_{l=1}^{n}\beta_l}\right) = \eta_1\left(\frac{P_0}{\sum_{l=1}^{k2}\beta_l}\right) \tag{17.68}$$

$$n \leq m$$

Fig. 17.4 Energy efficiency comparison curve

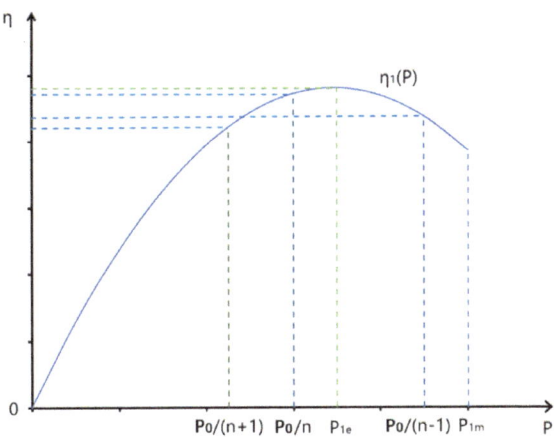

Fig. 17.5 Energy efficiency
comparison curve

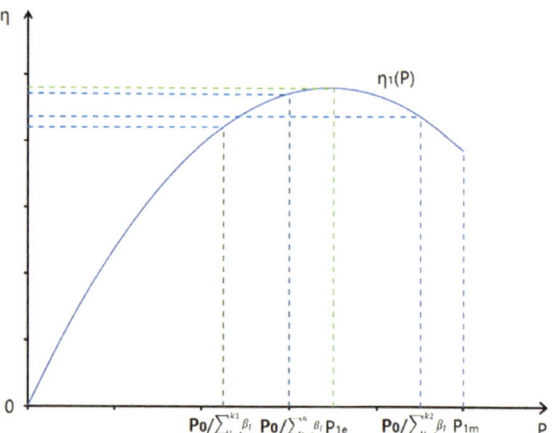

Then P_{02} is the optimal switching point between the operation of n transformers and the operation of k_2 transformers. When the total required power P_0 is less than P_{02}, the optimal number of transformers in operation is switched from n to k_2. If P_0 reduces until $P_0/\sum_{l=1}^{k_2}\beta_l = P_{1m}$, there isn't the point P_{02}, then $P_0/\sum_{l=1}^{k_2}\beta_l = P_{1m}$ is the switching point from n to k_2.

The analysis process is shown in Fig. 17.5.

k1 and k2 are any combination other than the optimal combination of n units this time, and also include other combinations of n units. Point $P_0/\sum_{l=1}^{k_1}\beta_l$ is the point closest to $P_0/\sum_{l=1}^{n}\beta_l$ to the left of point $P_0/\sum_{l=1}^{n}\beta_l$. Point $P_0/\sum_{l=1}^{k_2}\beta_l$ is the point closest to $P_0/\sum_{l=1}^{n}\beta_l$ to the right of point $P_0/\sum_{l=1}^{n}\beta_l$. If all transformers are identical, we have $k1 = n + 1$ and $k2 = n - 1$.

Chapter 18
Optimal Dispatch of Power Grid Load

Many power plants export electricity to the public grid, delivering it to factories and homes (Fig. 18.1). We should optimize the distribution and scheduling of the load rate of each line to minimize the power loss of the transmission system. The amount of energy wasted in the power grid transmission process is very considerable, and many scientists, scholars, students and engineers are engaged in research in this field.

Many optimization methods require the establishment of a power grid model. However, since accurate models of actual power grids are difficult to establish, these optimization methods will quickly become difficult to apply in actual power grids given the complexity of the models and algorithms, the curse of dimensionality, and the computational time.

An approach that avoids these complications would be highly advantageous, especially for business and industry.

18.1 Energy Consumption in a Power Grid

There are n transmission lines to supply power in an area, the power lines in parallel. The total apparent power required is S_t, VA, the line voltage U of each transmission line is identical. The total apparent current is I_0, the resistance and current of i-th line are R_i and I_i, respectively, the total energy consumption W_t of n transmission lines is expressed as

$$W_t = 3 \sum_{i=1}^{n} I_i^2 R_i = 3I_0^2 \sum_{i=1}^{n} \left(\frac{I_i}{I_0} \right)^2 R_i = 3I_0^2 \sum_{i=1}^{n} \theta_i^2 R_i \qquad (18.1)$$

$$I_0 = \sum_{i=1}^{n} I_i$$

© The Author(s) 2024
F. Yao and Y. Yao, *Efficient Energy-Saving Control and Optimization for Multi-Unit Systems*, https://doi.org/10.1007/978-981-97-4492-3_18

Fig. 18.1 Power grid

$$S_t = \sqrt{3}UI_0$$

where θ_i represents the load rate of the i-th transmission line, expressed as

$$\sum_{i=1}^{n} \theta_i = 1 \tag{18.2}$$

$$\theta_i = \frac{I_i}{I_0}$$

If the resistance of each line is the same as R_0, W_t of the n transmission lines is expressed as

$$W_t = 3I_0^2 R_0 \sum_{i=1}^{n} \theta_i^2 \tag{18.3}$$

18.2 Optimal Dispatching of Power Grids Composed of Lines with the Identical Resistance

Theorem: If the resistance of each line is i R_0, for the optimization problem of W_t, the minimization of the total energy consumption of the power grid.

$$\min \quad W_t \tag{18.4}$$
$$s.t. I_i > 0$$
$$\sum_{i=1}^{n} I_i = I_0$$

is given by

$$\min W_t = 3I_0^2 R_0 \frac{1}{n} \tag{18.5}$$

The optimal point is

$$I_1 = I_2 = \ldots = I_k = \frac{I_0}{k} \tag{18.6}$$

That is

$$\theta_1 = \theta_2 = \ldots = \theta_k = \frac{1}{k} \tag{18.7}$$

Proof: We begin our inductive proof by considering the case where n = 2.

The constraint condition then becomes

$$I_1 + I_2 = I_0 \tag{18.8}$$

where

$$I_1 > 0 \tag{18.9}$$

$$I_2 > 0$$

The objective function W_t is expressed as

$$W_t = 3R_0(I_1^2 + I_2^2) \tag{18.10}$$

The optimal condition is given for

$$W_t{}'(I_1) = 0 \tag{18.11}$$

We have

$$4I_1 - 2I_0 = 0 \tag{18.12}$$

It is then easily verified that

$$I_1 = I_2 = \frac{I_0}{2} \tag{18.13}$$

is an optimal point, that is

$$\theta_1 = \theta_2 = \frac{1}{2} \tag{18.14}$$

We then check the second derivative,

$$W_t'' > 0 \tag{18.15}$$

Therefore, the optimal point is the unique minimum.

The minimal value of the total energy consumption of the power grid is

$$\min W_t = 3I_0^2 R_0 \frac{1}{2} \tag{18.16}$$

We then assume that this holds for n = k. The above conclusion is extended to the case of n = k, and the optimal point is then

$$I_1 = I_2 = \ldots = I_k = \frac{I_0}{k} \tag{18.17}$$

That is

$$\theta_1 = \theta_2 = \ldots = \theta_k = \frac{1}{k} \tag{18.18}$$

The minimal value of W_t is

$$\max W_t = 3I_0^2 R_0 \frac{1}{k} \tag{18.19}$$

Our inductive case is then given by n = k + 1. For the total energy consumption of the power grid we have

$$W_t = 3R_0 \left(\sum_{i=1}^{k} I_i^2 + I_{k+1}^2 \right) \tag{18.20}$$

and the minimum of the first item is

$$\max 3R_0 \sum_{i=1}^{k} I_i^2 = 3R_0 (I_0 - I_{k+1})^2 \frac{1}{k} \tag{18.21}$$

where

$$I_1 = I_2 = \ldots = I_k \tag{18.22}$$

The expression for W_t becomes

$$W_t = 3R_0\left(\frac{(I_0 - I_{k+1})^2}{k} + I_{k+1}^2\right) \tag{18.23}$$

The optimal condition is given for

$$W_t'(I_{k+1}) = 0 \tag{18.24}$$

The solution is

$$I_{k+1} = \frac{I_0}{k+1} \tag{18.25}$$

and

$$I_1 = I_2 = \ldots = I_k = \frac{I_0 - I_{k+1}}{k} = \frac{I_0}{k+1} \tag{18.26}$$

Therefore, the optimal point is then

$$I_1 = I_2 = \ldots = I_k = I_{k+1} = \frac{I_0}{k+1} \tag{18.27}$$

That is

$$\theta_1 = \theta_2 = \ldots = \theta_k = \theta_{k+1} = \frac{1}{k+1} \tag{18.28}$$

The minimal value of total energy consumption of the power grid is

$$\min W_t = 3R_0 I_0^2 \frac{1}{k+1} \tag{18.29}$$

Load distribution theorem: In a power grid which consists of n main lines that have the identical resistance, the optimal control method is to keep each line to have the identical current I_0/n.

18.3 Optimal Scheduling of a Power Grid with the Different Power Line

Theorem: If the resistances of all power lines are different, respectively R1, R2, ..., Rn, for the optimization problem of W_t, the minimization of the total energy consumption of the power grid.

$$\min \quad W_t \tag{18.30}$$
$$s.t. I_i > 0$$
$$\sum_{i=1}^{n} I_i = I_0$$

is given by

$$\min W_t = 3I_0^2(R_1\|R_2\| \ldots \|R_n) = 3I_0^2 \frac{\prod_{j=1}^{n} R_j}{\sum_{m=1}^{n}\left(\left(\prod_{j=1}^{n} R_j\right)/R_m\right)} \tag{18.31}$$

The optimal point is

$$I_i = \frac{\left(\prod_{j=1}^{n} R_j\right)/R_i}{\sum_{m=1}^{n}\left(\left(\prod_{j=1}^{n} R_j\right)/R_m\right)} I_0 \tag{18.32}$$

Proof: We begin our inductive proof by considering the case where n = 2.

The constraint condition then becomes

$$I_1 + I_2 = I_0 \tag{18.33}$$

where

$$I_1 > 0 \tag{18.34}$$

$$I_2 > 0$$

The objective function W_t is expressed as

$$W_t = 3\left(I_1^2 R_1 + I_2^2 R_2\right) \tag{18.35}$$

The optimal condition is given for

$$W_t{'}(I_1) = 0 \tag{18.36}$$

The solution is

$$I_1 = \frac{R_2}{R_1 + R_2} I_0 \tag{18.37}$$

$$I_2 = \frac{R_1}{R_1 + R_2} I_0$$

Checking the second derivative, we see

$$W_t'' > 0 \tag{18.38}$$

Thus, the optimal point is the unique minimum.

The minimal value of the total energy consumption of the power grid is

$$\max W_t = 3I_0^2 (R_1 \| R_2) = 3I_0^2 \frac{R_1 R_2}{R_1 + R_2} \tag{18.39}$$

We then assume that this holds for n = k. The above conclusion is readily extended to the case of n = k, and the optimal point is then

$$I_i = \frac{(\prod_{j=1}^{k} R_j)/R_i}{\sum_{m=1}^{k} \left((\prod_{j=1}^{k} R_j)/R_m \right)} I_0 \tag{18.40}$$

The minimal value of W_t is

$$\max W_t = 3I_0^2 \frac{\prod_{j=1}^{k} R_j}{\sum_{m=1}^{k} \left((\prod_{j=1}^{k} R_j)/R_m \right)} \tag{18.41}$$

Our inductive case is then given by n = k + 1. For the total energy consumption of the power grid we have

$$W_t = 3\left(\sum_{i=1}^{k} I_i^2 R_i + I_{k+1}^2 R_{k+1} \right) \tag{18.42}$$

and the minimum of the first item is

$$3 \sum_{i=1}^{k} I_i^2 R_i = 3(I_0 - I_{k+1})^2 \frac{\prod_{j=1}^{k} R_j}{\sum_{m=1}^{k} \left((\prod_{j=1}^{k} R_j)/R_m \right)} \tag{18.43}$$

The expression for W_t becomes

$$W_t = 3(I_0 - I_{k+1})^2 \frac{\prod_{j=1}^{k} R_j}{\sum_{m=1}^{k} \left(\left(\prod_{j=1}^{k} R_j \right)/R_m \right)} + 3I_{k+1}^2 R_{k+1} \qquad (18.44)$$

The optimal condition is given for

$$W_t'(I_{k+1}) = 0 \qquad (18.45)$$

Based on above conclusion of $n = 2$, the solution is

$$I_{k+1} = \frac{\left(\prod_{j=1}^{k+1} R_j \right)/R_{k+1}}{\sum_{m=1}^{k+1} \left(\left(\prod_{j=1}^{k+1} R_j \right)/R_m \right)} I_0 \qquad (18.46)$$

Therefore, the optimal point is then

$$I_i = \frac{\left(\prod_{j=1}^{k+1} R_j \right)/R_i}{\sum_{m=1}^{k+1} \left(\left(\prod_{j=1}^{k+1} R_j \right)/R_m \right)} I_0 \qquad (18.47)$$

and the minimal value of total energy consumption of the power grid is

$$\min W_t = 3I_0^2 \frac{\prod_{j=1}^{k+1} R_j}{\sum_{m=1}^{k+1} \left(\left(\prod_{j=1}^{k+1} R_j \right)/R_m \right)} \qquad (18.48)$$

18.4 Optimal Scheduling of a Direct Current Power Grid

Theorem: If not all resistances of direct current power lines are identical, that is, $R_1, R_2, \ldots, R_n$, respectively, are different, for the optimization problem of W_t, the minimization of the total energy consumption of the power grid.

$$\begin{aligned} \min \quad & W_t \\ s.t. I_i &> 0 \\ \sum_{i=1}^{n} I_i &= I_0 \end{aligned} \qquad (18.49)$$

is given by

$$\min W_t = 2I_0^2(R_1 \| R_2 \| \ldots \| R_n) = 2I_0^2 \frac{\prod_{j=1}^{n} \mathbf{R_j}}{\sum_{m=1}^{n}\left(\left(\prod_{j=1}^{n} \mathbf{R_j}\right)/\mathbf{R_m}\right)} \quad (18.50)$$

The optimal point is

$$\mathbf{I}_i = \frac{\left(\prod_{j=1}^{n} \mathbf{R_j}\right)/\mathbf{R_i}}{\sum_{m=1}^{n}\left(\left(\prod_{j=1}^{n} \mathbf{R_j}\right)/\mathbf{R_m}\right)} \mathbf{I}_0 \quad (18.51)$$

*The Proof is **Omitted**.*

18.5 Conclusion

By assuming a fixed value of the total apparent current, we propose an optimal control method that does not rely on an accurate model of the grid.

The proofs of optimal control and switching methods given in this chapter are mainly based on the minimum energy consumption rule. Therefore, this optimal method has the following characteristics:

(1) Including lined and unlined systems,
(2) No systematic mathematical model is required,
(3) High versatility.

Chapter 19
Efficiency Optimization of Scraper Conveyors and Paper Machines

Multiple motors are used to drive long scraper conveyors or the wire section of paper machines, mainly to make long chains and long belts evenly stressed. This type of system does not have the problem of optimal switching of the number of drive motors. However, the system has the problem of how to distribute the load so that the system operates efficiently.

19.1 A Scraper Conveyor Driven by Two Different Motors

In coal mines, many scraper conveyors are very long, some exceeding 300 m. Most scraper conveyors are driven by more than one motor at the same time to ensure uniform force on the scraper conveyor. Scraper conveyors driven by two different motors are common. It has a traditional structure as shown in Fig. 19.1.

In Fig. 19.1, two different motors drive the scraper conveyor to ensure uniform force on the long scraper conveyor. There is a way to accomplish the same task that consumes the least amount of energy. In other words, there is a way to maximize the overall efficiency. However, there has not yet been a common conclusion on the efficiency optimization of such systems [1].

To achieve minimum energy, we should determine how much load each motor in the system carries.

This issue is becoming increasingly important with the increasing awareness of environmental protection and the need for energy-saving systems. Currently, optimization research on various scraper conveyors has been extensively studied [2–9]. Almost all such optimization methods require the establishment of an accurate model of the system. However, since accurate models of actual scraper conveyors are difficult to establish, these optimization methods quickly become difficult to apply in real systems. Solutions that avoid these complications would be highly beneficial, especially for business and industry.

© The Author(s) 2024
F. Yao and Y. Yao, *Efficient Energy-Saving Control and Optimization for Multi-Unit Systems*, https://doi.org/10.1007/978-981-97-4492-3_19

Fig. 19.1 The traditional structure of a scraper conveyor

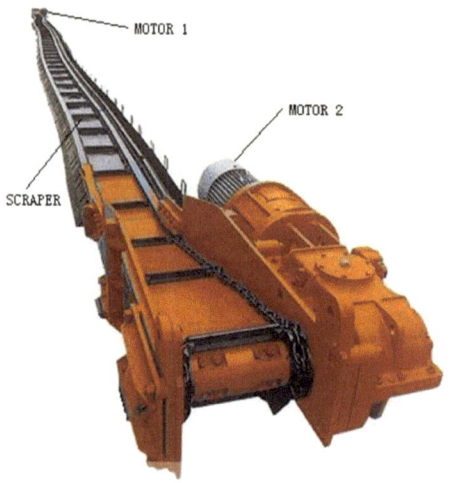

By assuming a fixed total load, we propose an optimal control method similar to traditional static optimization [10], which does not rely on an accurate model of the scraper conveyor and is only based on the shape of the efficiency function. By varying the total load, we also propose a method similar to traditional dynamic optimization [11], which only relies on the efficiency function of the device at constant output. These methods can be used for systems consisting of the same type of motor and gear and having the same energy efficiency function.

However, in industrial applications, there are still a large number of scraper conveyors, and all of their motors do not have the same model. To address the optimization problem of such systems, we extend the discussion presented in [10, 11]. This chapter defines the normalized energy efficiency function of the motor and the scraper conveyor driven by multiple motors with the same normalized energy efficiency function. We can find the functional expression of the scraper conveyor power consumption and its minimum value. When the normalized efficiency function of each motor is the same, our approach can still be adopted [10, 11]. The minimum power consumption and optimal switching point of this type of scraper conveyor are determined by using a normalized energy efficiency function.

19.1.1 Characteristics of a Scraper Conveyor

In general, the efficiency function of a motor at constant speed n_0 (rpm) is shown in Fig. 19.2. By exploiting the shape of the efficiency function, we are able to determine the total power consumption of a scraper conveyor.

In Fig. 19.2, M is torque (N m), and η (M) is the efficiency function, M_e is the torque at η_e which is the maximum efficiency. For a given motor, η_e and M_e are constants. η (M) can be considered approximately a concave non-negative function

Fig. 19.2 The efficiency function of a motor

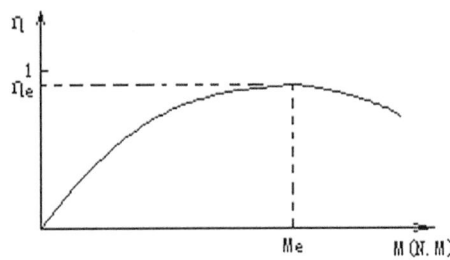

through the origin, and we have

$$M \geq 0$$
$$\eta(M) \geq 0 \qquad (19.1)$$
$$\eta(0) = 0$$

and

$$\eta''(M) < 0 \qquad (19.2)$$

Approximately, $\eta(M)$ becomes

$$\eta(M) = \sum_{i=0}^{\infty} a_i M^i = M \sum_{i=1}^{\infty} (a_i M^{i-1}) \approx M(a_1 + a_2 M) \geq 0 \qquad (19.3)$$

where

$$a_0 = 0 \qquad (19.4)$$

It follows that

$$a_1 + a_2 M > 0$$
$$a_1 > 0 \qquad (19.5)$$
$$a_2 < 0$$

The motor's power has the form

$$P(M) = \frac{Mn_0}{9550\eta(M)} = \frac{P_0}{\eta(M)} \qquad (19.6)$$

and

$$P_0 = \frac{Mn_0}{9550\eta} \qquad (19.7)$$

where $P(M)$ is the power consumption, M is the output torque, $\eta(M)$ is the efficiency at point (M, n_0), and P_0 is the ideal work.

If a scraper conveyor consists of two motors which are running, the total power consumption function has the form

$$P_t(M_t) = \frac{1}{9550} \sum_{i=1}^{2} \frac{n_i M_i}{\eta_i(M_i)} \tag{19.8}$$

and

$$M_t = \sum_{i=1}^{2} M_i \tag{19.9}$$

where $P_t(M)$ is the total power consumption function, M_i is the i-th motor's output torque, $\eta_i(M_i)$ is the ith motor's efficiency at point (M_i, n_i), M_t is the total load torque, and n_i is the speed of the i-th motor.

In order to ensure the long scraper conveyor stress evenly, we should have

$$M_i > 0 \tag{19.10}$$

If two motors are the same model in the scraper conveyor, their efficiency are the same, and the output speed of two motors is the same n_0. The $\eta_i(M_i)$ becomes $\eta(M_i)$, We have the following optimal conclusion [10].

The optimal control method is to keep

$$M_1 = M_2 = \frac{M_t}{2} \tag{19.11}$$

The minimal value of the total power consumption is

$$minP_t(M_t) = P_0 \frac{1}{\eta(\frac{M_t}{2})} \tag{19.12}$$

where P_0 is the ideal work, it has the form

$$P_0 = \frac{M_t n_0}{9550} \tag{19.13}$$

The total optimal efficiency is

$$max\eta_t(M_t) = \eta(\frac{M_t}{2}) \tag{19.14}$$

19.1.2 Scraper Conveyor with Similar Efficiency

We define the torque rate γ as

$$\gamma = \frac{M}{M_e} \tag{19.15}$$

We define $\eta_N(\gamma)$ as the normalized efficiency function of a motor. The normalized efficiency function $\eta_N(\gamma)$ has a shape shown in Fig. 19.3.

In Fig. 19.3, γ is a variable, $\eta_N(\gamma)$ is a function of γ, and $\eta_N(\gamma)$ has the following relationship with $\eta(M)$.

$$\eta(M) = \eta(\gamma M_e) = \eta_N(\gamma) \tag{19.16}$$

If the normalized efficiency functions of two different motors are the same, we have

$$\eta_{N1}(\gamma) = \eta_{N2}(\gamma) \tag{19.17}$$

We define motor 1 and motor 2 as motors with similar efficiency. Scraper conveyors using motors with similar efficiencies are called scraper conveyors with similar efficiencies.

Suppose γ_i is the i-th motor's torque rate and has the form

$$\gamma_i = \frac{M_i}{M_{ie}} \tag{19.18}$$

For a scraper conveyor with similar efficiency driven by two motors with the same output speed n_0, the total power consumption is of the form

$$P_t(M_t) = \frac{n_0}{9550} \sum_{i=1}^{2} \frac{M_i}{\eta_i(M_i)} = \frac{n_0}{9550} \sum_{i=1}^{2} \frac{\gamma_i M_{ie}}{\eta_N(\gamma_i)} \tag{19.19}$$

Fig. 19.3 The normalized efficiency function $\eta_N(\gamma)$

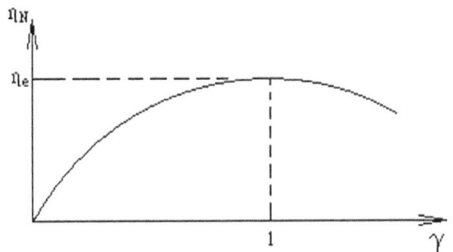

19.1.3 Optimal Control of the Scraper Conveyor with Similar Efficiencies

In this section, we derive the expression for the total power consumption of scraper conveyors with similar efficiencies and determine their minimum power consumption.

Theorem For the optimization problem of P_t (M_t), the minimization of the total power consumption.

$$minP_t(M_t)$$
$$s.t. \quad \gamma_i > 0, i = 1, 2 \tag{19.20}$$
$$\sum_{i=1}^{2} \gamma_i M_{ie} = M_t$$

is given by

$$minP_t(M_t) = \frac{P_0}{\eta_N\left(\frac{M_t}{M_{1e}+M_{2e}}\right)} \tag{19.21}$$

Proof There are two variables. We have

$$\gamma_1 M_{1e} + \gamma_2 M_{2e} = M_t \tag{19.22}$$

where

$$\gamma_1 > 0$$
$$\gamma_2 > 0 \tag{19.23}$$

The objective function P_t (M_t) is expressed as

$$P_t(M_t) = \frac{n_0}{9550}\left(\frac{\gamma_1 M_{1e}}{\eta_N(\gamma_1)} + \frac{\gamma_2 M_{2e}}{\eta_N(\gamma_2)}\right) \tag{19.24}$$

The optimal condition is given for

$$P_t'(\gamma_1) = 0 \tag{19.25}$$

We have

$$\frac{M_{1e}\eta_N(\gamma_1) - \gamma_1 M_{1e}\eta_N{'}(\gamma_1)}{\eta_N^2(\gamma_1)}$$

$$+\frac{-M_{1e}\eta_N\left(\frac{M_t-\gamma_1 M_{1e}}{M_{2e}}\right)+(M_t-\gamma_1 M_{1e})\frac{M_{1e}}{M_{2e}}\eta_N{}'\left(\frac{M_t-\gamma_1 M_{1e}}{M_{2e}}\right)}{\eta_N^2\left(\frac{M_t-\gamma_1 M_{1e}}{M_{2e}}\right)}=0 \qquad (19.26)$$

It is then easily verified that

$$\gamma_1 = \gamma_2 = \frac{M_t}{M_{1e}+M_{2e}} \qquad (19.27)$$

is an optimal point.

Checking the second derivative, we see

$$P_t{}''(\gamma_1) > 0 \qquad (19.28)$$

Therefore, the optimal point is the unique minimum.
The minimum of the total power consumption is then

$$minP_t = \frac{n_0 M_t}{9550}\frac{1}{\eta_N\left(\frac{M_t}{M_{1e}+M_{2e}}\right)} = \frac{P_0}{\eta_N\left(\frac{M_t}{M_{1e}+M_{2e}}\right)} \qquad (19.29)$$

where P_0 is the ideal work.

Remark If two motors have the same model, then the optimal point is.

$$\gamma_1 = \gamma_2 = \frac{M_t}{2M_e} \qquad (19.30)$$

$$M_1 = M_2 = \frac{M_t}{2}$$

The minimum of the total power consumption is

$$minP_t = \frac{n_0 M_t}{9550}\frac{1}{\eta_N\left(\frac{M_t}{2M_e}\right)} = \frac{P_0}{\eta\left(\frac{M_t}{2}\right)} \qquad (19.31)$$

19.1.4 Simulation

If a scraper conveyor is driven by two different motors, A and B, two motors run at the same speed 600 (rpm). The total torque is variable. The output torque of each motor is adjustable.

Motor A is smaller and its efficiency function with respect to torque M at speed 600 (rpm) is given by

$$\eta_1(M) = 0.0188M - 0.000094M^2 \tag{19.32}$$

where

$$M_{1e} = 100(N.M) \tag{19.33}$$

and

$$\eta_{1e} = 0.94 \tag{19.34}$$

Motor B is larger and its efficiency function with respect to torque M at speed 600 (r.p.m) is

$$\eta_2(M) = 0.0094M - 0.0000235M^2 \tag{19.35}$$

where

$$M_{2e} = 200(N.M) \tag{19.36}$$

and

$$\eta_{2e} = 0.94$$

We see that A and B have the same normalization efficiency function, given by the following

$$\eta_N(\gamma) = 1.88\gamma - 0.94\gamma^2 \tag{19.37}$$

According to our previous conclusion, the optimal control method is to keep motors A and B at the same torque rate

$$\gamma_1 = \gamma_2 = \frac{M_t}{M_{1e} + M_{2e}} = \frac{M_t}{300} \tag{19.38}$$

If the total load torque M_t is 360 (N.M), then

$$\begin{aligned} \gamma_1 = \gamma_2 &= \frac{M_t}{M_{1e} + M_{2e}} = 1.2 \\ M_1 &= \gamma_1 M_{1e} = 120 \\ M_2 &= \gamma_2 M_{2e} = 240 \end{aligned} \tag{19.39}$$

19.1.5 Conclusion

The proof of the optimal control method given in this chapter is mainly based on the characteristics of the efficiency function. The efficiency function can be approximately considered as a concave non-negative function passing through the origin. Therefore, the optimal method is independent of the linearity or nonlinearity of the system and does not require a state equation or transfer function of the system. The optimal method has the following characteristics:

(1) Includes linear and nonlinear systems,
(2) No systematic mathematical model is required,
(3) High versatility.

19.2 Wire Sections Driven by Three Different Motors in a Paper Machine

In the wire section of the paper machine, there is usually a very long felt belt. This kind of felt is often driven by multiple motors to ensure uniform felt tension. The structure of a piece of felt driven by three different motors is shown in Fig. 19.4.

In Fig. 19.4, three different motors drive a felt in a paper machine. There is a way to accomplish a task that consumes the least amount of energy, in other words, there is a way that maximizes overall energy efficiency.

The analysis method is basically the same as the idea in the previous section.

19.2.1 Optimal Control of the Wire Section with Similar Efficiencies

Theorem For the optimization problem of P_t (M_t), the minimization of the total power consumption.

Fig. 19.4 The structure of a felt driven by three motors

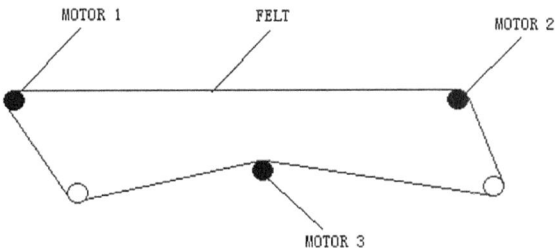

$$minP_t(M_t)$$

$$s.t. \quad \gamma_i > 0, i = 1, 2, 3$$

$$\sum_{i=1}^{3} \gamma_i M_{ie} = M_t$$

(19.40)

is given by

$$minP_t(M_t) = \frac{P_0}{\eta_N \left(\frac{M_t}{M_{1e}+M_{2e}+M_{3e}} \right)}$$

(19.41)

Proof We consider two kinds of situation.

(1) n = 2

There are two variables. We have

$$\gamma_1 M_{1e} + \gamma_2 M_{2e} = M_t$$

$$\gamma_1 > 0$$

$$\gamma_2 > 0$$

(19.42)

The objective function P_t is expressed as

$$P_t = \frac{n_0}{9550} \left(\frac{\gamma_1 M_{1e}}{\eta_N(\gamma_1)} + \frac{\gamma_2 M_{2e}}{\eta_N(\gamma_2)} \right)$$

(19.43)

The optimal condition is

$$P_t{}'(\gamma_1) = 0$$

(19.44)

We have

$$\frac{M_{1e}\eta_N(\gamma_1) - \gamma_1 M_{1e}\eta_N{}'(\gamma_1)}{\eta_N^2(\gamma_1)}$$

$$+ \frac{-M_{1e}\eta_N \left(\frac{M_t - \gamma_1 M_{1e}}{M_{2e}} \right) + (M_t - \gamma_1 M_{1e}) \frac{M_{1e}}{M_{2e}} \eta_N{}' \left(\frac{M_t - \gamma_1 M_{1e}}{M_{2e}} \right)}{\eta_N^2 \left(\frac{M_t - \gamma_1 M_{1e}}{M_{2e}} \right)} = 0$$

(19.45)

It is easy to see that

$$\gamma_1 = \gamma_2 = \frac{M_t}{M_{1e} + M_{2e}}$$

(19.46)

is an optimal point.

The minimal value of the total power consumption is

$$minP_t = \frac{n_0 M_t}{9550} \frac{1}{\eta_N \left(\frac{M_t}{M_{1e}+M_{2e}}\right)} = \frac{P_0}{\eta_N \left(\frac{M_t}{M_{1e}+M_{2e}}\right)} \tag{19.47}$$

P_0 is the ideal work.

(2) $n = 3$

There are three variables.

Based on known conditions, we have

$$\begin{aligned} \gamma_1 M_{1e} + \gamma_2 M_{2e} + \gamma_3 M_{3e} &= M_t \\ \gamma_1 &> 0 \\ \gamma_2 &> 0 \\ \gamma_3 &> 0 \end{aligned} \tag{19.48}$$

P_t expression becomes

$$P_t = \frac{n_0}{9550} \left(\frac{\gamma_1 M_{1e}}{\eta_N(\gamma_1)} + \frac{\gamma_2 M_{2e}}{\eta_N(\gamma_2)} + \frac{\gamma_3 M_{3e}}{\eta_N(\gamma_3)} \right) \tag{19.49}$$

Suppose γ_1 is the fixed optimal point, only γ_2 and γ_3 are variables. We have

$$\gamma_2 M_{2e} + \gamma_3 M_{3e} = M_t - \gamma_1 M_{1e} = constant \tag{19.50}$$

Based on the above conclusions at n = 2, the optimal point is at

$$\gamma_2 = \gamma_3 \tag{19.51}$$

Suppose that γ_2 is the fixed optimal point, only γ_1 and γ_3 are variables. We have

$$\gamma_1 M_{1e} + \gamma_3 M_{3e} = M_t - \gamma_2 M_{2e} = constant \tag{19.52}$$

According to the conclusion at n = 2, the optimal point is at

$$\gamma_1 = \gamma_3 \tag{19.53}$$

Similarly, assume that γ_3 is fixed optimal point, only γ_1 and γ_2 are variables. We have

$$\gamma_1 M_{1e} + \gamma_2 M_{2e} = M_t - \gamma_3 M_{3e} = constant \tag{19.54}$$

According to the conclusion at n = 2, the optimal point is at

$$\gamma_1 = \gamma_2 \tag{19.55}$$

Thus, we have the optimal points

$$\gamma_1 = \gamma_2 = \gamma_3 = \frac{M_t}{\sum_{i=1}^{3} M_{ie}} \tag{19.56}$$

The minimum of the total power consumption is

$$minP_t = \frac{P_0}{\eta_N\left(\frac{M_t}{\sum_{i=1}^{3} M_{ie}}\right)} \tag{19.57}$$

Remark: If three motors have the same model, then the optimal point is

$$\gamma_1 = \gamma_2 = \gamma_3 = \frac{M_t}{3M_e} \tag{19.58}$$

$$M_1 = M_2 = M_3 = \frac{M_t}{3}$$

The minimum of the total power consumption is

$$minP_t = \frac{n_0 M_t}{9550} \frac{1}{\eta_N\left(\frac{M_t}{3M_e}\right)} = \frac{P_0}{\eta\left(\frac{M_t}{3}\right)} \tag{19.59}$$

19.2.2 Conclusion

The proof of the optimal control method given in this chapter is mainly based on the characteristics of the energy efficiency function, which can be approximated as a concave non-negative function through the origin. Therefore, the optimal method is independent of the linearity or nonlinearity of the system and does not require a state equation or transfer function of the system.

In order to solve the optimization problem of general equipment, the authors have conducted extensive research [11–18]. The conclusions given in this chapter can also be applied to the optimization of pumps, fans, blowers, compressors, etc. We would like to thank Zhang Yanfang from Hebei Automation Company and Yao Bosheng from Beijing IAO Technology Development Company for their valuable help and suggestions during the development and experiment of this theory.

References

1. Yao F, Sun H (2012) Efficiency optimal control and dispatching method for General equipment. Machine Press, China
2. Kusumaningtyas I, Lodewijks G (2007) Toward intelligent power consumption optimization in long high-speed passenger conveyors. In: Intelligent transportation systems conference, 2007. ITSC 2007. IEEE, pp 597–602
3. Qichen Q, Hao T, Lei Z, Qi J (2013) The optimization control of single conveyor-serviced production station with variable service rate. In: Control conference (CCC), 2013 32nd Chinese, pp. 2180–2184
4. Zhang S, Tang Y (2011) Optimal control of operation efficiency of belt conveyor. In: Power and energy engineering conference (APPEEC), 2011 Asia-Pacific, pp 1–4
5. De Beauregard DM, Benthaus B, Conradi A, Kulig S (2013) Electromechanical model of an induction machine driven roller conveyor. In: 2012 13th international conference on optimization of electrical and electronic equipment (OPTIM). pp 1071–1078
6. Masoudinejad M, Feldhorst S, Javadian F, ten Hompel M (2013) Energy optimized speed regulation of permanent magnet synchronous motors for driving roller conveyors. In: 2013 IEEE 10th international conference on power electronics and drive systems (PEDS). pp 500–505
7. Lodewijks G (2012) Energy efficient use of belt conveyors in baggage handling systems. In: 2012 9th IEEE international conference on networking, sensing and control (ICNSC). pp 97–102
8. Pang Y, Lodewijks G (2011) Improving energy efficiency in material transport systems by fuzzy speed control. In: 2011 3rd IEEE International Symposium on Logistics and Industrial Informatics (LINDI). pp 159–164
9. Prasse C, Kamagaew A, Gruber S, Kalischewski K, Soter S, ten Hompel M (2011) Survey on energy efficiency measurements in heterogenous facility logistics systems. In: 2011 IEEE International Conference on Industrial Engineering and Engineering Management (IEEM). pp 1140–1144
10. Yao F, Sun H (2011) Optimal control for a common system. In: 2011 3rd IEEE international conference on information management and motorering (IEEE ICIME 2011), May 21–22, 2011, Zhengzhou, China
11. Yao F, Sun H (2011) Optimal switch for a common system. In: IEEE power motorering and automation conference (PEAM 2011), September 8–9, 2011 in Wuhan, China
12. Yao F, Sun H (2011) Optimal control and switch in high speed train. In: The 3rd international conference on computational and information sciences (ICCIS2011), Chengdu, China, from October 21 to 23, 2011
13. Yao F, Sun H (2011) Optimal control in variable-speed pumping stations. In: The 2011 IEEE international conference on mechatronics and automation (ICMA 2011), August 7–10, 2011, Beijing, China
14. Yao F, Sun H (2011) Optimal switch in variable-speed pumping stations. In: 2011 IEEE 2nd international conference on computing, control and industrial motorering (CCIE 2011), August 20–21, 2011, Wuhan, China
15. Yao F, Sun H (2011) Optimal control in variable-speed fan stations. In: IEEE the 18th international conference on industrial monitoring and monitoring management, September 3–5, 2011. Chang Chun, P.R. China
16. Yao F, Sun H (2011) Optimal switch in variable-speed fan stations. In: 2011 international conference of renewable energy sources and environmental materials, May 20–22, 2011, Shanghai, China
17. Yao FL, Zhang YF (2009) Electrical energy saving control method and practice. China Electric Power Press, Beijing
18. Yao FL, Sun H (2011) Frequency-converter and energy-saving control crash course. Publishing House of Electronics Industry

Chapter 20
Optimal Control of Electric Transportation Vehicles Such as High-Speed Rail

Many electric transportation vehicles such as locomotives, subways, electric cars, ships, etc. are jointly driven by multiple electric motors (Fig. 20.1). In these occasions and fields, there are optimization problems of how multiple motors or multiple engines output power and how to determine the number of operating units.

20.1 Total Power Consumption of High-Speed Train Drive System

The work performed by transport vehicles is mainly to overcome friction. Take the EMU as an example. The carriages have the same speed. How to distribute the output of the motors in each carriage to minimize the total energy consumption?

The total force required by the EMU is F_t, the speed V_0 of each power unit is equal, the output force F_i of each carriage is part of the total force, and the energy consumption expression of the EMU is:

$$P_t = k3 \sum_{i=1}^{n} \frac{F_i V_i}{\eta_i(F_i, V_i)} = k3 F_t V_0 \sum_{i=1}^{n} \frac{F_i}{F_t} \times \frac{1}{\eta_i(F_i)} \tag{20.1}$$

where k3 is a constant.

The expression for energy consumption written as torque and speed is

$$P_t = k3 M_t n_0 \sum_{i=1}^{n} \frac{M_i}{M_t} \frac{1}{\eta(M_i, n_0)} = k3 M_t n_0 \sum_{i=1}^{n} \theta_i \frac{1}{\eta(\theta_i M_t, n_0)}$$

$$= P_0 \sum_{i=1}^{n} \theta_i \frac{1}{\eta(\theta_i M_t)} \tag{20.2}$$

© The Author(s) 2024
F. Yao and Y. Yao, *Efficient Energy-Saving Control and Optimization for Multi-Unit Systems*, https://doi.org/10.1007/978-981-97-4492-3_20

Fig. 20.1 Electric transportation vehicles

Fig. 20.2 The efficiency curve of the first power unit

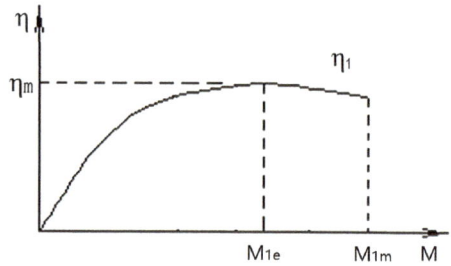

where

$$\theta_i = \frac{M_i}{M_t} \tag{20.3}$$

$$P_0 = k3M_t n_0$$

The total torque M_t required by the EMU is the total load, and the rotation speed of each power unit is equal to n_0.

The efficiency curve of the first drive unit (including motor and reducer) at the speed n_0 is shown in Fig. 20.2.

In Fig. 20.2, M1m is the maximum output torque of the first power unit, M_{1e} is the optimal output torque of the first power unit, and M_{1e} corresponds to the maximum efficiency η_m.

20.2 Optimized Load Distribution Method for Drive Motors of High-Speed Trains

Assuming that M_t and n_0 remain unchanged, let's find the optimal control method for the EMU. Since n_0 is a constant, the operating efficiency of the speed-regulated motor of each power unit only considers the torque variable.

Consider minimizing the total power consumption of the drive system

$$\min P_t$$
$$s.t.\, \theta_i > 0, \quad i = 1, 2, \ldots n$$
$$\sum_{i=1}^{n} \theta_i = 1 \tag{20.4}$$

This problem can also be written as

$$\min P_0 \sum_{i=1}^{n} \theta_i \frac{1}{\eta_i(\theta_i M_0)}$$
$$s.t.\, \theta_i > 0, i = 1, 2, \ldots n \tag{20.5}$$
$$\sum_{i=1}^{n} \theta_i = 1$$

We consider three cases:

(1) $n = 2$

The system has two variables and has

$$\theta_1 + \theta_2 = 1$$
$$\theta_1 > 0 \tag{20.6}$$
$$\theta_2 > 0$$

The objective function P_t can be expressed as

$$P_t = P_0 \left(\theta_1 \frac{1}{\eta_1(\theta_1 M_t)} + \theta_2 \frac{1}{\eta_2(\theta_2 M_t)} \right)$$
$$= P_0 \left(\theta_1 \frac{1}{\eta_1(\theta_1 M_t)} + (1 - \theta_1) \frac{1}{\eta_2((1 - \theta_1)M_t)} \right) \tag{20.7}$$

The optimization condition is

$$P_{t'}(\theta_1) = 0 \tag{20.8}$$

We have

$$\frac{1}{\eta_1(\theta_1 M_t)} - \theta_1 \frac{\eta_1{'}(\theta_1 M_t)M_t}{\eta_1(\theta_1 M_t)^2} - \frac{1}{\eta_2((1 - \theta_1)M_t)} + (1 - \theta_1) \frac{\eta{'}_2((1 - \theta_1)M_t)M_t}{\eta_2((1 - \theta_1)M_t)^2} = 0 \tag{20.9}$$

Since all the motors are variable speed motors and the identical model, it is easy to see that

$$\theta_1 = \frac{1}{2}$$

$$\eta_{1'}\left(\frac{M_t}{2}\right) = \eta_{2'}\left(\frac{M_t}{2}\right) \tag{20.10}$$

$$\eta_1\left(\frac{M_t}{2}\right) = \eta_2\left(\frac{M_t}{2}\right)$$

is an optimization point. Then, we have

$$\theta_2 = 1 - \theta_1 = \theta_1 = \frac{1}{2} = \frac{1}{n} \tag{20.11}$$

That is, the optimal control method is to keep

$$M_1 = M_2 = \frac{M_t}{2} = \frac{M_t}{n} \tag{20.12}$$

The minimum of the total power dissipation is

$$minP_t = P_0 \frac{1}{\eta_1\left(\frac{M_t}{2}\right)} = P_0 \frac{1}{\eta_1\left(\frac{M_t}{n}\right)} \tag{20.13}$$

(2)   n = 3

The system has three variables, based on known conditions, we have

$$\theta_1 + \theta_2 + \theta_3 = 1 \tag{20.14}$$

$$\theta_1 > 0$$

$$\theta_2 > 0$$

$$\theta_3 > 0$$

The P_t expression becomes

$$P_t = P_0\left(\theta_1 \frac{1}{\eta_1(\theta_1 M_t)} + \theta_2 \frac{1}{\eta_2(\theta_2 M_t)} + \theta_3 \frac{1}{\eta_3(\theta_3 M_t)}\right) \tag{20.15}$$

Assuming that θ_1 is fixed and is an optimization point, and only θ_2 and θ_3 are variables, then we have

$$\theta_2 + \theta_3 = 1 - \theta_1 = constant \tag{20.16}$$

Based on the conclusion of n = 2 above, we have

$$\theta_2 = \theta_3 \tag{20.17}$$

to be the optimal point.

Assuming that θ_2 is fixed and is an optimization point, and only θ_1 and θ_3 are variables, then we have

$$\theta_1 + \theta_3 = 1 - \theta_2 = constant \tag{20.18}$$

According to the conclusion of n = 2 above, we have

$$\theta_1 = \theta_3 \tag{20.19}$$

to be the optimal point.

Similarly, assuming that θ_3 is fixed and is an optimization point, and only θ_1 and θ_2 are variables, then we have

$$\theta_1 + \theta_2 = 1 - \theta_3 = constant \tag{20.20}$$

According to the conclusion of n = 2 above, we have

$$\theta_1 = \theta_2 \tag{20.21}$$

to be is the optimal point.

So, we have the optimization point

$$\theta_1 = \theta_2 = \theta_3 = \frac{1}{3} = \frac{1}{n} \tag{20.22}$$

The optimal control method is to keep

$$M_1 = M_2 = M_3 = \frac{M_t}{3} = \frac{M_t}{n} \tag{20.23}$$

The minimum of the total power dissipation is

$$minP_t = P_0 \frac{1}{\eta_1\left(\frac{M_t}{3}\right)} = P_0 \frac{1}{\eta_1\left(\frac{M_t}{n}\right)} \tag{20.24}$$

(3) n = k

The system has n variables, and the above conclusion can be extended to the case of n = k, the optimal point is

$$\theta_1 = \theta_2 = \ldots = \theta_k = \frac{1}{k} \tag{20.25}$$

The optimal control method is to keep

$$M_1 = M_2 = \ldots = M_k = \frac{M_t}{k} \tag{20.26}$$

The minimum of the total power dissipation is

$$minP_t = P_0 \frac{1}{\eta_1\left(\frac{M_t}{k}\right)} \tag{20.27}$$

The optimal overall efficiency is

$$max\eta_t(M_t) = \eta_1\left(\frac{M_t}{k}\right) \tag{20.28}$$

20.3 The Optimal Number of Operating Motors for High-Speed Trains

Assume that the EMU has m cars of the identical type, and all cars are driven by motors that can be adjusted in speed. The speed of each car is the same as n_0, and the total torque is M_t. There are n car motors running with speed regulation, and n is less than is equal to m, the output torque of the i-th car motor is M_i, and the total power consumption is the minimum, there is

$$minP_t = P_0 \frac{1}{\eta_1\left(\frac{M_t}{n}\right)}$$
$$n \leq m \tag{20.29}$$
$$\frac{M_t}{n} \leq M_{1m}$$

When the total torque M_t changes due to wind resistance, uphill, downhill, etc., we consider 2 cases:

1. M_t/n is equal to or less than m_{1e}, which is the torque corresponding to the highest efficiency point, as shown in Fig. 20.3.
   In Fig. 20.3, we have

$$max\left(\eta_1\left(\frac{M_t}{n-1}\right), \eta_1\left(\frac{M_t}{n}\right), \eta_1\left(\frac{M_t}{n+1}\right)\right) = \eta_1\left(\frac{M_t}{n}\right) \tag{20.30}$$

Fig. 20.3 M_t/n is equal to or less than M_{1e}

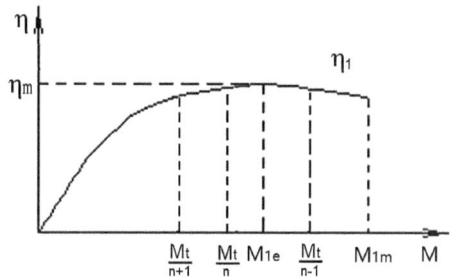

When M_t increases, M_t/n also increases. If $M_t/n > M_{1e}$ and $\eta_1(M_t/n) < \eta_1 (M_t/(n+1))$, then the optimal number of operating units becomes $n + 1$, and we should increase the number of running motors from n to $n + 1$, there are

$$max\left(\eta_1\left(\frac{M_t}{n-1} \right), \eta_1\left(\frac{M_t}{n} \right), \eta_1\left(\frac{M_t}{n+1} \right) \right) = \eta_1\left(\frac{M_t}{n+1} \right) \qquad (20.31)$$

The optimal switching point is

$$\eta_1\left(\frac{M_t}{n} \right) = \eta_1\left(\frac{M_t}{n+1} \right) \qquad (20.32)$$

When M_t decreases, M_t/n also decreases. If $M_t/n > M_{1e}$ and $\eta_1(M_t/n) < \eta_1(M_t/(n-1))$, then the optimal number of running units becomes $n - 1$. We should reduce the number of running motors from n to $n - 1$, and then we have

$$max\left(\eta_1\left(\frac{M_t}{n-1} \right), \eta_1\left(\frac{M_t}{n} \right), \eta_1\left(\frac{M_t}{n+1} \right) \right) = \eta_1\left(\frac{M_t}{n-1} \right) \qquad (20.33)$$

The optimal switching point is

$$\eta_1\left(\frac{M_t}{n} \right) = \eta_1\left(\frac{M_t}{n-1} \right) \qquad (20.34)$$

2. M_t n is greater than M_{1e}, as shown in Fig. 20.4
   In Fig. 20.4, there are

$$max\left(\eta_1\left(\frac{M_t}{n-1} \right), \eta_1\left(\frac{M_t}{n} \right), \eta_1\left(\frac{M_t}{n+1} \right) \right) = \eta_1\left(\frac{M_t}{n} \right) \qquad (20.35)$$

When M_t increases, M_t/n also increases. If $\eta_1(M_t/n) < \eta_1(M_t/(n+1))$, then the optimal number of running units becomes $n + 1$, and we should increase the number of running motors from n to $n + 1$, which yields

Fig. 20.4 M_t/n is greater than M_{1e}

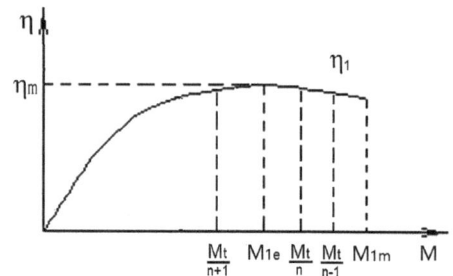

$$max\left(\eta_1\left(\frac{M_t}{n-1}\right), \eta_1\left(\frac{M_t}{n}\right), \eta_1\left(\frac{M_t}{n+1}\right)\right) = \eta_1\left(\frac{M_t}{n+1}\right) \tag{20.36}$$

The optimal switching point is

$$\eta_1\left(\frac{M_t}{n}\right) = \eta_1\left(\frac{M_t}{n+1}\right) \tag{20.37}$$

When M_t decreases, M_t/n also decreases, if $\eta_1(M_t/n) < \eta_1(M_t/(n-1))$, then the optimal number of running units becomes $n-1$, and we should reduce the number of running motors from n to $n-1$, which yields

$$max\left(\eta_1\left(\frac{M_t}{n-1}\right), \eta_1\left(\frac{M_t}{n}\right), \eta_1\left(\frac{M_t}{n+1}\right)\right) = \eta_1\left(\frac{M_t}{n-1}\right) \tag{20.38}$$

The optimal switching point is

$$\eta_1\left(\frac{M_t}{n}\right) = \eta_1\left(\frac{M_t}{n-1}\right) \tag{20.39}$$

Chapter 21
Energy Efficiency Optimization of Heating and Cooling Systems

In a heat source plant, there are many boilers. In the cooling system, there are many chillers. Whether it is a cooling system or a heating system, they are all thermal energy supply systems. There is a minimum energy consumption to provide constant thermal energy.

21.1 Energy Efficiency of a Boiler or a Chiller

The energy efficiency of a boiler or chiller under constant output temperature difference Δt_0 (°C) is shown in Fig. 21.1. For the convenience of the following description, we refer to boilers or chillers collectively as equipment or devices.

In Fig. 21.1, Q is the flow rate (tons/hour), η is the efficiency, and Q_e is the flow rate at maximum efficiency. We have

$$
\begin{aligned}
Q &\geq 0 \\
\eta(Q) &\geq 0 \\
\eta(0) &= 0 \\
\eta''(Q) &< 0
\end{aligned}
\tag{21.1}
$$

Approximately, we have

$$
\eta(Q) = \sum_{i=0}^{\infty} a_i Q^i = Q \sum_{i=1}^{\infty} (a_i Q^{i-1}) \approx Q(a_1 + a_2 Q)
\tag{21.2}
$$

where

F. Yao and Y. Yao, *Efficient Energy-Saving Control and Optimization for Multi-Unit Systems*, https://doi.org/10.1007/978-981-97-4492-3_21

Fig. 21.1 The energy
efficiency function of a
boiler or a chiller

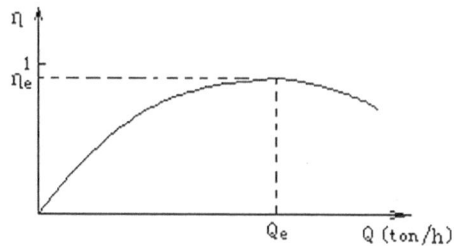

$$a_1 + a_2 Q > 0$$
$$a_1 > 0 \qquad\qquad (21.3)$$
$$a_2 < 0$$

The equipment's output energy is P(Q), which can be expressed as

$$P(Q) = \frac{k \Delta t_0 Q}{\eta(Q)} \qquad\qquad (21.4)$$

where k is a constant related to heat capacity.

21.2 Optimal Control of Heating Systems Composed of Equipment of the Identical Type

A thermal energy supply system has n devices of the identical model. The system is used to supply constant temperature Δt_0 water. The total water flow is Q_t. The flow of the i-th device is Q_i. Q_i is greater than zero. The water flow of all devices is variable. We have

$$Q_i > 0$$
$$\sum_{i=1}^{n} Q_i = Q_t \qquad\qquad (21.5)$$

$\eta (Q_i)$ is the energy efficiency of the i-th device at point $(Q_i, \Delta t_0)$, η is the total energy efficiency of the thermal energy supply system, and P_t is the total energy consumption of the system.

$$P_t = k \Delta t \sum_{i=1}^{n} Q_i \frac{1}{\eta(Q_i)} \qquad\qquad (21.6)$$

Using the contents of the previous chapters, it is not difficult to prove the following conclusions:

The optimal control method is to keep

$$Q_1 = Q_2 = \ldots = Q_n = \frac{Q_t}{n} \qquad (21.7)$$

The minimum value of total energy consumption is

$$minP_t = P_0 \frac{1}{\eta(\frac{Q_t}{n})} \qquad (21.8)$$

The ideal work P_0 is

$$P_0 = kQ_t \Delta t_0 \qquad (21.9)$$

The overall optimal efficiency is

$$max\eta_t(Q_t) = \eta(\frac{Q_t}{n}) \qquad (21.10)$$

21.3 The Energy Efficiency Similarity of Different Equipment

We define the load rate γ as

$$\gamma = \frac{Q}{Q_e} \qquad (21.11)$$

We call $\eta_N(\gamma)$ as the normalized energy efficiency function of an equipment. The Normalized energy efficiency function $\eta_N(\gamma)$ has a shape shown in Fig. 21.2.

In Fig. 21.2, γ is the variable and η_N is the efficiency. η_N and η have the following relationship.

Fig. 21.2 The normalized efficiency function η_N (γ)

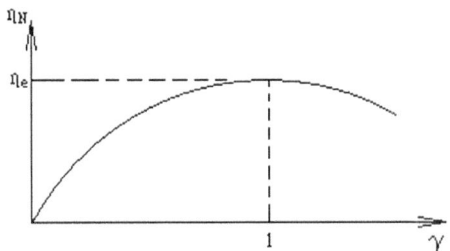

$$\eta(Q) = \eta(\gamma Q_e) = \eta_N(\gamma) \tag{21.12}$$

If the normalized efficiency functions of two different devices are identical, we have

$$\eta_{N1}(\gamma) = \eta_{N2}(\gamma) \tag{21.13}$$

We call them efficiency-similar devices. A thermal energy supply system containing efficiency-similar devices is called an efficiency-similar heating system.

Let γ_i be the load rate of the i-th device, and its form is:

$$\gamma_i = \frac{Q_i}{Q_{ie}} \tag{21.14}$$

For an efficiency-similar heat energy supply system with n-unit running equipment, the total energy consumption has the form

$$P_t = k \Delta t_0 \sum_{i=1}^{n} \frac{Q_i}{\eta_i(Q_i)} = k \Delta t_0 \sum_{i=1}^{n} \frac{\gamma_i Q_{ie}}{\eta_N(\gamma_i)} \tag{21.15}$$

21.4   Optimal Control of an Efficiency-Similar Heating System

When the load rate of each running equipment is greater than zero, consider the minimization of the total energy consumption

$$\begin{aligned} &\min P_t \\ &s.t. \ \gamma_i > 0, i = 1, 2, \ldots n \\ &\sum_{i=1}^{n} \gamma_i Q_{ie} = Q_t \end{aligned} \tag{21.16}$$

We consider three kinds of situation.

(1) n = 2

There are two variables.

We have

$$\begin{aligned} &\gamma_1 Q_{1e} + \gamma_2 Q_{2e} = Q_t \\ &\gamma_1 > 0 \\ &\gamma_2 > 0 \end{aligned} \tag{21.17}$$

The objective function P_t is expressed as

$$P_t = k\Delta t_0 \left(\frac{\gamma_1 Q_{1e}}{\eta_N(\gamma_1)} + \frac{\gamma_2 Q_{2e}}{\eta_N(\gamma_2)} \right) \tag{21.18}$$

The optimal condition is

$$P_t'(\gamma_1) = 0 \tag{21.19}$$

We have

$$\frac{Q_{1e}\eta_N(\gamma_1) - \gamma_1 Q_{1e}\eta_N{}'(\gamma_1)}{\eta_N^2(\gamma_1)} + \frac{-Q_{1e}\eta_N(\frac{Q_t-\gamma_1 Q_{1e}}{Q_{2e}}) + (Q_t - \gamma_1 Q_{1e})\frac{Q_{1e}}{Q_{2e}}\eta_N{}'(\frac{Q_t-\gamma_1 Q_{1e}}{Q_{2e}})}{\eta_N^2(\frac{Q_t-\gamma_1 Q_{1e}}{Q_{2e}})} = 0 \tag{21.20}$$

It is easy to see that

$$\gamma_1 = \gamma_2 = \frac{Q_t}{Q_{1e} + Q_{2e}} \tag{21.21}$$

to be an optimal point.

The minimum value of total energy consumption is

$$minP_t = kQ_t\Delta t \frac{1}{\eta_N(\frac{Q_t}{Q_{1e}+Q_{2e}})} = \frac{P_0}{\eta_N(\frac{Q_t}{Q_{1e}+Q_{2e}})} \tag{21.22}$$

P_0 is the ideal work.

(2) n = 3

There are three variables.

Based on known conditions, we have

$$\gamma_1 Q_{1e} + \gamma_2 Q_{2e} + \gamma_3 Q_{3e} = Q_t$$
$$\gamma_1 > 0$$
$$\gamma_2 > 0 \tag{21.23}$$
$$\gamma_3 > 0$$

P_t expression becomes

$$P_t = k\Delta t_0 \left(\frac{\gamma_1 Q_{1e}}{\eta_N(\gamma_1)} + \frac{\gamma_2 Q_{2e}}{\eta_N(\gamma_2)} + \frac{\gamma_3 Q_{3e}}{\eta_N(\gamma_3)} \right) \tag{21.24}$$

Assume that γ_1 is a fixed optimal point and only γ_2 and γ_3 are variables. We have

$$\gamma_2 Q_{2e} + \gamma_3 Q_{3e} = Q_t - \gamma_1 Q_{1e} = constant \tag{21.25}$$

According to the conclusion drawn with n = 2, the optimal point is

$$\gamma_2 = \gamma_3 \tag{21.26}$$

Assume that γ_2 is a fixed optimal point and only γ_1 and γ_3 are variables. We have

$$\gamma_1 Q_{1e} + \gamma_3 Q_{3e} = Q_t - \gamma_2 Q_{2e} = constant \tag{21.27}$$

According to the conclusion at n = 2, the optimal point is at

$$\gamma_1 = \gamma_3 \tag{21.28}$$

Similarly, assume that γ_3 is fixed optimal point and only γ_1 and γ_2 are variables. We have

$$\gamma_1 Q_{1e} + \gamma_2 Q_{2e} = Q_t - \gamma_3 Q_{3e} = constant \tag{21.29}$$

According to the conclusion at n = 2, the optimal point is at

$$\gamma_1 = \gamma_2 \tag{21.30}$$

We have the optimal points

$$\gamma_1 = \gamma_2 = \gamma_3 = \frac{Q_t}{\sum_{i=1}^{3} Q_{ie}} \tag{21.31}$$

The minimum value of total energy consumption is

$$minP_t = \frac{P_0}{\eta_N \left(\frac{Q_t}{\sum_{i=1}^{3} Q_{ie}} \right)} \tag{21.32}$$

(3) n = k

There are n variables.

The above conclusion can be extended to the situation at n = k, the optimal point is

$$\gamma_1 = \gamma_2 = ... = \gamma_k = \frac{Q_t}{\sum_{i=1}^{k} Q_{ie}} \tag{21.33}$$

The minimum value of total energy consumption is

$$minP_t = \frac{P_0}{\eta_N\left(\frac{Q_t}{\sum_{i=1}^{k} Q_{ie}}\right)} \tag{21.34}$$

21.5 Optimal Number of Running Units

If n is the optimal and all equipment are the identical model, there must be

$$\frac{P_0}{\eta_N\left(\frac{Q_t}{\sum_{i=1}^{n-1} Q_{ie}}\right)} \geq \frac{P_0}{\eta_N\left(\frac{Q_t}{\sum_{i=1}^{n} Q_{ie}}\right)} \leq \frac{P_0}{\eta_N\left(\frac{Q_t}{\sum_{i=1}^{n} Q_{ie}}\right)} \tag{21.35}$$

Namely

$$\eta_N\left(\frac{Q_t}{\sum_{i=1}^{n-1} Q_{ie}}\right) \leq \eta_N\left(\frac{Q_t}{\sum_{i=1}^{n} Q_{ie}}\right) \geq \eta_N\left(\frac{Q_t}{\sum_{i=1}^{n+1} Q_{ie}}\right) \tag{21.36}$$

If n is the optimal and all devices have the identical energy efficiency, there must be

$$\frac{P_0}{\eta_N\left(\frac{Q_t}{\sum_{i=1}^{n1} Q_{ie}}\right)} \geq \frac{P_0}{\eta_N\left(\frac{Q_t}{\sum_{i=1}^{n} Q_{ie}}\right)} \tag{21.37}$$

That is

$$\eta_N\left(\frac{Q_t}{\sum_{i=1}^{n1} Q_{ie}}\right) \leq \eta_N\left(\frac{Q_t}{\sum_{i=1}^{n} Q_{ie}}\right) \tag{21.38}$$

where n1 is the operating number of devices in any combination.

21.6 Optimal Switch of Efficiency-Similar Heating System

Assume that there are M devices in a thermal energy supply system, n devices are running, and n is equal to or less than M. If n is optimal, the total energy consumption is the minimum.

Fig. 21.3 η_N (γ) curve when $\gamma(Q_t,n)$ is less than 1

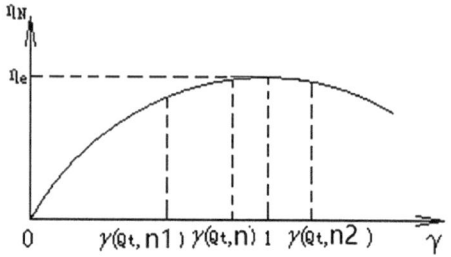

$$minP_t = \frac{P_0}{\eta_N\left(\dfrac{Q_t}{\sum_{i=1}^{n} Q_{ie}}\right)}$$

$$n \le M$$

$$\eta_N\left(\frac{Q_t}{\sum_{i=1}^{n1} Q_{ie}}\right) \le \eta_N\left(\frac{Q_t}{\sum_{i=1}^{n} Q_{ie}}\right) \ge \eta_N\left(\frac{Q_t}{\sum_{i=1}^{n2} Q_{ie}}\right) \qquad (21.39)$$

$$\gamma(Q_t, n) = \frac{Q_t}{\sum_{i=1}^{n} Q_{ie}}$$

n1 and n2 are the running number of equipment in any combinations.

When Q_t changes, we consider three situations.

(1) The γ (Q_t, n) is less than 1.

η_e is the maximum efficiency, and 1 is the load rate at η_e, as shown in Fig. 21.3. In Fig. 21.3, we have

$$max(\eta_N(Q_t, n1), \eta_N(Q_t, n), \eta_N(Q_t, n1)) = \eta_N(Q_t, n) \qquad (21.40)$$

When Q_t increases, the load rate $\gamma(Q_t,n)$ increases, and $\eta_N(Q_t,n)$ and $\eta_N(Q_t,n1)$ also increase. However, when $\gamma(Qt,n) > 1$, Q_t increases, $\eta_N(Q_t, n1))$ still increases, but $\eta_N(Q_t, n)$ decreases. When $\eta_N(Q_t, n) < \eta_N(Q_t, n1)$, n1 is optimal, we should change the number of running units from n to n1, we have

$$max(\eta_N(Q_t, n1), \eta_N(Q_t, n), \eta_N(Q_t, n2)) = \eta_N(Q_t, n1) \qquad (21.41)$$

The optimal switch point is

$$\eta_N(Q_t, n) = \eta_N(Q_t, n1) \qquad (21.42)$$

When Q_t decreases, the load rate $\gamma(Q_t,n)$ decreases, and $\eta_N(Q_t, n)$ decreases also. However, $\eta_N(Q_t, n2)$ increases. When $\eta_N(Q_t, n) < \eta_N(Q_t, n2)$, n2 is the optimal, we should change the number of running units from n to n2, we have

$$max(\eta_N(Q_t, n1), \eta_N(Q_t, n), \eta_N(Q_t, n2)) = \eta_N(Q_t, n2) \qquad (21.43)$$

Fig. 21.4 η_N (γ) curve when
$\gamma(Q_t,n)$ is greater than 1

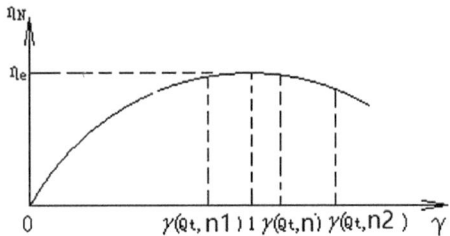

The optimal switch point is

$$\eta_N(Q_t, n) = \eta_N(Q_t, n2) \tag{21.44}$$

(2)   The $\gamma(Q_t,n)$ is greater than 1, as shown in Fig. 21.4.

In Fig. 21.4, we have

$$max(\eta_N(Q_t, n1), \eta_N(Q_t, n), \eta_N(Q_t, n2)) = \eta_N(Q_t, n) \tag{21.45}$$

When Q_t increases, the load rate γ (Q_t, n) increases, and $\eta_N(Q_t, n1)$ increases also, however $\eta_N(Q_t, n)$ decreases. When $\eta_N(Q_t, n) < \eta_N(Q_t, n1)$, n1 is the optimal, we should change the number of running units from n to n1, we have

$$max(\eta_N(Q_t, n1), \eta_N(Q_t, n), \eta_N(Q_t, n2)) = \eta_N(Q_t, n1) \tag{21.46}$$

The optimal switch point is

$$\eta_N(Q_t, n) = \eta_N(Q_t, n1) \tag{21.47}$$

When Q_t decreases, the load rate $\gamma(Q_t, n)$ decreases, $\eta_N(Q_t, n)$ and $\eta_N(Q_t, n2)$ both increase. When $\gamma(Q_t, n) < 1$, Q_t decreases, η_N (Q_t, n2) still increases, however $\eta_N(Q_t, n)$ decreases. When $\eta_N(Q_t, n) < \eta_N(Q_t, n2)$, n2 is the optimal, we should change the number of running units from n to n2, we have

$$max(\eta_N(Q_t, n1), \eta_N(Q_t, n), \eta_N(Q_t, n2)) = \eta_N(Q_t, n2) \tag{21.48}$$

The optimal switch point is

$$\eta_N(Q_t, n) = \eta_N(Q_t, n2) \tag{21.49}$$

(3)   The $\gamma(Q_t, n)$ is equal to 1.

When Q_t increases, the load rate $\gamma(Q_t,n)$ increases, $\eta_N(Q_t, n1)$ increase also, however $\eta_N(Q_t, n)$ decreases. When $\eta_N(Q_t, n) < \eta_N(Q_t, n1)$, n1 is the optimal, we should change the number of running units from n to n1, we have

$$max(\eta_N(Q_t, n1), \eta_N(Q_t, n), \eta_N(Q_t, n2)) = \eta_N(Q_t, n1) \qquad (21.50)$$

The optimal switch point is

$$\eta_N(Q_t, n) = \eta_N(Q_t, n1) \qquad (21.51)$$

When Q_t decreases, the load rate $\gamma(Q_t, n)$ decreases, $\eta_N(Q_t, n)$ decreases also, however $\eta_N(Q_t, n2)$ increases. When $\eta_N(Q_t, n) < \eta_N(Q_t, n2)$, n2 is the optimal, we should change the number of running units from n to n2, we have

$$max(\eta_N(Q_t, n1), \eta_N(Q_t, n), \eta_N(Q_t, n2)) = \eta_N(Q_t, n2) \qquad (21.52)$$

The optimal switch point is

$$\eta_N(Q_t, n) = \eta_N(Q_t, n2) \qquad (21.53)$$

21.7   Conclusion

The proof of the optimal control and switching method given in this chapter is mainly based on the characteristics of the energy efficiency function, which can be approximated as a concave non-negative function through the origin. The optimal method has the following characteristics:

(1) Includes linear and nonlinear systems,
(2) No systematic mathematical model is required,
(3) High versatility.

Chapter 22
Efficiency Optimization of Wind Power Hydrogen Production System

Due to the involvement of chemical reactions, fission reactions and other factors, the energy output by some systems is not all converted from the energy consumed by the system. Such systems can only maximize the output as the optimal goal of the overall energy efficiency of the system. For example, when electrolyzing water to produce hydrogen, you cannot say that the hydrogen energy produced is completely converted from the electrical energy of the electrolyzer, because the hydrogen in the water itself has a certain amount of energy. See Chap. 2.

In order to achieve the dual-carbon goal of mankind, all countries are vigorously developing green energy, and wind power is one of them (Fig. 22.1).

Due to the large randomness of wind energy, it is difficult to dispatch wind power to the grid. In order not to waste the electric energy generated by wind power, people transmit the wind power to the hydrogen production station. There are multiple hydrogen production machines in the hydrogen production station, and the hydrogen production machines convert the electric energy into hydrogen energy and store it. For the identical wind power, arrange the operating number of hydrogen generators and the load of each hydrogen generator to produce the most hydrogen. This is a multi-machine energy efficiency optimization system.

22.1 Energy Efficiency Curve of Hydrogen Generator

Due to the large amount of power generated by the wind farm, the power input of the hydrogen generator is measured in MW. The energy efficiency curve of a hydrogen generator is shown in Fig. 22.2.

In Fig. 22.2, W is the electric energy input of the hydrogen generator, η is the energy efficiency of the hydrogen generator. W_m is the maximum load input of the device, η_e is the highest energy efficiency of the device, W_e is the optimal energy input when the device has the highest operating energy efficiency.

F. Yao and Y. Yao, *Efficient Energy-Saving Control and Optimization for Multi-Unit Systems*, https://doi.org/10.1007/978-981-97-4492-3_22

Fig. 22.1 Wind farm

Fig. 22.2 Energy efficiency
curve of hydrogen generator

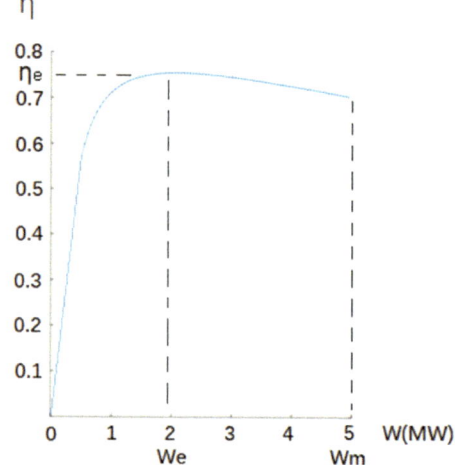

Below we refer to the hydrogen generator as device for short, and the hydrogen
production station as system.

22.2 Optimal Load Distribution Theorem for Hydrogen Production System

Assuming that there are m devices in the hydrogen production station, the total
input power is W_t, the electric energy input by the i-th device is W_i, and the overall
efficiency of the system is η_t, then the overall energy efficiency expression of the
system is

$$
\begin{aligned}
\eta_t &= \sum_{i=1}^{m} \frac{W_i}{W_t} \eta_i(W_i) \\
s.t. \sum_{i=1}^{m} W_i &= W_t > 0 \\
W_{imax} &\geq W_i > 0
\end{aligned}
\tag{22.1}
$$

Assuming that n is optimal, W_t is constant, consider the following three situations:

(1) $n = 2$

The system has two variables W_1 and W_2

$$
\begin{aligned}
W_1 + W_2 &= W_t \\
W_1 &> 0 \\
W_2 &> 0
\end{aligned}
\tag{22.2}
$$

The objective function can be expressed as

$$
\eta_t = \frac{W_1}{W_t}\eta_1(W_1) + \frac{W_2}{W_t}\eta_2(W_2)
\tag{22.3}
$$

The optimization condition is

$$
\eta_t{}'(W_1) = 0
\tag{22.4}
$$

According to known conditions

$$
W_2 = W_t - W_1
\tag{22.5}
$$

We have

$$
\left(\eta_1(W_1) + W_1\eta_1'(W_1) - \eta_2(W_2) - W_2\eta_2'(W_2)\right)/W_t = 0
\tag{22.6}
$$

If two devices are devices with the identical energy efficiency, then

$$
\eta_2(W) = \eta_1(W)
\tag{22.7}
$$

and it is easy to see that

$$
\begin{aligned}
W_1 &= \frac{W_t}{2} \\
W_2 &= W_1 = \frac{W_t}{2}
\end{aligned}
\tag{22.8}
$$

is an optimization point.
  The optimal control method is to keep

$$
W_1 = W_2 = \frac{W_t}{n}
\tag{22.9}
$$

The maximum overall energy efficiency of system is

$$
\max\eta_{t2}(W_t) = \eta_1\left(\frac{W_t}{n}\right)
\tag{22.10}
$$

If two devices are similar energy efficiency devices, then

$$\eta_2(W_2) = \eta_1\left(\frac{W_1}{\beta_2}\right) \tag{22.11}$$

and it is easy to see that

$$W_1 = \frac{W_t}{1+\beta_2}$$
$$W_2 = \beta_2 W_1 \tag{22.12}$$

is an optimization point.

The optimal control method is to keep

$$W_1 = \frac{W_t}{1+\beta_2}$$
$$W_2 = \beta_2 W_1 \tag{22.13}$$

The maximum value of the overall operating energy efficiency $\eta_{t2}(W_t)$ of the system is

$$\max\eta_{t2}(W_t) = \eta_1\left(\frac{W_t}{1+\beta_2}\right) \tag{22.14}$$

Since the shape of the overall efficiency curve of the system is the same as that of a single device, the second derivative is also less than zero. The overall energy efficiency is a maximum value.

(2) $n = 3$

The system has three variables W_1, W_2 and W_3

$$W_1 + W_2 + W_3 = W_t$$
$$W_1 > 0$$
$$W_2 > 0 \tag{22.15}$$
$$W_3 > 0$$

η_t expression is

$$\eta_t = \frac{W_1}{W_t}\eta_1(W_1) + \frac{W_2}{W_t}\eta_2(W_2) + \frac{W_3}{W_t}\eta_3(W_3) \tag{22.16}$$

If the three devices have the identical energy efficiency, assuming that W_3 is the optimal point and fixed, W_1 and W_2 are variables, leading to

$$W_1 + W_2 = W_t - W_3 \tag{22.17}$$

Based on the conclusion of n = 2, we have

$$W_2 = W_1 \tag{22.18}$$

Similarly, assuming that W_2 is the optimal point and has been fixed, we have

$$W_3 = W_1 \tag{22.19}$$

Assuming that W_1 is the optimal point and it has been fixed, we have

$$W_3 = W_2 \tag{22.20}$$

where

$$W_1 = W_2 = W_3 = \frac{W_t}{3} \tag{22.21}$$

is an optimization point.
The optimal control method is to keep

$$W_1 = W_2 = W_3 = \frac{W_t}{n} \tag{22.22}$$

The overall maximum operating energy efficiency $\eta_{t3}(W_t)$ of the system is

$$\max \eta_{t3}(W_t) = \eta_1 \left(\frac{W_t}{3} \right) \tag{22.23}$$

If the three devices have similar energy efficiency, assuming that W_3 is the optimal point and fixed, based on the conclusion of n = 2, we have

$$W_2 = \beta_2 W_1 \tag{22.24}$$

Similarly, assuming that W_2 is the optimal point and has been fixed, we have

$$W_3 = \beta_3 W_1 \tag{22.25}$$

where

$$\begin{aligned} W_1 &= \frac{W_t}{1+\beta_2+\beta_3} \\ W_2 &= \beta_2 W_1 \\ W_3 &= \beta_3 W_1 \end{aligned} \tag{22.26}$$

is an optimization point.
The optimal control method is to keep

$$W_1 = \frac{W_t}{\sum_{i=1}^{3} \beta_i}$$
$$W_2 = \beta_2 W_1 \tag{22.27}$$
$$W_3 = \beta_3 W_1$$

The maximum value of the overall operating energy efficiency $\eta_{t3}(W_t)$ of the system is

$$\max \eta_{t3}(W_t) = \eta_1 \left(\frac{W_t}{\sum_{i=1}^{3} \beta_i} \right) \tag{22.28}$$

(3) n = k

If k devices are devices with the identical energy efficiency, the above conclusion is extended to the case of n = k, and the optimal point is

$$W_1 = W_2 = \cdots = W_k = \frac{W_t}{k} \tag{22.29}$$

The optimal control method is to keep

$$W_1 = W_2 = \cdots = W_k = \frac{W_t}{k} \tag{22.30}$$

The maximum value of the overall operating energy efficiency $\eta_{tk}(W_t)$ of the system is

$$\max \eta_{tk}(W_t) = \eta_1 \left(\frac{W_t}{k} \right) \tag{22.31}$$

If k devices are devices with similar energy efficiency, the above conclusion is extended to the case of n = k, and the optimal point is

$$W_1 = \frac{W_t}{\sum_{i=1}^{k} \beta_i}$$
$$W_2 = \beta_2 W_1 \tag{22.32}$$
$$\cdots$$
$$W_k = \beta_k W_1$$

The optimal control method is to keep

$$W_1 = \frac{W_t}{\sum_{i=1}^{k} \beta_i}$$
$$W_2 = \beta_2 W_1 \tag{22.33}$$
$$\cdots$$
$$W_k = \beta_k W_1$$

The maximum value of the overall operating energy efficiency $\eta_{tk}(W_t)$ of the system is

$$\max\eta_{tk}(W_t) = \eta_1\left(\frac{W_t}{\sum_{i=1}^{k}\beta_i}\right) \tag{22.34}$$

Whether it is device with the identical energy efficiency or device with similar energy efficiency, their optimal load distribution methods have one thing in common, that is, the energy efficiency of all operating device is identical, that is,

$$\eta_1(W_1) = \eta_2(W_2) = \cdots = \eta_n(W_n) \tag{22.35}$$

Optimal load distribution theorem (Yao Theorem 1): The optimal load distribution method is to keep the operating energy efficiency of each operating device equal.

$$\eta_1(W_1) = \eta_2(W_2) = \cdots = \eta_n(W_n) \tag{22.36}$$

22.3 Optimal Switching Theorem for Hydrogen Production System

The optimal methods of load distribution obtained above are all obtained under the assumption that n is already optimal, but is the n optimal? We analyze two cases.

The highest overall energy efficiency of system is

$$\max\eta_{tn}(W_t) = \eta_1\left(\frac{W_t}{n}\right) \tag{22.37}$$

For the same W_t, there are $(n-1)$ running devices with the identical energy efficiency, and the highest overall energy efficiency of system is

$$\max\eta_{t(n-1)}(W_t) = \eta_1\left(\frac{W_t}{n-1}\right) \tag{22.38}$$

For the same W_t, there are $(n+1)$ running devices with the identical energy efficiency, and the highest overall energy efficiency of the system is

$$\max\eta_{t(n+1)}(W_t) = \eta_1\left(\frac{W_t}{n+1}\right) \tag{22.39}$$

where

Fig. 22.3 Energy efficiency comparison curve

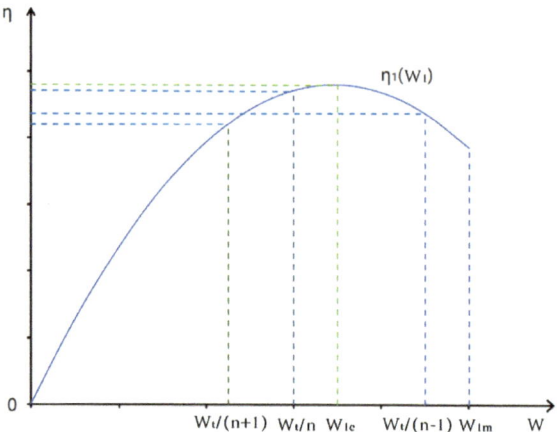

$$\frac{W_t}{n-1} > \frac{W_t}{n} > \frac{W_t}{n+1} \tag{22.40}$$

On the $\eta_1(\ )$ energy efficiency curve, the $W_t/(n-1)$ point is on the right side of the W_t/n point, and the $W_t/(n+1)$ point is on the left side of the W_t/n point.

If the following formula holds

$$\eta_1\left(\frac{W_t}{n}\right) = max\left\{\eta_1\left(\frac{W_t}{n-1}\right), \eta_1\left(\frac{W_t}{n}\right), \eta_1\left(\frac{W_t}{n+1}\right)\right\} \tag{22.41}$$

we say that n is truly optimal.

As shown in Fig. 22.3, $\eta_1(W_t/n)$ has the highest operating energy efficiency, and the W_t/n point is closer to the W_{1e} point than $W_t/(n-1)$ and $W_t/(n+1)$,

In the above discussion, we have always regarded W_t as an invariable constant. In practical applications, W_t changes with the size of the air volume.

When W_t increases, $\eta_1(W_t/(n+1))$ also increases, then the following conditions are met

$$\eta_1\left(\frac{W_t}{n+1}\right) = \eta_1\left(\frac{W_t}{n}\right) \tag{22.42}$$

The switching point is reached, if W_t continues to increase, then

$$\eta_1\left(\frac{W_t}{n+1}\right) > \eta_1\left(\frac{W_t}{n}\right) \tag{22.43}$$

It should be switched the number of running devices from n to n + 1.

Similarly, when W_t decreases, $\eta_1(W_t/(n-1))$ increases, when the following conditions are met

$$\eta_1\left(\frac{W_t}{n-1}\right) = \eta_1\left(\frac{W_t}{n}\right) \tag{22.44}$$

The switching point is reached, if W_t continues to decrease, then

$$\eta_1\left(\frac{W_t}{n-1}\right) > \eta_1\left(\frac{W_t}{n}\right) \tag{22.45}$$

should switch the number of running devices from n to n-1.

Due to the limitation of W_{1m}, when W_t increases, W_t/n also increases until $W_t/n = W_{1m}$, which is still not satisfied

$$\eta_1\left(\frac{W_t}{n+1}\right) = \eta_1\left(\frac{W_t}{n}\right) \tag{22.46}$$

If W_t/n continues to increase, the device will be overloaded, and it is necessary to force switching to (n + 1) devices at the point $W_t/n = W_{1m}$.

The total value of the input energy is W_t, and there are n devices with similar energy efficiency running, and the maximum value of the overall operating energy efficiency $\eta_{tn}(W_t)$ of system is

$$\max\eta_{tn}(W_t) = \eta_1\left(\frac{W_t}{\sum_{i=1}^{n}\beta_i}\right) \tag{22.47}$$

For the same W_t, there are k devices with similar energy efficiency operating, k is any feasible combination except n, and the highest overall operating energy efficiency of system is

$$\max\eta_{tk}(W_t) = \eta_1\left(\frac{W}{\sum_{i=1}^{k}\beta_i}\right) \tag{22.48}$$

If the number of running units n is optimal, it must satisfy

$$\eta_1\left(\frac{W_t}{n\sum_{i=1}^{n}\beta_i}\right) = max\left\{\eta_1\left(\frac{W_t}{\sum_{i=1}^{n}\beta_i}\right), \eta_1\left(\frac{W_t}{\sum_{i=1}^{k}\beta_i}\right), \cdots\right\} \tag{22.49}$$

When W_t increases, the condition is satisfied

$$\eta_1\left(\frac{W_t}{\sum_{i=1}^{n}\beta_i}\right) = \eta_1\left(\frac{W_t}{\sum_{i=1}^{k_1}\beta_i}\right) \tag{22.50}$$

The switching point has been reached. If W_t continues to increase, it should switch the number of operating devices from n to k_1.

When W_t decreases, the condition is satisfied

$$\eta_1\left(\frac{W}{\sum_{i=1}^{n}\beta_i}\right) = \eta_1\left(\frac{W_t}{\sum_{i=1}^{k_2}\beta_i}\right) \tag{22.51}$$

The switching point has been reached. If W_t continues to decrease, it should switch the number of operating devices from n to k_2.

k1 and k2 are any combination other than the optimal combination of n units this time, and also include other combinations of n units.

If there is no point of equal energy efficiency, the optimal switching point is at the maximum load point of the equipment.

Optimal switching theorem (Yao Theorem 2): The optimal switching point for the number of operating units is at the point of equal efficiency or at the maximum output point of the devices.

$$\eta_1\left(\frac{P_t}{\sum_{i=1}^{n}\beta_i}\right) = \eta_1\left(\frac{P_t}{\sum_{i=1}^{k_2}\beta_i}\right) or \eta_1\left(\frac{P_t}{\sum_{i=1}^{n}\beta_i}\right) = \eta_{1m} \tag{22.52}$$

22.4 Engineering Optimal Versus Theoretical Optimal

In actual engineering, because the electrolytic cell needs several hours to start up, there is energy loss in preheating. Thus it can only achieve engineering optimization, not theoretical optimization.

Chapter 23
Energy Efficiency Optimization of Pumping Stations and Fan Stations

Many occasions and fields have requirements for transporting liquids and gases, so pumps and fans are widely used in chemical industry, pharmaceuticals, power plants, steel mills, papermaking, oil refining, water conservancy, water supply, drainage and irrigation, sewage, secondary water supply, hospitals, office buildings, shopping malls, hotels, hotels and other occasions (Fig. 23.1). The total electricity consumption of pumps and fans accounts for 30–35% of the world's total electricity consumption.

In the field of energy saving for pumps and fans, speed control devices have developed rapidly, especially frequency conversion speed control technology has become relatively mature. At present, many internationally renowned companies, such as Siemens, ABB, Schneider, Fuji, Toshiba, Sanken, General Electric, AB, etc., have launched water pump fan type inverters. In terms of energy-saving calculation of water pumps and fans, many multinational companies have also launched some tools, such as ABB's "PumpSave" and "FanSave" software, and Siemens' "SinaSave" software. However, there are still many problems in terms of versatility, accuracy and energy-saving thoroughness.

The energy-saving transformation of some pumping stations and fan stations is far from the expected results, which not only damages the reputation of energy-saving equipment manufacturers, but also makes users dissatisfied. Such cases include large multinationals and listed companies. Some pumping stations and fan stations have increased power consumption after energy-saving renovations. The reason is unknown, and the two sides started arguing. Since there are many ways to save energy, and each has its own advantages and disadvantages, some users simply think that they can use the equipment of the company that promises the largest energy-saving ratio. Turning technical issues such as energy-saving ratio into business issues that can be negotiated. Many strange phenomena have occurred. The promised power saving ratio has broken the law of conservation of energy, and the two parties are still negotiating seriously.

There are different opinions on energy-saving methods, and the vast number of energy-saving workers feel clouded. Some experts' articles are well written, but the

F. Yao and Y. Yao, *Efficient Energy-Saving Control and Optimization for Multi-Unit Systems*, https://doi.org/10.1007/978-981-97-4492-3_23

Fig. 23.1 Application of pump and fan

conclusion is only correct for the water pump fan station they experimented with. If generalized, there will be problems. Some energy-saving methods are very good, but they are only suitable for specific working conditions. The following problems are what you have to face:

Where are inverters suitable for application?

Is there no potential for energy saving if the inverter is already used?

In fact, many special cases can also illustrate some general trends:

(1) When the frequency converter works at 50 Hz, it is better to run directly at power frequency to save power. Using methods such as series resistance speed regulation and liquid viscosity speed regulation, the efficiency is close to 100% when running close to the rated speed, which is more energy-saving than a frequency converter;

(2) When multiple pumps are running, the variable frequency pump that only maintains pressure but does not output water wastes 100% of energy. Turning off the pump at this time will not affect the output pressure and flow of the pump station;

(3) Even if the frequency converter is used and has a good power-saving effect, if there is no optimal operation method, it may still cause a large waste of electric energy.

23.1 The Total Energy Consumption of a Fan Station

The energy consumed by a fan is

$$W = k_8 \frac{QP_0}{\eta(Q)} \tag{23.1}$$

where W is the energy consumed by the fan, k_8 is a constant, Q is the flow rate of the fan, and η is the energy efficiency of the fan, P_0 is the total pressure provided by the fan.

A fan station has m fans, and the total pressure P_0 provided by each fan is equal. The total energy consumption of this fan station is

$$W_t = \sum_{i=1}^{m} W_i = k_8 P_0 \sum_{i=1}^{m} Q_i \frac{1}{\eta(Q_i)} \tag{23.2}$$

where Q_i is the output flow of the i-th fan, η_i is the energy efficiency of the i-th fan, and W_i is the energy consumed by the i-th fan. W_t is the total energy consumption of the fan station.

23.2 Total Energy Consumption of a Pumping Station

The energy consumed by a pump is

$$W = \frac{QH_0}{367\eta(Q)} \tag{23.3}$$

where W is the energy consumed by the pump, Q is the flow rate of the pump, and η is the energy efficiency of the pump, H_0 is the full head provided by the pump.

A pumping station has m pumps, and the full head H_0 provided by each pump is equal. The total energy consumption of this pumping station is

$$W_t = \sum_{i=1}^{m} W_i = \frac{H_0}{367} \sum_{i=1}^{m} Q_i \frac{1}{\eta_i(Q_i)} \tag{23.4}$$

where Q_i is the output flow of the i-th pump, η_i is the energy efficiency of the i-th pump, and W_i is the energy consumed by the i-th pump, W_t is the total energy consumption of the pumping station.

23.3   Unification of Energy Efficiency Optimization for Pumping Stations and Fan Stations

Comparing the energy consumption expressions of the water pump station and the fan station, the changing parts are the same. Only one type of energy efficiency optimization method needs to be solved to generalize to the other type. Below we only analyze the energy efficiency optimization of pumping stations.

For a pumping station that outputs a constant full head H0, the efficiency of the i-th centrifugal variable speed pump is shown in Fig. 23.2.

In Fig. 23.2, Q is flow (m³/hour), and η is the efficiency, Q_{ie} is the flow rate at the maximum efficiency η_{ie}, Q_{im} is the maximum flow rate of the pump.

$$Q \geq 0 \tag{23.5}$$

$$\eta_i(Q) \geq 0$$

$$\eta_i(0) = 0$$

$$\eta_i''(Q) < 0$$

Approximately, we have

$$\eta_i(Q) = \sum_{i=0}^{\infty} a_i Q^i = Q \sum_{i=1}^{\infty} \left(a_i Q^{i-1} \right) \tag{23.6}$$

$$\sum_{i=1}^{\infty} \left(a_i Q^{i-1} \right) > 0$$

$$a_0 = 0$$

Fig. 23.2 Centrifugal pump efficiency function

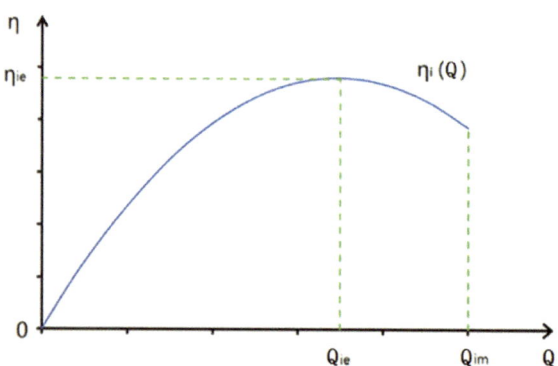

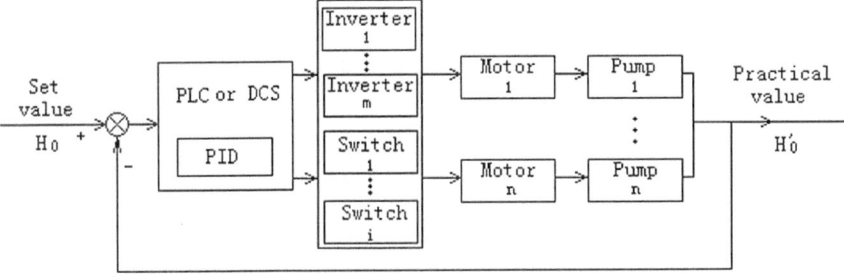

Fig. 23.3 Typical single closed-loop control block diagram

The minimum energy consumption of the pumping station is

$$min W_t = min\left(\frac{H_0}{367}\sum_{i=1}^{m}Q_i\frac{1}{\eta_i(Q_i)}\right)$$
$$s.t. \sum_{i=1}^{m}Q_i = Q_t > 0 \tag{23.7}$$
$$Q_{imx} \geq W_i > 0$$

where Q_t is the total flow rate of the pumping station.

23.4 Typical Single Closed-Loop PID Control Method

A typical single closed-loop control of a pumping station is shown in Fig. 23.3.

In Fig. 23.3, the typical PID control method is to measure the actual value H'$_0$ to be controlled, compare H'$_0$ with the set value H$_0$, and feed the error back to the PID. The PID changes a manipulated variable to control the frequency converter and adjust the speed of the motor and pump. When H'$_0$ is less than H$_0$, PID increases the pump speed. If H'$_0$ is still less than H$_0$ when the pump speed reaches the maximum speed, PLC or DCS will turn on another pump. When H'$_0$ is greater than H$_0$, PID reduces the pump speed. If H'$_0$ is still greater than H$_0$ when the pump speed reaches the minimum speed, PLC or DCS will shut down a pump.

This control method does not consider the energy consumption of the pump station, so it cannot achieve minimum power consumption.

23.5 Optimal Control of Pumping Stations Composed of Identical Energy Efficiency Pumps

A pumping station has n water pumps of the identical model running. The pumping station is used to transport water at a constant full head H$_0$. The total flow rate is Q$_t$. The flow rate of the i-th water pump is Q$_i$. Q$_i$ is greater than zero. The speed of all pumps is variable, and the full head of the i-th pump remains at H_0, we have

$$Q_i > 0 \tag{23.8}$$

$$\sum_{i=1}^{n} Q_i = Q_t$$

$\eta(Q_i)$ is the i-th pump efficiency at the point (Q_i, H_0), η_t is the total efficiency of the pumping station, W_t is the total power consumption of the pumping station.

$$W_t = \frac{H_0}{367} \sum_{i=1}^{n} Q_i \frac{1}{\eta(Q_i)} \tag{23.9}$$

Based on the discussion in the previous chapters, we know that the optimal control method is to keep

$$Q_1 = Q_2 = \ldots = Q_n = \frac{Q_t}{n} \tag{23.10}$$

This is "*Same pump, Same load*", it is equivalent to "*Same pump, Same speed*", "*Same pump, Same frequency*" and "*Same pump, Same efficiency*".

The minimum value of the total power consumption is

$$minW_t = W_0 \frac{1}{\eta(\frac{Q_t}{n})} \tag{23.11}$$

The ideal work P_0 is

$$W_0 = \frac{Q_t H_0}{367} \tag{23.12}$$

The overall optimal efficiency is

$$max\eta_t(Q_t) = \eta\left(\frac{Q_t}{n}\right) \tag{23.13}$$

23.6 Pumping Station Composed of Pumps of Similar Efficiency

We define the load rate γ as

$$\gamma = \frac{Q}{Q_e} \tag{23.14}$$

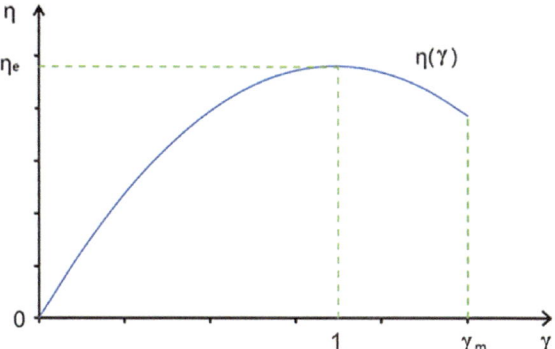

Fig. 23.4 The normalized efficiency function $\eta_N(\gamma)$

We call $\eta_N(\gamma)$ as the normalized efficiency function of a pump. The normalized efficiency function $\eta_N(\gamma)$ has a shape shown in Fig. 23.4.

In Fig. 23.4, γ is a variable and η_N is the efficiency. η_N and η have the following relationship.

$$\eta(Q) = \eta(\gamma Q_e) = \eta_N(\gamma) \tag{23.15}$$

If the normalized efficiency functions of two different pumps are identical, we have

$$\eta_{N1}(\gamma) = \eta_{N2}(\gamma) \tag{23.16}$$

We refer to them as pumps with similar efficiencies. A pumping station composed of pumps with similar efficiencies is called a similar efficiency pumping station.

Assume that γ_i is the load rate of the i-th pump, and its form is

$$\gamma_i = \frac{Q_i}{Q_{ie}} \tag{23.17}$$

For a pumping station consisting of n pumps with similar efficiencies, the total power consumption is of the form

$$W_t = \frac{H_0}{367} \sum_{i=1}^{n} \frac{Q_i}{\eta_i(Q_i)} = \frac{H_0 Q_t}{367} \sum_{i=1}^{n} \left(\frac{1}{Q_t} \frac{\gamma_i Q_{ie}}{\eta_N(\gamma_i)} \right) = W_0 \frac{1}{\eta_t} \tag{23.18}$$

where W_0 is the ideal work, η_t is the overall energy efficiency of the pumping station.

$$W_0 = \frac{H_0 Q_t}{367} \tag{23.19}$$

$$\eta_t = 1 \left/ \sum_{i=1}^{n} \left(\frac{1}{Q_t} \frac{\gamma_i Q_{ie}}{\eta_N(\gamma_i)} \right) \right.$$

23.7   Energy Efficiency Optimization of a Similar Efficiency Pumping Station

When the flow rate of each running pump is greater than zero, consider the minimization of the total power consumption

$$min W_t \tag{23.20}$$

$$s.t. \gamma_i > 0, i = 1, 2, \ldots n$$

$$\sum_{i=1}^{n} \gamma_i Q_{ie} = Q_t$$

We consider three kinds of situation.

(1) $n = 2$

There are two variables.
   We have

$$\gamma_1 Q_{1e} + \gamma_2 Q_{2e} = Q_t \tag{23.21}$$

$$\gamma_1 > 0$$

$$\gamma_2 > 0$$

The objective function P_t is expressed as

$$W_t = \frac{H_0}{367} \left(\frac{\gamma_1 Q_{1e}}{\eta_N(\gamma_1)} + \frac{\gamma_2 Q_{2e}}{\eta_N(\gamma_2)} \right) \tag{23.22}$$

The optimal condition is

$$W_t'(\gamma_1) = 0 \tag{23.23}$$

We have

$$\frac{Q_{1e}\eta_N(\gamma_1) - \gamma_1 Q_{1e}\eta_N'(\gamma_1)}{\eta_N^2(\gamma_1)}$$
$$+ \frac{-Q_{1e}\eta_N\left(\frac{Q_t - \gamma_1 Q_{1e}}{Q_{2e}}\right) + (Q_t - \gamma_1 Q_{1e})\frac{Q_{1e}}{Q_{2e}}\eta_N'\left(\frac{Q_t - \gamma_1 Q_{1e}}{Q_{2e}}\right)}{\eta_N^2\left(\frac{Q_t - \gamma_1 Q_{1e}}{Q_{2e}}\right)} = 0 \tag{23.24}$$

It is easy to see that

$$\gamma_1 = \gamma_2 = \frac{Q_t}{Q_{1e} + Q_{2e}} \tag{23.25}$$

is an optimal point.

The minimum value of the total power consumption is

$$\min W_t = \frac{H_0 Q_t}{367} \frac{1}{\eta_N \left(\frac{Q_t}{Q_{1e} + Q_{2e}}\right)} = \frac{W_0}{\eta_N \left(\frac{Q_t}{Q_{1e} + Q_{2e}}\right)} \tag{23.26}$$

The maximum energy efficiency of the pumping station is

$$\max \eta_{t2} = \eta_N \left(\frac{Q_t}{Q_{1e} + Q_{2e}}\right) = \eta_N \left(\frac{Q_t}{\sum_{i=1}^{2} Q_{ie}}\right) \tag{23.27}$$

(2) $n = 3$

There are three variables.

Based on known conditions, we have

$$\gamma_1 Q_{1e} + \gamma_2 Q_{2e} + \gamma_3 Q_{3e} = Q_t \tag{23.28}$$

$$\gamma_1 > 0$$

$$\gamma_2 > 0$$

$$\gamma_3 > 0$$

W_t expression becomes

$$W_t = \frac{H_0}{367} \left(\frac{\gamma_1 Q_{1e}}{\eta_N(\gamma_1)} + \frac{\gamma_2 Q_{2e}}{\eta_N(\gamma_2)} + \frac{\gamma_3 Q_{3e}}{\eta_N(\gamma_3)} \right) \tag{23.29}$$

Assume that γ_1 is a fixed optimal point and only γ_2 and γ_3 are variables. We have

$$\gamma_2 Q_{2e} + \gamma_3 Q_{3e} = Q_t - \gamma_1 Q_{1e} = constant \tag{23.30}$$

Based on the above conclusions at $n = 2$, the optimal point is at

$$\gamma_2 = \gamma_3 \tag{23.31}$$

Assume that γ_2 is a fixed optimal point and only γ_1 and γ_3 are variables. We have

$$\gamma_1 Q_{1e} + \gamma_3 Q_{3e} = Q_t - \gamma_2 Q_{2e} = constant \tag{23.32}$$

According to the conclusion at n = 2, the optimal point is at

$$\gamma_1 = \gamma_3 \tag{23.33}$$

Similarly, assume that γ_3 is a fixed optimal point and only γ_1 and γ_2 are variables. We have

$$\gamma_1 Q_{1e} + \gamma_2 Q_{2e} = Q_t - \gamma_3 Q_{3e} = constant \tag{23.34}$$

According to the conclusion at n = 2, the optimal point is at

$$\gamma_1 = \gamma_2 \tag{23.35}$$

So that we have the optimal points

$$\gamma_1 = \gamma_2 = \gamma_3 = \frac{Q_t}{\sum_{i=1}^{3} Q_{ie}} \tag{23.36}$$

The minimum value of the total power consumption is

$$minP_t = \frac{P_0}{\eta_N \left(\frac{Q_t}{\sum_{i=1}^{3} Q_{ie}}\right)} \tag{23.37}$$

The maximum overall energy efficiency of the pumping station is

$$max\eta_{t3} = \eta_N \left(\frac{Q_t}{\sum_{i=1}^{3} Q_{ie}}\right) \tag{23.38}$$

(3) n = k

There are k variables.

The above conclusion can be extended to the situation at n = k, the optimal point is

$$\gamma_1 = \gamma_2 = \ldots = \gamma_k = \frac{Q_t}{\sum_{i=1}^{k} Q_{ie}} \tag{23.39}$$

This is "*Same load rate, Same efficiency* ".

The minimum value of the total power consumption is

$$minW_t = \frac{P_0}{\eta_N \left(\frac{Q_t}{\sum_{i=1}^{k} Q_{ie}}\right)} \tag{23.40}$$

The maximum overall energy efficiency of the pumping station is

$$\max \eta_{tk} = \eta_N \left(\frac{Q_t}{\sum_{i=1}^{k} Q_{ie}} \right) \tag{23.41}$$

23.8 Optimal Number of Running Pumps

For pumping stations composed of the identical pumps, if n is the optimal, there must be

$$\frac{P_0}{\eta_N \left(\frac{Q_t}{\sum_{i=1}^{n-1} Q_{ie}} \right)} \geq \frac{P_0}{\eta_N \left(\frac{Q_t}{\sum_{i=1}^{n} Q_{ie}} \right)} \leq \frac{P_0}{\eta_N \left(\frac{Q_t}{\sum_{i=1}^{n+1} Q_{ie}} \right)} \tag{23.42}$$

where

$$\eta_N \left(\frac{Q_t}{\sum_{i=1}^{n-1} Q_{ie}} \right) \leq \eta_N \left(\frac{Q_t}{\sum_{i=1}^{n} Q_{ie}} \right) \geq \eta_N \left(\frac{Q_t}{\sum_{i=1}^{n+1} Q_{ie}} \right) \tag{23.43}$$

For pumping stations composed of the pumps with similar energy efficiency, if n is the optimal, we have

$$\frac{P_0}{\eta_N \left(\frac{Q_t}{\sum_{i=1}^{n1} Q_{ie}} \right)} \geq \frac{P_0}{\eta_N \left(\frac{Q_t}{\sum_{i=1}^{n} Q_{ie}} \right)} \tag{23.44}$$

where

$$\eta_N \left(\frac{Q_t}{\sum_{i=1}^{n1} Q_{ie}} \right) \leq \eta_N \left(\frac{Q_t}{\sum_{i=1}^{n} Q_{ie}} \right) \tag{23.45}$$

n1 is any combination other than this n combination.

23.9 Optimal Switch of the Pumping Station with the Identical Model Pumps

Assuming that n is optimal, the total power consumption is the smallest

$$\min W_t = \frac{W_0}{\eta_N \left(\frac{Q_t}{\sum_{i=1}^{n} Q_{ie}} \right)} \tag{23.46}$$

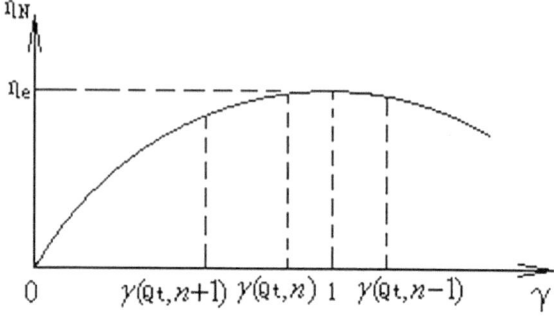

Fig. 23.5 η_N (γ) curve of the pumping station with the identical pumps when $\gamma(Q_t, n)$ is less than 1

$$n \leq m$$

$$\eta_N \left(\frac{Q_t}{\sum_{i=1}^{n-1} Q_{ie}} \right) \leq \eta_N \left(\frac{Q_t}{\sum_{i=1}^{n} Q_{ie}} \right) \geq \eta_N \left(\frac{Q_t}{\sum_{i=1}^{n+1} Q_{ie}} \right)$$

$$\gamma(Q_t, n) = \frac{Q_t}{\sum_{i=1}^{n} Q_{ie}}$$

When Q_t changes, we consider three situations.

(1) The $\gamma(Q_t, n)$ is less than 1.

η_e is the maximum efficiency, and 1 is the load rate at η_e, as shown in Fig. 23.5. In Fig. 23.5, we have

$$\max(\eta_N(Q_t, n-1), \eta_N(Q_t, n), \eta_N(Q_t, n+1)) = \eta_N(Q_t, n) \qquad (23.47)$$

When Q_t increases, the load rate $\gamma(Q_t, n)$ increases, $\eta_N(Q_t, n)$ and $\eta_N(Q_t, n + 1)$ increase also. However when $\gamma(Q_t, n) > 1$, Q_t continues to increase, $\eta_N(Q_t, n + 1)$ still increases, $\eta_N(Q_t, n)$ decreases. When $\eta_N(Q_t, n) < \eta_N(Q_t, n + 1)$, $n + 1$ is the optimal, we should increase the number of running pumps from n to n + 1, leading to

$$\max(\eta_N(Q_t, n-1), \eta_N(Q_t, n), \eta_N(Q_t, n+1)) = \eta_N(Q_t, n+1) \qquad (23.48)$$

The optimal switch point is

$$\eta_N(Q_t, n) = \eta_N(Q_t, n+1) \qquad (23.49)$$

When Q_t decreases, the load rate $\gamma(Q_t, n)$ decreases, $\eta_N(Q_t, n)$ decreases also, however $\eta_N(Q_t, n-1)$ increases. When $\eta_N(Q_t, n) < \eta_N(Q_t, n-1)$, n-1 is the optimal, we should decreases the number of running pumps from n to n-1, which yields,

Fig. 23.6 η_N (γ) curve of the pumping station with the identical pumps when $\gamma(Q_t, n)$ is greater than 1

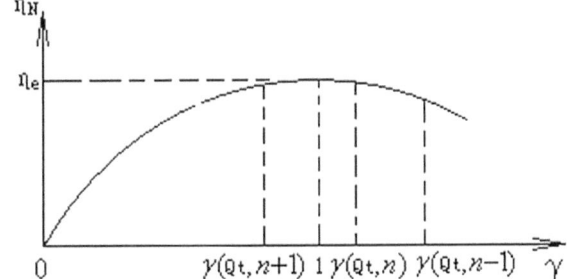

$$\max(\eta_N(Q_t, n-1), \eta_N(Q_t, n), \eta_N(Q_t, n+1)) = \eta_N(Q_t, n-1) \qquad (23.50)$$

The optimal switch point is

$$\eta_N(Q_t, n) = \eta_N(Q_t, n-1) \qquad (23.51)$$

(2) The $\gamma(Q_t, n)$ is greater than 1, as shown in Fig. 23.6.

In Fig. 23.6, we have

$$\max(\eta_N(Q_t, n-1), \eta_N(Q_t, n), \eta_N(Q_t, n+1)) = \eta_N(Q_t, n) \qquad (23.52)$$

When Q_t increases, the load rate γ (Q_t, n) increases, and $\eta_N(Q_t, n+1)$ increases also, however $\eta_N(Q_t, n)$ decreases. When $\eta_N(Q_t, n) < \eta_N(Q_t, n+1)$, $n+1$ is the optimal, we should increase the number of running pumps from n to $n+1$, and we have

$$\max(\eta_N(Q_t, n-1), \eta_N(Q_t, n), \eta_N(Q_t, n+1)) = \eta_N(Q_t, n+1) \qquad (23.53)$$

The optimal switch point is

$$\eta_N(Q_t, n) = \eta_N(Q_t, n+1) \qquad (23.54)$$

When Q_t decreases, the load rate $\gamma(Q_t, n)$ decreases, $\eta_N(Q_t, n)$ and $\eta_N(Q_t, n-1)$ both increase. When $\gamma(Q_t, n) < 1$, Q_t continues to decrease, $\eta_N(Q_t, n-1)$ still increases, however $\eta_N(Q_t, n)$ decreases. When $\eta_N(Q_t, n) < \eta_N(Q_t, n-1)$, n-1 is the optimal, we should decreases the number of running pumps from n to n-1, and we have

$$\max(\eta_N(Q_t, n-1), \eta_N(Q_t, n), \eta_N(Q_t, n+1)) = \eta_N(Q_t, n-1) \qquad (23.55)$$

The optimal switch point is

$$\eta_N(Q_t, n) = \eta_N(Q_t, n-1) \qquad (23.56)$$

(3)   The $\gamma(Q_t, n)$ is equal to 1

When Q_t increases, the load rate $\gamma(Q_t, n)$ increases, $\eta_N(Q_t, n + 1)$ increase also, however $\eta_N(Q_t, n)$ decreases. When $\eta_N(Q_t, n) < \eta_N(Q_t, n + 1)$, $n + 1$ is the optimal, we should increase the number of running pumps from n to $n + 1$, which leads to,

$$\max(\eta_N(Q_t, n - 1), \eta_N(Q_t, n), \eta_N(Q_t, n + 1)) = \eta_N(Q_t, n + 1) \qquad (23.57)$$

The optimal switch point is

$$\eta_N(Q_t, n) = \eta_N(Q_t, n + 1) \qquad (23.58)$$

When Q_t decreases, the load rate $\gamma(Q_t, n)$ decreases, $\eta_N(Q_t, n)$ decreases also, however $\eta_N(Q_t, n-1)$ increases. When $\eta_N(Q_t, n) < \eta_N(Q_t, n-1)$, $n-1$ is the optimal, we should decreases the number of running pumps from n to $n-1$, and we have

$$\max(\eta_N(Q_t, n - 1), \eta_N(Q_t, n), \eta_N(Q_t, n + 1)) = \eta_N(Q_t, n - 1) \qquad (23.59)$$

The optimal switch point is

$$\eta_N(Q_t, n) = \eta_N(Q_t, n - 1) \qquad (23.60)$$

This is "*Same efficiency Switching*".

23.10   Optimal Switch of a Similar Efficiency Pumping Station

Assuming that n is optimal, the total power consumption is the smallest and the overall operating energy efficiency is the highest.

$$\min W_t = \frac{W_0}{\eta_N\left(\frac{Q_t}{\sum_{i=1}^{n} Q_{ie}}\right)} \qquad (23.61)$$

$$n \leq m$$

$$\eta_N\left(\frac{Q_t}{\sum_{i=1}^{n1} Q_{ie}}\right) \leq \eta_N\left(\frac{Q_t}{\sum_{i=1}^{n} Q_{ie}}\right)$$

$$\gamma(Q_t, n) = \frac{Q_t}{\sum_{i=1}^{n} Q_{ie}}$$

n1 is any combination other than this n combination.

Fig. 23.7 η_N (γ) curve of the similar efficiency pumping station when $\gamma(Q_t, n)$ is less than 1

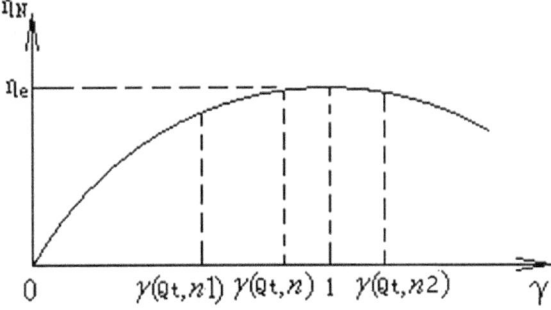

When Q_t changes, we consider three situations.

(1) The $\gamma(Q_t, n)$ is less than 1

η_e is the maximum efficiency, and 1 is the load rate at η_e, as shown in Fig. 23.7. In Fig. 23.7, we have

$$max(\eta_N(Q_t, n1), \eta_N(Q_t, n), \eta_N(Q_t, n2)) = \eta_N(Q_t, n) \qquad (23.62)$$

When Q_t increases, the load rate $\gamma(Q_t, n)$ increases, $\eta_N(Q_t, n)$ and $\eta_N(Q_t, n1)$ increase also. However when $\gamma(Q_t, n) > 1$, Q_t continues to increase, $\eta_N(Q_t, n1)$ still increases, $\eta_N(Q_t, n)$ decreases. When $\eta_N(Q_t, n) < \eta_N(Q_t, n1)$, n1 is the optimal, we should change the number of running pumps from n to n1, which leads to,

$$max(\eta_N(Q_t, n1), \eta_N(Q_t, n), \eta_N(Q_t, n2)) = \eta_N(Q_t, n1) \qquad (23.63)$$

The optimal switch point is

$$\eta_N(Q_t, n) = \eta_N(Q_t, n1) \qquad (23.64)$$

When Q_t decreases, the load rate $\gamma(Q_t, n)$ decreases, $\eta_N(Q_t, n)$ decreases also, however $\eta_N(Q_t, n2)$ increases. When $\eta_N(Q_t, n) < \eta_N(Q_t, n2)$, n-1 is the optimal, we should decrease the number of running pumps from n to n2, and we have

$$max(\eta_N(Q_t, n1), \eta_N(Q_t, n), \eta_N(Q_t, n2)) = \eta_N(Q_t, n2) \qquad (23.65)$$

The optimal switch point is

$$\eta_N(Q_t, n) = \eta_N(Q_t, n2) \qquad (23.66)$$

(2) The $\gamma(Q_t, n)$ is greater than 1, as shown in Fig. 23.8

In Fig. 23.8, we have

Fig. 23.8 η_N (γ) curve of the similar efficiency pumping station when $\gamma(Q_t, n)$ is greater than 1

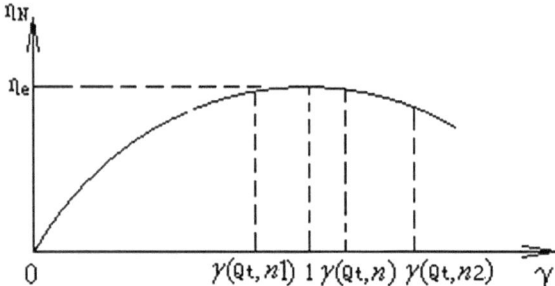

$$max(\eta_N(Q_t, n1), \eta_N(Q_t, n), \eta_N(Q_t, n2)) = \eta_N(Q_t, n) \tag{23.67}$$

When Q_t increases, the load rate $\gamma(Q_t, n)$ increases, $\eta_N(Q_t, n+1)$ increases also, however $\eta_N(Q_t, n)$ decreases. When $\eta_N(Q_t, n) < \eta_N(Q_t, n1)$, n1 is the optimal, we should change the number of running pumps from n to n1, and we have

$$max(\eta_N(Q_t, n1), \eta_N(Q_t, n), \eta_N(Q_t, n2)) = \eta_N(Q_t, n1) \tag{23.68}$$

The optimal switch point is

$$\eta_N(Q_t, n) = \eta_N(Q_t, n1) \tag{23.69}$$

When Q_t decreases, the load rate $\gamma(Q_t, n)$ decreases, $\eta_N(Q_t, n)$ and $\eta_N(Q_t, n2)$ both increase. When $\gamma(Q_t, n) < 1$, Q_t continues to decrease, $\eta_N(Q_t, n2)$ still increases, however $\eta_N(Q_t, n)$ decreases. When $\eta_N(Q_t, n) < \eta_N(Q_t, n2)$, n2 is the optimal, we should change the number of running pumps from n to n-1, leading to

$$max(\eta_N(Q_t, n1), \eta_N(Q_t, n), \eta_N(Q_t, n2)) = \eta_N(Q_t, n2) \tag{23.70}$$

The optimal switch point is

$$\eta_N(Q_t, n) = \eta_N(Q_t, n2) \tag{23.71}$$

(3) The $\gamma(Q_t, n)$ is equal to 1

When Q_t increases, the load rate $\gamma(Q_t, n)$ increases, $\eta_N(Q_t, n1)$ increase also, however $\eta_N(Q_t, n)$ decreases. When $\eta_N(Q_t, n) < \eta_N(Q_t, n1)$, n1 is the optimal, we should change the number of running pumps from n to n1, which yields,

$$max(\eta_N(Q_t, n1), \eta_N(Q_t, n), \eta_N(Q_t, n2)) = \eta_N(Q_t, n1) \tag{23.72}$$

The optimal switch point is

$$\eta_N(Q_t, n) = \eta_N(Q_t, n1) \qquad (23.73)$$

When Q_t decreases, the load rate $\gamma(Q_t, n)$ decreases, $\eta_N(Q_t, n)$ decreases also, however $\eta_N(Q_t, n2)$ increases. When $\eta_N(Q_t, n) < \eta_N(Q_t, n2)$, n2 is the optimal, we should change the number of running pumps from n to n2, and we have

$$max(\eta_N(Q_t, n1), \eta_N(Q_t, n), \eta_N(Q_t, n2)) = \eta_N(Q_t, n2) \qquad (23.74)$$

The optimal switch point is

$$\eta_N(Q_t, n) = \eta_N(Q_t, n2) \qquad (23.75)$$

This is "*Same efficiency Switching*".

23.11 Pressure Measurement of Pump and Fan

Take the water supply pipe network as an example, as shown in Fig. 23.9. If the resistance loss in the pipe network is not considered and there is no external energy input, according to the law of conservation of energy, the energy of any two cross sections 1–1 and 2–2 in the pipe network should be equal, that is, the pressure energy, potential energy, and kinetic energy of the liquid on the two cross sections should be equal.

In Fig. 23.9, p1, v_1, h_1, and s_1 respectively represent the pressure (pa, N/m^2), water velocity (m/s), height (m), and area (m^2) of the 1–1 cross section, while p2, v_2, h_2, and s_2 respectively represent the pressure, water velocity, height, and area of the 2–2 cross section. The pressure F_1 (N) of cross section 1–1 and the pressure F_2 of cross section 2–2 are calculated as follow

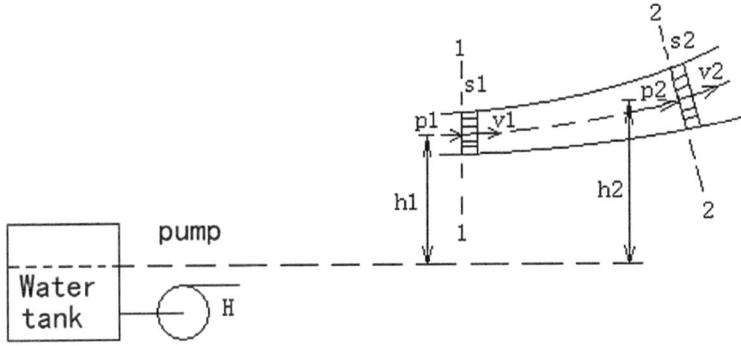

Fig. 23.9 Bernoulli's equation

$$F_1 = p_1 \times s_1 \tag{23.76}$$

$$F_2 = p_2 \times s_2$$

Calculate the volumetric flow rate Q as follow

$$Q = s_1 \times v_1 = s_2 \times v_2 \tag{23.77}$$

where Q is volumetric flow rate, m^3/s.

Calculate the mass flow rate as follow

$$m = \gamma \times Q \tag{23.78}$$

where m is the mass of the volumetric flow rate Q, and m is expressed in kg, γ is the density of the fluid in kg/m^3. Unit time $\Delta t = 1$ s, the pressure energy at cross sections 1–1 and 2–2 in the pipeline causes changes in potential energy and kinetic energy as follow

$$(p_1 \times s_1) \times v_1 \times \Delta t - (p_2 \times s_2) \times v_2 \times \Delta t = (mgh_2 - mgh_1) + \left(\frac{mv_2^2}{2} - \frac{mv_1^2}{2} \right) \tag{23.79}$$

where h_1 and h_2 are heights, m. Both sides of Eq. (23.79) are divided by mg, and the terms with the same base standard on both sides are merged. The relationship between pressure energy, potential energy, and kinetic energy per unit weight is obtained as follows:

$$\frac{p_1}{\gamma g} + h_1 + \frac{v_1^2}{2g} = \frac{p_2}{\gamma g} + h_2 + \frac{v_2^2}{2g} \tag{23.80}$$

where h_1 and h_2 are the positional water heads, $p_1 / (\gamma\ g)$ and $p_2 / (\gamma\ g)$ is the pressure head (also known as static pressure), and $v_1^2/(2\ g)$ and $v_2^2/(2\ g)$ are the kinetic head (also known as dynamic pressure). Equation (23.80) is the famous Bernoulli equation. It represents the relationship between pressure energy, potential energy, and kinetic energy that can be converted into each other without external work and without resistance loss.

In the case of $h_1 = h_2$, an increase in kinetic energy will lead to a decrease in pressure energy. Similarly, a decrease in kinetic energy will lead to an increase in pressure energy. For example, if the pipeline becomes thicker, the flow rate will decrease, the dynamic pressure will decrease, and the static pressure will increase. If the pipeline becomes thinner, the flow rate will increase, the dynamic pressure will increase, and the static pressure will decrease. A fire water gun is an example of increasing dynamic pressure.

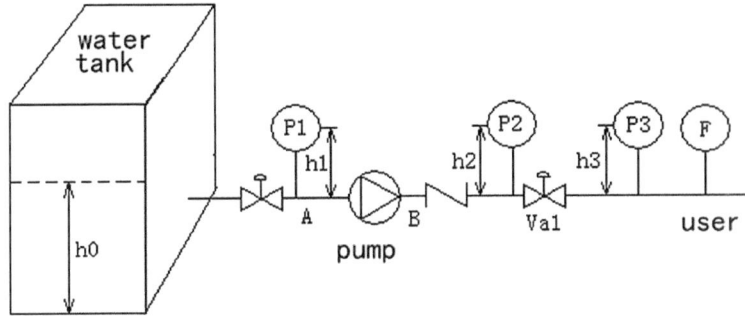

Fig. 23.10 Measurement of parameters in a pump station

The total head H provided by the water pump (m, calculated from the inlet water surface) is converted into pressure energy, potential energy, and kinetic energy in the pipeline network without considering losses, leading to

$$H = \frac{p_1}{\gamma g} + h_1 + \frac{v_1^2}{2g} = \frac{p_2}{\gamma g} + h_2 + \frac{v_2^2}{2g} \tag{23.81}$$

It must be noted that the concept of head in the pump characteristic curve and pipeline network characteristic curve includes three parts: pressure energy, potential energy, and kinetic energy. When implementing energy-saving control for actual pumping stations, we are concerned about the operating parameters from the pump inlet to the pump outlet. We need to have a comprehensive understanding of the pressure, flow rate, and liquid level situation in this section, as shown in Fig. 23.10.

In Fig. 23.10, h0, h1, h2, and h3 represent the liquid level height of the water tank, the height from the pump inlet pressure gauge to the pump shaft, the height from the pump outlet pressure gauge to the pump shaft, and the height from the pipe network pressure gauge to the pump shaft. P_1, P_2, and P_3 represent the water pump inlet pressure, water pump outlet pressure, and pipe network pressure, respectively. F represents the flow rate of the pump station.

When the liquid level h_0 of the water tank is higher than the center of the water pump shaft, if the inlet pipeline of the water pump is shorter, the resistance loss is not significant, and P_1 is positive pressure. When the liquid level of the water tank is lower than the center of the water pump shaft, or if the inlet pipeline of the pump is longer, the resistance loss is significant, resulting in P_1 being negative pressure. P_1 is static pressure.

When using a pressure gauge to measure static pressure, the actual pressure value at the measure point is the displayed value on the pressure gauge, plus the additional pressure added by the height of the pressure measuring element on the pressure gauge. Therefore, the actual static pressure at the inlet of the water pump is equal to $P_1/(\gamma g) + h_1$, the actual static pressure at the outlet of the water pump is equal to $P_2/(\gamma g) + h_2$. The actual static pressure of the pipeline network is equal to $P_3/(\gamma g) + h_3$.

Everyone must pay attention that the height of the pressure measuring element on the pressure gauge and the height of the pressure gauge display are sometimes not the same, unless they are integrated. The electrical signals measured by the pressure measuring element can be transmitted to far places to display.

Pressure p (pa), height h (m) and density γ (kg/m^3), their conversion relationship is as follows:

$$h = \frac{p}{\gamma g} \tag{23.82}$$

Due to density γ different, the same pressure value results in different heights. For water, the density $\gamma = 1000$ kg/m^3, 1Mpa pressure value is equivalent to a height of 102 m water, 1 kg/cm^2 pressure corresponds to a height of 10 m water, 1 standard atmospheric pressure (ATM) corresponds to a height of 10.33 m water, and 1 bar corresponds to a height of 10.2 m water. For mercury liquids, the same pressure is much lower when converted to a height of mercury. On the contrary, for oil with lower density, it is higher.

$$\begin{aligned} &1 \text{ (kg m)} = 9.8 \text{ (J)} \\ &1 \text{ (kW h)} = 367, 170 \text{ (kg m)} \end{aligned} \tag{23.83}$$

The flow rate of the pump is divided by the cross-sectional area of the pipeline to obtain the average flow velocity in the pipeline, and then the dynamic pressure is calculated. The total head of a pump is equal to the total pressure at the pump outlet (static pressure + dynamic pressure + h_2) minus the total pressure at the pump inlet (static pressure + dynamic pressure + h_1). When the diameter difference between the inlet and outlet of the pump is not significant, it can be approximated that the inlet and outlet dynamic pressures are equal. The total head of the pump is approximately equal to the static pressure at the outlet minus the static pressure at the inlet. In order to reduce resistance loss, the flow rate of pipelines designed according to specifications is generally not too high, so the proportion of dynamic pressure $v_2^2/(2 g)$ is not too large and can sometimes be ignored.

For the fan system, the dynamic pressure value already accounts for a significant proportion of the total pressure and cannot be ignored anymore, due to the density γ of the gas very small, the pressure formed by the installation height h of the pressure gauge is γgh, the numerical value is very small and can be ignored, so only static pressure and dynamic pressure are considered in the fan system.

The static pressure in the fan system can be directly measured using a pressure gauge or pressure transmitter, as shown in Fig. 23.11.

The measurement of dynamic pressure in a fan system is usually obtained by subtracting the static pressure from the total pressure. The total pressure and static pressure can be directly measured using a pitot tube, as shown in Fig. 23.12.

In Fig. 23.12, point A is located at the center of the pipeline, directly facing the direction of the incoming air. The pressure tap at point A measures the total pressure in the pipeline, while point B is located at the center of the pipeline. The pressure

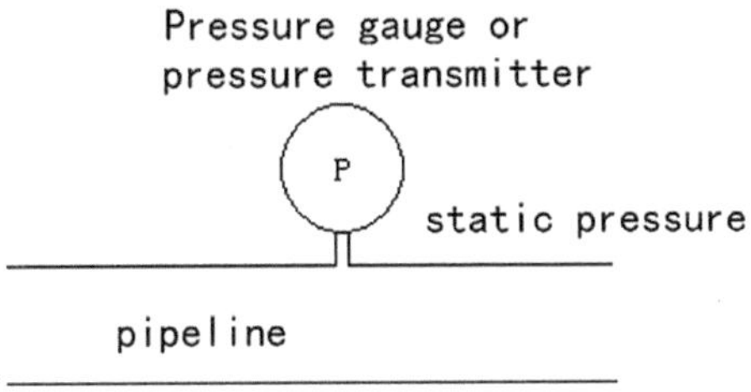

Fig. 23.11 Measurement of
static pressure in fan systems

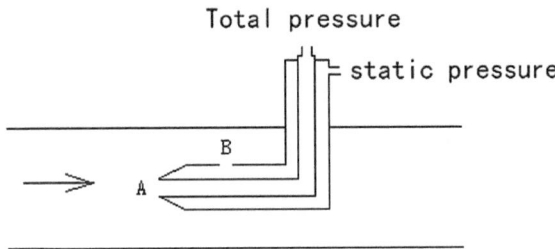

Fig. 23.12 Static pressure
and total pressure

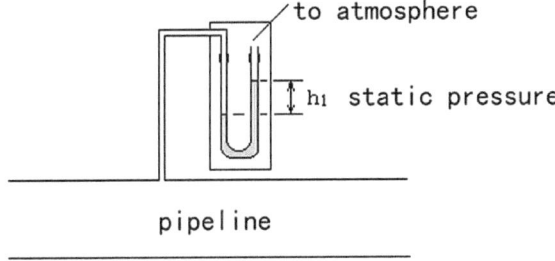

tap at point B is perpendicular to the direction of the incoming air, and measures the static pressure in the pipeline. The dynamic pressure is equal to the total pressure minus static pressure. The total pressure and static pressure are simultaneously fed into the two pressure taps of the differential pressure transmitter to directly measure the dynamic pressure.

For ventilation fans, the pressure is not high, and sometimes the static pressure h_1, full pressure h, and dynamic pressure h_2 can be directly measured on site using a U-shaped tube, as shown in Figs. 23.13, 23.14, and 23.15.

Use pressure sensors to measure the pressure in pipelines and containers.

Fig. 23.13 Measuring static
pressure with a U-tube

Fig. 23.14 Measure full pressure using a U-tube

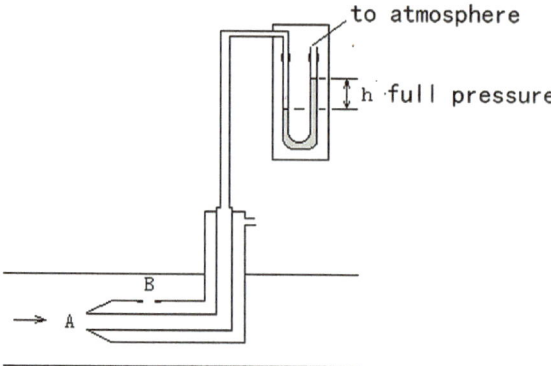

Fig. 23.15 Using a U-shaped tube to measure dynamic pressure

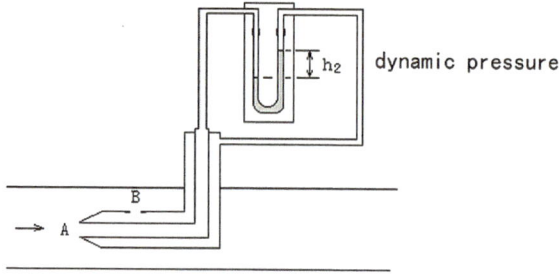

23.12   Practical Case of Energy Saving in a Pumping Station

The above theory and two theorems have been successfully applied to nearly a thousand pumping stations in the fields of building secondary water supply, urban water supply, steel, petrochemical, pharmaceutical and other fields.

A pumping station with a daily water supply of 400,000 tons is shown in Fig. 23.16. It has three speed-regulating centrifugal pumps of the identical model.

Fig. 23.16 The water delivery pump station of the water plant

Fig. 23.17 Flow-efficiency
curve Q-η

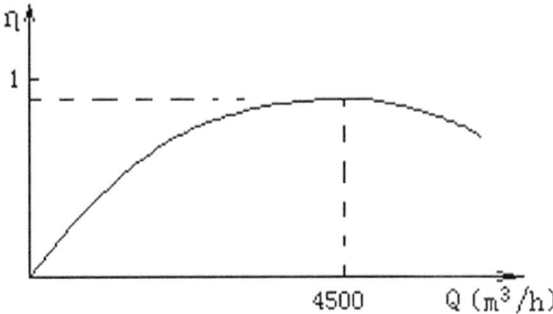

Fig. 23.18 Flow-Head
Curve Q-H

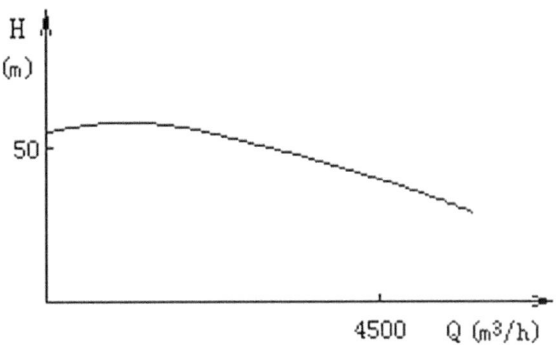

The rated head of the pump is 39 (m), the rated flow is 4500 (m³/h), the rated speed n_0 is 980 (r/min). The flow-efficiency curve of the pump at the rated speed is shown in Fig. 23.17.

The flow-head curve H(Q) of the pump at the rated speed n_0 is shown in Fig. 23.18.

The pumping station pressurizes the water to a fixed height H_0, H_0 is equal to 20 (m).

Based on the above conclusions, "*Same pump, Same frequency*" and "*Same efficiency Switching*", referred to as "*4S*" technology, there is a curve as shown in Fig. 23.19.

The optimal switching point Q_{12} of 1 pump and 2 pumps is at $Q_t = 4050$(m3/hour), which leads to

$$Q_{12} = 4050 \, (\text{m}^3/\text{h}) \tag{23.84}$$

When the total flow Q_t is less than Q_{12}, use 1 water pump and keep $Q_1 = Q_t$; when the total flow Q_t is greater than Q_{12}, use 2 water pumps and keep $Q_1 = Q_2 = 0.5Q_t$.

The optimal switching point Q_{23} of 2 water pumps and 3 water pumps is at $Q_t = 7254$(m3/hour), and we have

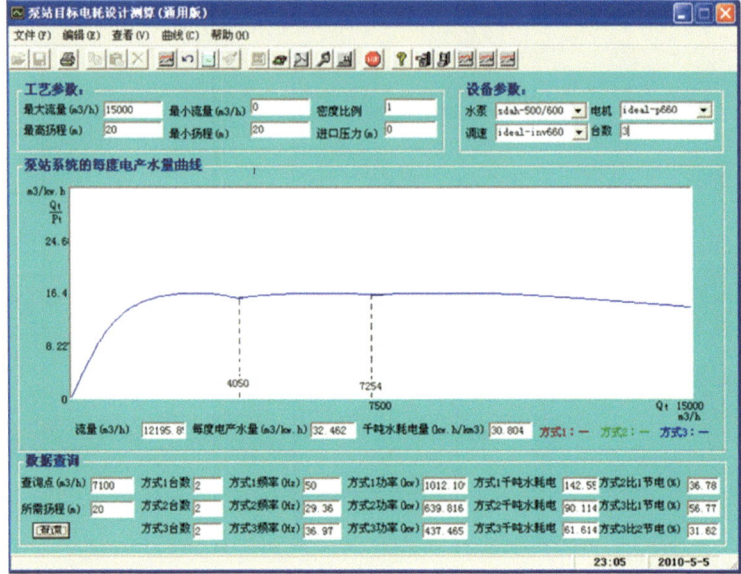

Fig. 23.19 Optimal running curve

$$Q_{23} = 7254(m^3/h) \tag{23.85}$$

When the total flow Q_t is less than Q_{23}, use 2 water pumps and keep $Q_1 = Q_2 = 0.5Q_t$; when the total flow Q_t is greater than Q_{23}, use 3 water pumps and keep $Q_1 = Q_2 = Q_3 = Q_t/3$.

For example, if $Q_t = 7100$ (m3/hour), optimal control is achieved with 2 pumps, and the optimal control method is to keep

$$Q_1 = Q_2 = \frac{Q_t}{2} = 3550(m^3/h) \tag{23.86}$$

The minimum total power consumption is

$$minP_t = \frac{Q_t H_0}{367\eta_1(\frac{Q_t}{2})} = 437.465(kW) \tag{23.87}$$

Similarly, when $Q_t = 11,700$ (m3/hour), 3 pumps are used to achieve optimal control, and the optimal control method is to keep

$$Q_1 = Q_2 = Q_3 = \frac{Q_t}{3} = 3900(m^3/h) \tag{23.88}$$

The minimum total power consumption is

Fig. 23.20 Yao curve

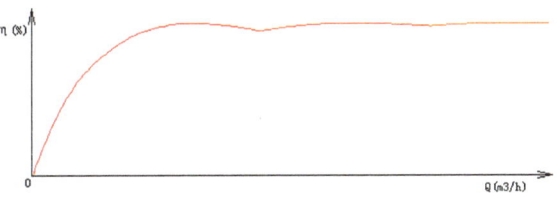

$$minP_t = \frac{Q_t H_0}{367 \eta_1 (\frac{Q_t}{3})} = 738.124(kW) \qquad (23.89)$$

This case was successfully applied in May 1999.

23.13 Judgment Method for Optimal Efficiency Operation of Pumping Stations

Judgment method: For a pumping station composed of the identical type of pump, measure the operating efficiency curve within the full flow range when H = 0.5H$_e$. H$_e$ is the rated head of the pump. If the pumping station operates with optimal energy efficiency, the curve has no jump point and the switching points are between n and n + 1 or n and n-1, as shown in Fig. 23.20; if the pump station operating efficiency curve jumps up and down, it is definitely not optimal.

We call the optimal operating efficiency curve of H = 0.5H$_e$ as the *Yao curve*.

The optimal operating efficiency curve when the operating head of the pumping station satisfies 0 < H < 0.7 h is in line with the above rules and can also be used as a *Yao curve*.

23.14 Stability of Pumps and Fans at Rated Speed

For some pumps and fans with low specific speed, their Q-H curves often appear in hump shape, while for pumps and fans with high specific speed, their Q-H curves often appear in saddle shape, as shown in Figs. 23.21 and 23.22.

When such a pump or fan is connected to the pipe network, there may be more than one operating point, resulting in unstable operation.

Take the hump-shaped Q-H characteristic curve of a pump as an example, as shown in Fig. 23.23.

In Fig. 23.23, curve 1 is the Q-H curve of the pump at the rated speed ne, and curve 2 is the characteristic curve of the water supply network. The head of the pump reaches the highest point H$_m$ at point B in the figure, and the pump flow rate corresponding to the highest head H$_m$ is Q$_m$. For such a pump and pipe network, there are two operating points, A and C. If the system works at point A, due to interference,

Fig. 23.21 Q-H curves with
a hump

Fig. 23.22 Q-H curves with
a saddle

Fig. 23.23 Working status
of pumps and fan with hump
in Q-H curve

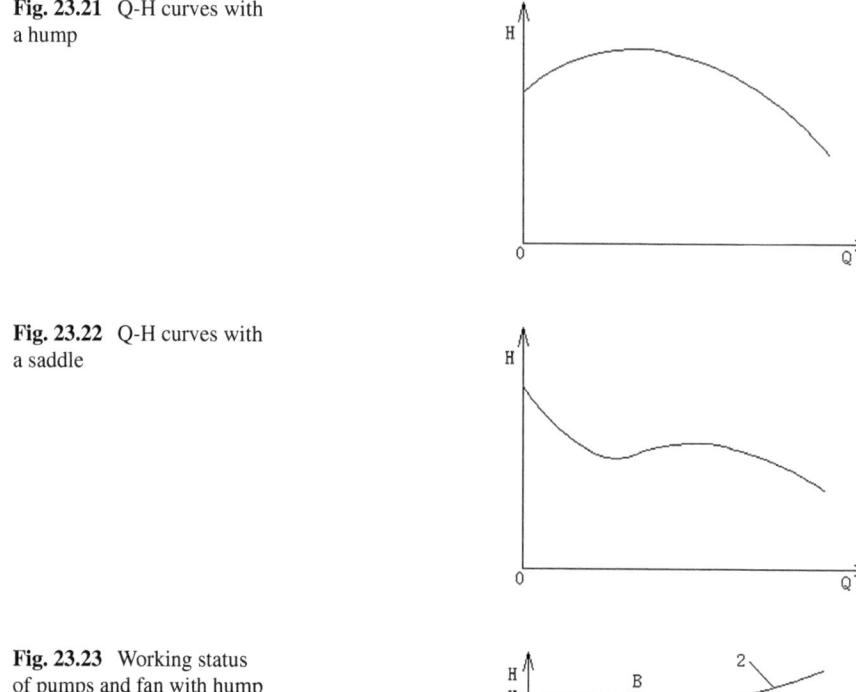

the working point A moves to the left. The head of the pump on the left side of point A is less than the head required by the pipe network, the flow rate decreases, and the working point continues to move to the left until the flow rate of the pump drops to zero. If the disturbance causes the working point A to move to the right, the head of the pump on the right side of A will be higher than the pressure required by the pipe network, the flow rate of the pump will increase. The working point will continue to move to the right until it reaches point C, and the system will be stable down, so the working point A is unstable, and it cannot return to the original working position after being disturbed.

Let's take a look at the working conditions of point C. Suppose that after the disturbance occurs, the working point C moves to the left. The head of the pump on the left side of point C is higher than the head required by the pipe network, so the flow increases, and the working point moves to the right again, return to point C. If the interference causes the working point C to move to the right, the lift of the pump on the right side of point C will be lower than the pressure required by the pipe network. The flow rate of the pump will decrease, the working point will move to the left, and return to point C, the system automatically stabilized, so the working point C is stable.

In fact, corresponding to the pump shown in Fig. 23.23, the working area on the left side of Q_m is an unstable area, and the working area on the right side of Q_m is a stable area. When the system works in an unstable region, the system is very likely to become unstable.

For the pump whose Q-H characteristic curve is a saddle shape, the analysis of its stability is the same as the above process.

Next, we take a type of pump with a container (water storage, oil storage, etc.) at the outlet as an example to analyze the instability of pump operation. Assume that the operating system of the pump is as shown in Fig. 23.24, and the Q-H characteristic curve of the pump is in the shape of a hump.

In Fig. 23.24, the water pump supplies water to the water tank, and the water flowing out of the water tank is supplied to the user. For the water pump, the difference between the liquid level of the water tank and the liquid level of the inlet pool is both the geodesic height Hg of the pipe network and the water outlet pressure of the water pump. It must be greater than H_g, and the pump can send out water, the characteristic equation of the pipe network is:

$$H = H_g + K \times Q^2 \tag{23.90}$$

If the water tank is not far from the water pump, the second part $(K \times Q^2)$ is smaller, and the characteristic curve of the pipe network is relatively flat. As shown in Fig. 23.25, when the water tank is very close to the water pump, the pipe network characteristic curve is approximately a horizontal straight line.

In Fig. 23.25, Curve 1 is the Q-H characteristic curve of the water pump, and Curve 2, Curve 3 and Curve 4 are the characteristic curves of the pipe network at different tank liquid levels.

When the liquid level of the water tank is $H_g = H1$, the characteristic curve of the pipe network is curve 2. At this time, the user's water consumption is Q_d, the user's water consumption is greater than Q_m, and the working point is stable at point D. When the user's water consumption increases, the water tank's liquid level will

Fig. 23.24 A water tank in the pipeline

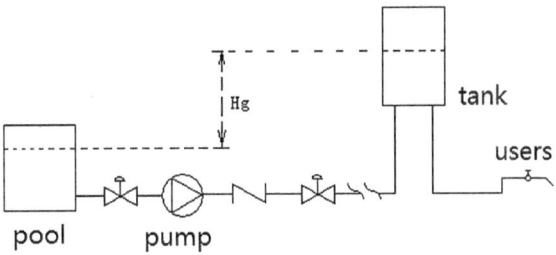

pool   pump

Fig. 23.25 Working status
of a pump with hump in Q-H
curve

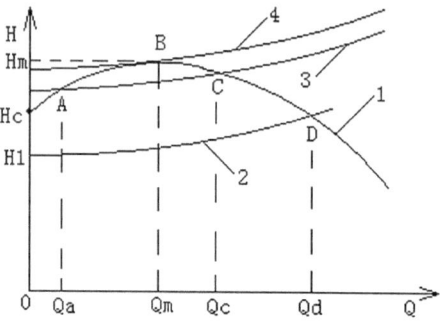

decrease, the characteristic curve of the pipe network will move downward, and the flow rate of the pump will increase. The network can realize self-stabilizing operation.

When the water consumption of the user is Q_m, the liquid level of the water tank reaches the highest point, and the characteristic curve of the pipe network moves up to the highest point, which is curve 4. At this time, if the water consumption of the user decreases, the characteristic of the pipe network moves from point B to the left, and the pump characteristic curve is separated from the characteristic curve of the pipe network. The head of the pump is less than the pressure of the water in the water tank, and the water flows back. Because there is a check valve in front of the pump, the water cannot flow back, and the flow rate of the pump drops to zero. As the user uses water, the liquid level of the water tank gradually decreases. When the liquid level of the water tank drops below the closing head H_c of the pump, the head of the pump is greater than the pressure of the water tank, and the pump starts to discharge water. Since the water consumption of the user is small, the pump's flow rate is larger than the user's water consumption, the liquid level of the water tank rises, and the characteristic curve of the pipe network moves up. The pump will oscillate back and forth in the unstable range from H_c to H_m. The amplitude of the oscillation depends on the size of the user's flow rate. When the distance is small, the oscillation period will be very short.

For such a system, only when the user's water consumption is greater than Q_m, the system can achieve stable water supply, and the liquid level of the water tank will be stable at a certain level.

For the fan system, there is usually no check valve at the outlet of the fan, so when the shock occurs, the gas backflow phenomenon will appear, which is called the surge phenomenon. The division of the unstable area of the fan is the same as the analysis process of the above-mentioned water pump.

The pipe network characteristic equation of the gas path is generally:

$$P = K_1 \times Q^2 \tag{23.91}$$

The characteristic curve of the pipe network is shown in Fig. 23.26.

In Fig. 23.26, the curve is the Q-H characteristic curve of the fan. For the fan system, in most cases, as long as there is wind pressure, there will be flow in the

Fig. 23.26 Q-P
characteristic curve of gas
path

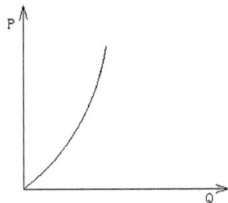

pipeline, so the shape of the pipe network characteristics of the air supply system is basically a quadratic curve passing through the origin $P_g = 0$.

In fact, there are also some fan systems, only when the wind pressure is greater than a certain value, there is flow in the pipeline. For such pipelines, $P_g \neq 0$, the pipe network characteristic curve is not a quadratic parabola passing through the origin, so what kind of pipeline has this characteristic? For example, the pipe network with a water tank filter in the air circuit, and the pipe network with a constant pressure opening valve, that is the case. Take the pipe network with a water tank filter in the air circuit as an example, as shown in Fig. 23.27.

The shape of the characteristic curve of the pipe network is shown in Fig. 23.28. This type of pipe has flow rate only when the pressure is greater than P_g.

Next, we use an example of a type of fan with a large container at the outlet to analyze the unstable operation of the fan. Assume that the fan system is shown in Fig. 23.29, and the Q-H characteristic curve of the fan in the figure is a hump shape.

Fig. 23.27 A water tank in
the fan pipeline

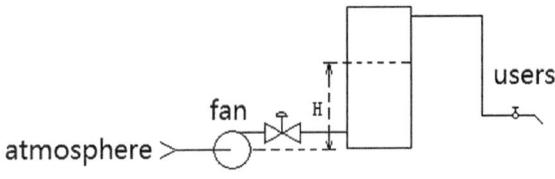

Fig. 23.28 The
characteristic curve of the
pipe network with water tank

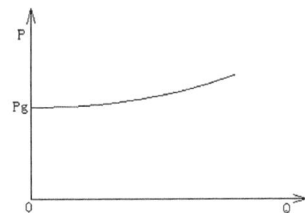

Fig. 23.29 A large container
or a thick pipe network in the
fan pipeline

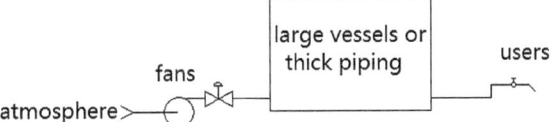

Fig. 23.30 Working status
of fan with hump in Q-P
curve

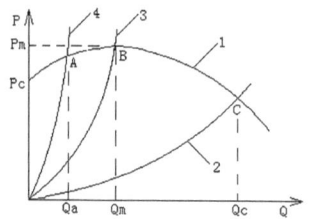

In Fig. 23.29, the fan supplies air to the user through a large container (thicker pipeline). Because the large container has the function of storing gas, the pressure of the pipe network is greater than the outlet pressure of the fan when the fan is powered off for a short time. If the fan outlet valve is not closed, air from the large container will flow back to the fan.

The operation analysis of the fan's flow-pressure characteristic curve and the pipe network characteristic curve is shown in Fig. 23.30.

In Fig. 23.30, curve 1 is the Q-P characteristic curve of the fan. When the gas consumption of the user is Q_c, the characteristic curve of the pipe network is curve 2. Since the gas consumption of the user is greater than Q_m, the working point is stable at point C. When the air volume increases, the pressure of the pipe network decreases and the flow rate of the fan increases. When the air consumption of the user decreases, the pressure of the pipe network increases and the flow rate of the fan decreases. The fan and pipe network can realize self-stabilizing operation.

When the gas consumption of the user is Q_m, the output pressure of the fan reaches the highest. At this time, if the gas consumption of the user decreases and the pressure of the pipe network increases again, the output pressure of the fan will be lower than that in the pipe network (also a large container). The large container supplies gas to the user, and the gas in the large container also flows back to the fan. The fan has a negative flow rate, and the operation curve enters the left side of the P coordinate axis. With the gas consumption, the pressure in the large container drops rapidly. When the pressure in the large container drops below the closing head P_c of the fan, the pressure of the fan is greater than the pressure in the large container, and the fan starts to supply air to the outside. Due to the small air consumption of the user, the flow rate of the fan is larger than the user's air consumption, and the pressure in the large container rises, the fan will oscillate back and forth in the unstable range from P_c to P_m, and the oscillation range depends on the size of the user's flow. When the distance between P_c and P_m is small, the oscillation cycle will be accelerated. For such a system, only when the user's gas consumption is greater than Q_m, the system can achieve stable gas supply.

The stability analysis of the fan and pipe network curve of the saddle-shaped Q-P is shown in Fig. 23.31.

In Fig. 23.31, curve 1 is the Q-P characteristic curve of the fan. When the gas consumption of the pipe network is at point Q_a, the pipe network curve is curve 2, and the operating point is A. The system can run stably. The analysis process is as before. When the pipe network When the gas consumption of the network is at

Fig. 23.31 Pipeline
behavior with saddle Q-P
curve

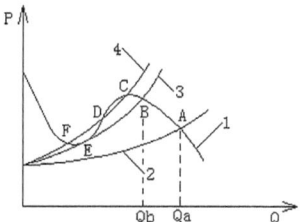

Q_b, the working point is B, and the pipe network curve is curve 3. There will be two intersection points B and E between the pipe network curve and the Q-P characteristic curve. When the gas consumption of the pipe network is less than Q_b such as curve as shown in 4, the pipe network curve of the system and the Q-P characteristic curve of the fan will have three intersection points C, D and F. When the gas flow rate of the pipe network is less than Q_b, the system will easily enter an unstable operation state, and the analysis process is omitted.

When the pipe network curve intersects with the Q-P characteristic curve of the fan, as shown in Fig. 23.32, the system is likely to enter an unstable operation state, and the analysis process is omitted.

In Fig. 23.32, curve 1 is the Q-P characteristic curve of the fan, and curve 2 is the pipe network curve.

When the pump or fan does not adjust the speed, the stability judgment and stability guarantee measures are summarized as follows:

(1) When the flow head Q-H characteristic curve of the pump or the flow pressure Q-P characteristic curve of the fan are in the shape of a hump or a saddle, such a pump or fan may have unstable operation. It is the most fundamental factor affecting the stable operation. If possible, choose a pump or fan with a flow head (pressure) characteristic curve without hump and saddle shape.

(2) If it is not possible to choose a pump or fan without hump and saddle shape on the flow head (pressure) characteristic curve, select the operating point to make the system run in the stable area on the right side of the highest head (highest pressure) point, and avoid working at the highest head (highest pressure) point to the unstable area on the left. We can change the characteristic curve of the pipe network, change the vane angle of the inlet deflector to change the flow head (pressure) characteristic curve of the pump or fan, or use a bypass valve or other measures to increase the flow rate.

Fig. 23.32 Q-P curve with
saddle and pipe network
curve

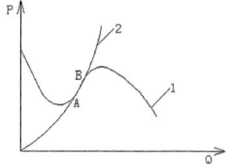

23.15   Stability Criteria After Speed Adjustment of Pumps and Fans

For pumps or fans with hump and saddle-shaped characteristic curves of flow head (pressure), the problem that one frequency corresponds to two flow points will also appear after the implementation of frequency conversion speed regulation, so if no targeted measures are taken, there will also be problems Control instability, as described in the automatic control theory in a large number of pages, stability is the primary problem of the control system.

For the pump or fan with hump and saddle shape on the flow-head (pressure) characteristic curve, the system is set to automatically control according to the constant pressure H_s, as shown in Fig. 23.33.

In Fig. 23.33, curve 1 is the pipe network characteristics required for constant pressure, and the required constant pressure value is Hs. Curve 2 is the connecting curve of the highest point of the Q-H characteristic curve after speed regulation. Curve 2 intersects the constant pressure line 1 at Q_m, and curves 3, 4, 5, and 6 are the Q-H characteristic curves of the pump or fan at four speeds of n1, n2, n3, and n4 respectively.

It can be seen from Fig. 23.33 that when the water consumption Q of the pipe network is greater than Q_m, the larger the flow rate of the pipe network (Q3 to Q4), the higher the speed of the pump or fan (n3 to n2). From the principle of automatic control, this is a normal feedback control system, and general controllers or regulators can control this system very well. When the flow rate Q of the pipe network is less than Q_m, the flow rate of the pipe network is smaller (Q2 to Q1), and the higher the required speed (n3 to n2). This adjustment method is opposite to the above normal situation, the automatic control system has the problem of control law uncertainty. Q_m is the critical unstable point of the system, and the right side of Q_m is the stable operation area, the left side of Q_m is the unstable operation area.

Assume that the system automatically controls pressure according to the characteristics of the pipe network under different flow rates, as shown in Fig. 23.34.

Fig. 23.33 Stability of constant pressure speed regulation

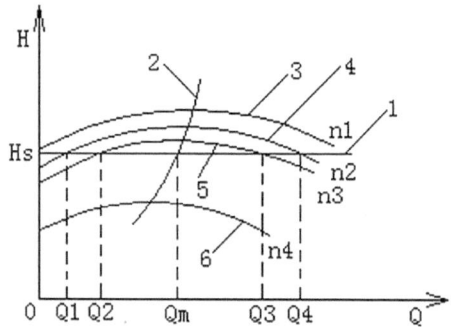

Fig. 23.34 Stability after
speed change

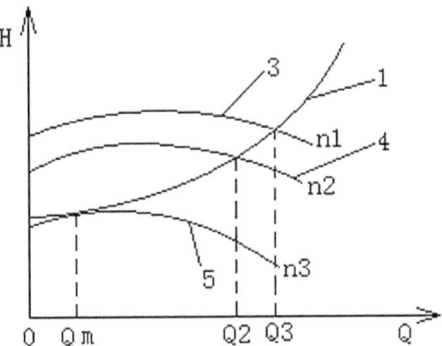

In Fig. 23.34, curve 1 is the characteristic curve of the pipe network, and curves 3, 4, and 5 are the Q-H characteristic curves of the pump or fan at three speeds of n1, n2, and n3.

It can be seen from Fig. 23.34 that the characteristics of the pipe network are tangent to curve 5 at Q_m. When the water consumption Q of the pipe network is greater than Q_m, the greater the flow rate of the pipe network (Q2 to Q3), the higher the speed of the pump or fan (n2 to n1). From the principle of automatic control, this is a normal feedback control system, and general controllers or regulators can control this system very well. When the flow rate Q of the pipe network is less than Q_m, the pipe network has smaller flow rate, the higher the required speed. This adjustment method is contrary to the above normal situation. The automatic control system has control uncertainty problems. Q_m is the critical unstable point of the system, and the right side of Q_m is the stable operation area. The left side of Q_m is the unstable operation area.

The occurrence of these unstable phenomena is mainly due to the fact that the flow-head (pressure) characteristic curve of the pump or fan is not like the normal flow-head (pressure) characteristic curve, and a hump appears.

Although the flow head (pressure) characteristic curve of some water pump fans has no hump, its top changes gently in a wide range. In such a system, a small change in the speed in the low flow area will cause a large change in flow, which will also give Control stability brings problems. At this time, it is necessary to improve the accuracy of control output, reduce the range of output change, and not use sensitive speed signals as control targets. Use the flow signal to limit the range of frequency changes.

Stability criteria (Yao's stability criteria): Assuming that Q_m and P_m are the flow rate and head corresponding to the highest hump point of the pump or fan flow head curve, and H_{sv} is the full head (or full pressure) set by the process, and the pump or fan is centrifugal, the necessary condition for the stable operation of the pump or fan after speed regulation is

$$Q > Q_m \times \sqrt{\frac{H_{SV}}{H_m}} \qquad (23.92)$$

Note: This oscillation is not an oscillation caused by the mechanical resonance frequency of the pump (for example, a long-axis water pump), but a controlled oscillation caused by the shape of the pump characteristic curve.

23.16   Overload Problems of Pumps and Fans

For the pump or fan with low specific speed, the flow rate-shaft power Q-N characteristic curve, the greater the flow rate, the greater the shaft power. If the outlet pressure is very low or the outlet is damaged, the flow rate will be too large, and the motor driving the pump or fan is prone to overload and heat generation, exceeding the rated power of the matched motor, the motor is overheated, the alarm trips, and it burns out, as shown in Fig. 23.35.

In Fig. 23.35, when the flow rate is greater than Q_m, the shaft power of the pump or fan is greater than N_m, and the output power of the motor driving the pump or fan is close to the rated power. If the flow rate is larger, the motor will be overloaded. To avoid overload, we need to use current transmitter monitors the running current of the motor, and when it is found that the running current is too large, the valve is turned off a little. To ensure that the flow rate of the pump or fan will not be too large.

For the pump or fan with high specific speed, the flow rate-shaft power Q-N characteristic curve, the smaller the flow rate, the greater the shaft power. So for this kind of pump or fan, if a valve is used to control the outlet flow rate, it must be noted that if the valve opening is too small, the flow rate is too small or zero, which will cause the shaft power of the pump or fan to rise too fast, exceeding the rated power of the motor, and the motor will overheat, and the alarm will trip, as shown in Fig. 23.36.

In Fig. 23.36, when the flow rate is less than Q_m, the shaft power of the pump or fan is greater than N_m, and the output power of the motor driving the pump or

Fig. 23.35 Q-N flow rate-shaft power characteristic curve (1)

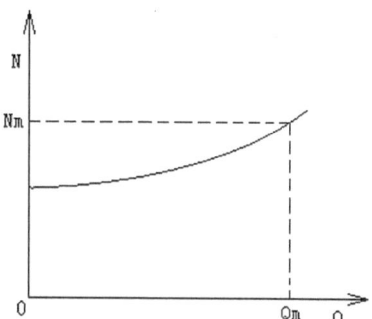

Fig. 23.36 Q-N flow
rate-shaft power
characteristic curve (2)

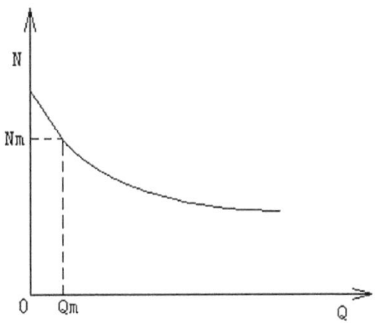

fan is close to the rated power. If the flow rate decreases further, the motor will be overloaded. In order to avoid the overload problem, measures need to be taken to ensure that the flow rate control valve of the pump or fan is not closed or closed too small, and a current transmitter is used to monitor the operating current of the motor. When the operating current is found to be too large, the valve is opened a little.

23.17   Mechanical Resonance Problem of Pumps and Fans

For oscillations caused by the mechanical resonance frequency of pumps and fans (such as long-axis pumps), dangerous high and low speed limit zones should be set to avoid operating in this range.

For example, for the ACS510 inverter, set the codes 2501 (CRIT SPEED SEL), 2502 (CRIT SPEED LO), and 2503 (CRIT SPEED HI).

23.18   Other Issues that Need Attention

Most water supply systems consist of multiple water pumps connected in parallel. A small amount of water dripping from the pump shaft can ensure heat dissipation at the shaft end. The pressure at the center of a centrifugal pump is low, and air will be sucked into the pump through the shaft end. Air accumulation in the pump will cause the pump to be unable to supply water, or the flow rate of the pump will be reduced. Gradually reduce until no water flows out. This can easily happen when the water level in the pumping station's inlet pool is lower than the height of the pump shaft. This should attract enough attention. At this point, beginners don't need to panic. Note the minimum fluid level in the inlet tank.

23.19   Conclusion

There is no need to establish an accurate mathematical model of a pumping station, based on the characteristics of the energy efficiency function, this chapter presents a constrained, nonlinear, integer-real-number hybrid energy efficiency optimization method for pumping stations.

This optimization method includes two theorems: optimal load distribution theorem and optimal switching theorem.

Optimal load distribution theorem: The optimal load distribution method of a pumping station is to keep the operating energy efficiency of each operating pump equal, i.e., ***Yao Theorem 1***.

$$\eta_N(\gamma_1) = \eta_N(\gamma_2) = \cdots = \eta_N(\gamma_n) \tag{23.93}$$

Optimal switching theorem: The optimal switching point for the number of operating pumps is at the point of equal efficiency or at the maximum output point of the pumps, i.e., ***Yao Theorem 2***.

$$\eta_N\left(\frac{Q_t}{\sum_{i=1}^n Q_{ie}}\right) = \eta_N\left(\frac{Q_t}{\sum_{i=1}^k Q_{ie}}\right) or \frac{Q_t}{\sum_{i=1}^n Q_{ie}} = \gamma_{1m} \tag{23.94}$$

We call above optimization method as the ***Quantum Optimization Method of Pumping Stations***.

We call above theory as the ***Energy Efficiency Predictive Theory of Pumping Stations***.

Acknowledgements In order to solve the energy efficiency optimization problem of multi-unit system, the author has conducted long-term research. Thanks to Yanfang Zhang from Hebei Automation Company and Bosheng Yao from Beijing IAO Technology Development Company for their valuable help and advice during the development and experiment of this theory.

Chapter 24
Other Commonly Used Energy-Saving Operation Methods

24.1 Reactive Power Compensation of Motors

24.1.1 Reactive Current and Reactive Power

Some of the electrical equipment used in the industry are resistive loads, such as resistance heating furnaces, reactor heating rods, chemical reaction vessel heating jackets, etc., while more electrical equipment is electric motors, transformers, etc. For resistive electrical loads, the phase of the current in the line is the same as that of the working voltage. For inductive electrical loads, due to the limitation of power in the circuit, the current on the inductance cannot change abruptly, so the phase of the current in the inductive line lags behind working voltage 90°.

Taking a three-phase motor as an example, assuming that the motor is in a Y connection, the phase current is I, the line current I_i is equal to the phase current I, the phase voltage is U, the line voltage U_i is equal to U, and the phase current I lags φ behind the phase angle of the phase voltage U, we decompose the phase current I in the line into two parts I_1 and I_2, as shown in Fig. 24.1.

In Fig. 24.1, the current I_1 is in the same direction as the phase voltage U, I_1 is the active current, I_1 and U generate single-phase active power P_1, and the active power P_1 is converted into mechanical energy and heat energy, as shown in Eq. (24.1).

$$P_1 = U \times I_1 = U \times I \times COS\varphi \qquad (24.1)$$

In Fig. 24.2, the current I_2 is perpendicular to the phase voltage U, and I_2 is a reactive current, which is caused by the coil inductance L. I_2 and U generate single-phase reactive power P_2. The P_2 is used for the connection between the electric field and the magnetic field although it does not do work, this energy conversion is necessary for the operation of the motor, so the reactive current is also necessary in the motor. The reactive power P_2 is shown as follows

© The Author(s) 2024
F. Yao and Y. Yao, *Efficient Energy-Saving Control and Optimization for Multi-Unit Systems*, https://doi.org/10.1007/978-981-97-4492-3_24

Fig. 24.1 Decomposition of current

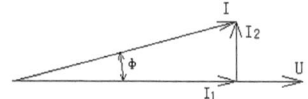

$$P_2 = U \times I_2 = U \times I \times \sin \varphi \tag{24.2}$$

When the motor is in no-load operation, the active current component of the motor is very small, and the current of the motor is basically the reactive excitation current. At this time, $\cos_\varphi$ tends to 0, and $\sin_\varphi$ tends to 1.

Seen from the power supply side, the total capacity (also called apparent power) transmitted by the line is S, as shown in Eq. (24.3).

$$S = \sqrt{3} \times U_i \times I_i \tag{24.3}$$

In Eq. (24.3), Ui is the line voltage of the three-phase motor, and I_i is the line current of the three-phase motor.

The three-phase total active power of the motor is P, which is also the power consumption of the electric motor we often say, such as Eq. (24.4).

$$P = 3 \times U \times I_1 = 3 \times U \times I \times COS\varphi = \sqrt{3} \times U_i \times I_i \times COS\varphi = S \times COS\varphi \tag{24.4}$$

The three-phase total reactive power of the motor is Q, such as Eq. (24.5).

$$Q = 3 \times U \times I_2 = 3 \times U \times I \times \sin \varphi = \sqrt{3} \times U_i \times I_i \times \sin \varphi = S \times \sin \varphi \tag{24.5}$$

The reactive current I_2 is necessary to establish a rotating magnetic field in the motor, but the reactive current in the transmission line will generate heat loss $I_2{}^2 \times R$, and these losses will be wasted. The reactive current flowing through the transformer will also occupy the capacity of the power supply transformer, and this part of the capacity does not do work, which reduces the power supply capacity of the transformer.

24.1.2 Compensation of Reactive Current and Reactive Power

Generally, the reactive current needs to be compensated in the line. Most of the compensation measures are carried out by connecting capacitors in parallel, because the current on the capacitor is ahead of the voltage by 90° and can be offset by the current on the inductor lagging behind the voltage by 90°. The relationship of current after connecting capacitors in parallel is shown in Fig. 24.2.

Fig. 24.2 Compensation of
reactive current

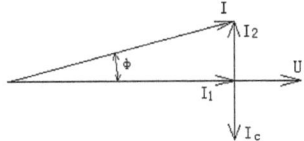

Fig. 24.3 Current before
and after compensation

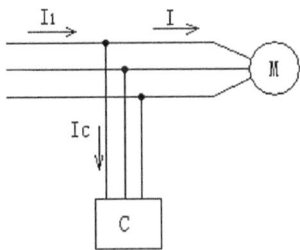

In Fig. 24.2, if the current I_c of the compensation capacitor is equal to the amplitude of I_2, but the phase is opposite, $I_c = -I_2$, then the current in the power supply line after compensation changes from the original I to I_1, and Fig. 24.3 supplies power to the motor current representation before and after line compensation.

In Fig. 24.3, before compensation, the current in the motor power supply line is I, and after compensation, the current in the line is reduced to I_1, and the heat loss is reduced from $I^2 \times R$ to $I_2{}^2 \times R$. In actual compensation measures, 100% compensation is generally not performed, that is, the power factor $\cos_\varphi$ after compensation is not required to be 1, and of course overcompensation cannot occur.

If the power factor before compensation is $\cos\varphi_1$, and the power factor required after compensation is $\cos\varphi_2$, assuming that the active power of the load is P_a (kW), the capacity Q_a (kvar) to be compensated is shown in Eq. (24.6).

$$Q_a = \sqrt{3} \times U_i \times I_i \times \sin\varphi_1 - \sqrt{3} \times U_i \times I_i \times \sin\varphi_2 = P_a \times (\tan\varphi_1 - \tan\varphi_2)$$

$$= P_a \times \left(\sqrt{\frac{1}{\cos^2 \varphi_1} - 1} - \sqrt{\frac{1}{\cos^2 \varphi_2} - 1} \right) \tag{24.6}$$

When performing compensation according to Eq. (24.6), it is necessary to measure the actual active power Pa of the electric load and the power factor $\cos\varphi1$ before compensation, and then calculate the compensation capacity Q_a.

Under rated operating conditions, the relationship between the actual active power Pa, the rated output power Pe of the electrical load, and the rated efficiency ηe is shown in Eq. (24.7).

$$P_a = \frac{P_e}{\eta_e} \tag{24.7}$$

24.1.3 *Reactive Power Compensation of Motors*

If the distance between the motor and the power distribution room is relatively close, we can adopt the method of centralized compensation in the power distribution room. For the situation far away from the power distribution room, the reactive current of the motor will cause loss in the line, and this part of the loss cannot be distributed in the power distribution room. It is compensated by reactive power compensation in the room, so it is necessary to perform local compensation next to the motor.

According to the requirements of reactive power in-situ compensation for small and medium-sized motors, it is generally required to compensate 90% of the no-load current of the motor. Several commonly used calculation methods are introduced below.

(1) Compensate according to Eq. (24.6), and the power factor $\cos\varphi2$ after compensation is generally required to be 0.95–0.97.
(2) In order to avoid overcompensation, it is generally compensated according to 90% of the no-load current, and the compensation capacity Q_a (kvar) is calculated according to Eq. (24.8).

$$Q_a = 0.9 \times \sqrt{3} \times U_i \times I_0 \times 0.001 \qquad (24.8)$$

In the Eq. (24.24–24.8), Ui is the rated voltage of the motor, and I0 is the no-load current of the motor, which can be found out from the samples of the motor.

Compensation capacity Qa calculated according to Eq. (24.7), and then calculated power factor $\cos\varphi2$ after compensation according to Eq. (24.6) and Eq. (24.7), pay attention Compensation has occurred.

(3) Calculate the compensation capacity according to the data on the motor nameplate, such as Eq. (24.9).

$$Q_a = (2.12.7) \times \left(1 - \cos^2\varphi_e\right) \times I_e \times U_e \times 0.001 \qquad (24.9)$$

In the Eq. (24.8–24.9), $\cos\varphi e$ is the rated power factor, I_e is the rated current of the motor and U_e is the rated voltage of the motor, the coefficient in front of the 2-pole motor is 2.1. The coefficient in front of the 4-pole motor is 2.4, and the coefficient in front of the 6-pole motor is 2.55, and the coefficient in front of the 8-pole and 10-pole motors is 2.7.

Issues to be aware of:

(1) We cannot simply calculate the power by multiplying the ratio of the actual current to the rated current of the motor by the rated power. Unless the actual current of the motor is near the rated current. The no-load current of the motor is not too small.
(2) The compensation methods mentioned above are all calculated for fixed-speed motors. If a frequency converter is used for speed regulation, compensation is not required.

(3) Capacitor switching of the capacitor compensation device on the power supply side of the inverter may lead to a sudden drop in voltage, sometimes resulting in a malfunction of the inverter and a protective shutdown.

(4) The compensation capacity calculated by Eq. (24.6) is suitable for other power lines and loads.

24.2 Reduce Voltage and Save Electricity When the Motor is Lightly Loaded

(1) For some fixed-speed motors with large load changes, they often work under no-load, light-load or even power generation conditions, such as various industrial conveyors, port belt conveyors, coal mine belt conveyors, cement belt conveyors, steel rolling equipment, machine tools, grinding machines, punching machines, polishing machines, cutting machines, compressors, oil well pumping units, forging machines, presses, concrete molding machines, rubber molding machines, etc.

If the load is not allowed to reduce the speed when it is lightly loaded, you can consider reducing the voltage of the motor to save power, such as some belt conveyors. In the state of power generation, it may be necessary to take measures to feedback electric energy to the grid. If the equipment is allowed to change the speed, such as the dust removal fan in a steel mill, you can consider speed regulation and power saving measures.

Because factors such as voltage fluctuations and equipment overload must be considered when selecting a motor, its power margin usually has a certain ratio. A certain amount of electric energy is wasted. Therefore, when the motors of these devices are under light load or no load, appropriately reducing their operating voltage can reduce their loss and improve the operating efficiency of the motor.

Reducing the voltage of a light-load or no-load constant-speed motor can reduce the excitation current and iron loss of the motor, thereby producing a power-saving effect. The way to reduce the voltage can be through transformer step-down, thyristor step-down or series partial voltage reactance implemented in the form of a device. The power supply voltage of the motor should not be reduced too much, otherwise the torque of the motor will drop too much. Under the condition of constant load, it may cause the slip rate to rise and the rotor current to rise. The total consumption of the motor will increase instead if the rate at which the copper loss rises is increased.

Ways to reduce voltage by using inductance coils and transformers:

(1) Three inductance coils are connected in series at the three input terminals of the three-phase motor (similar to the governor of a household electric fan). There are two types of adjustable and non-adjustable inductance of the inductance coil for voltage reduction.

(2) The stepwise reduction of voltage can be realized by changing the position of the tap connected to the transformer coil, or by stepless adjustment of the self-coupling voltage regulating transformer output tap position to continuously and smoothly adjust the output voltage. The method of using series inductors and transformers to divide voltage has no harmonic pollution, but the disadvantage is that it is bulky and the adjustment speed is slow by switching contacts.

The method of using the thyristor to step down the voltage is to change the conduction angle of the bidirectional thyristor on the three-phase power line, so as to achieve the purpose of reducing the voltage. The power saver in this way can also easily realize the soft start and Soft stop function. The disadvantage of this method is that there is harmonic interference to the power grid, and the advantage is that the adjustment speed is fast and the volume is small.

Figure 24.4 shows two implementations of voltage reduction. U, V, and W are connected to the input voltage of the power line. By detecting the magnitude of the motor current and the phase angle between the current and voltage, the load rate of the motor can be judged. When the electric motor is in light-load operation, for the series inductance mode, open the contact of KM1, and connect the inductance L1, L2, L3 to the power supply circuit of the motor in series. Due to the voltage drop of the inductance, the working voltage on the motor is reduced. For thyristor step-down mode, open the contact of KM1, control the conduction angle of bidirectional thyristor SCR1, SCR2, SCR3, and change the average effective voltage value on the motor.

What needs to be reminded is that if the variable load equipment that does not require the motor to run at a constant speed, such as hydraulic presses, injection molding machines, dust removal fans in steel mills, etc., when maintaining, cooling or venting, it is to ensure the oil pressure and not stop the machine. This power-saving method is far less effective than the method of frequency conversion speed regulation and speed reduction.

2. When the $\triangle$-connected motor running at a constant speed is in a light-load working state, the slip of the motor is less than the rated value, and the rotor current is not large. At this time, the line voltage of the winding can be reduced to 1/2 of

Fig. 24.4 Two ways of step-down regulation

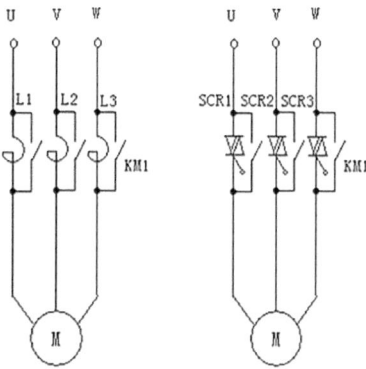

the original by changing the Y connection, so that the iron loss and the no-load current decrease. Because the torque of the motor is proportional to the square of the voltage, the torque decreases to 1/3 of the original, the slip increases slightly, and the rotor current increases slightly, but the total motor losses are reduced. However, the definition of light load means that when the motor with $\triangle$ connection is changed to Y connection, the total loss of the motor must be reduced. The load rate when the losses in the two ways are equal is the critical load rate $\beta \triangle Y$, and the critical load rate of the $\triangle$ connected motor can be calculated according to the relevant parameters of the $\triangle$ connected motor are obtained as follows.

$$\beta_{\triangle Y} = \sqrt{\frac{\frac{2}{3} \times P_{Fe} + 0.75 \times P_{0Cu}}{2 \times \left[\left(\frac{1}{\eta_e} - 1\right) \times P_e - P_0\right]}} \qquad (24.10)$$

In Eq. (24.10), P_{Fe} is the iron loss of the motor (kW), P_{0Cu} is the no-load copper loss of the motor (kW), P_e is the rated power of the motor (kW), P_0 is the no-load loss of the motor (kW), η is the rated efficiency of the motor (kW).

When the actual load rate of the $\triangle$connection motor is greater than $\beta \triangle Y$, the motor remains in $\triangle$connection, and when the actual load rate of the $\triangle$connection motor is less than $\beta \triangle Y$, the motor switches to Y connection.

If there is no variable load equipment that requires the motor to run at a constant speed, the Y-$\triangle$ conversion method may not be as effective as the frequency conversion speed regulation and speed reduction method to save electricity.

3. For light-load motors with variable frequency operation, at the same frequency, the iron loss of the motor can be reduced by reducing the output voltage of the frequency converter. Because the product of voltage and current has decreased to realize the energy-saving operation of the motor, many inverters provide energy-saving operation mode options. When the load is likely to be in light-load operation, the "energy-saving" operation mode should be selected.

24.3 Power-Saving Control of Equipment Such as Hydraulic Presses, Injection Molding Machines, and Dust Removal Fans

In the industry, there are also many loads that are under load changes. However, due to the characteristics of the equipment itself, when the equipment does not need to provide liquid flow, the power consumption of the equipment increases instead, such as injection molding machines, hydraulic machines and other loads. The hydraulic oil pump will have a large load change under different working conditions. Since most of the oil pumps that transport hydraulic oil are positive displacement oil pumps, the characteristics of this oil pump are that the oil output is proportional to the speed, and the oil output is small, and the pressure is small. Therefore, in the standby and pressure maintenance sections of this type of oil pump, although the oil consumption decreases, the output pressure rises and the power consumption increases. The oil exceeding the safety pressure returns through the relief valve, resulting in a lot of waste of electric energy. Since the load allows the oil pump to reduce the speed of the oil pump while maintaining the oil pressure during the pressure maintaining stage, the frequency conversion transformation can have a very good power saving effect.

Due to the impact of the smelting cycle, the dust removal fan in the steel plant needs a large air volume during the oxygen blowing and steelmaking stages (about half the time) for the production of a furnace of steel. After the converter taps steel (about half the time), the required air volume is immediately reduced, and the fan is often in this alternate working mode. The fan adjusts the air volume by the damper at the light load stage, which wastes a lot of electric energy. This type of load can be adjusted in speed. In the blowing stage, the fan works at a higher speed level. When the load is light, the motor speed is reduced to a lower level, which can produce a significant power saving effect. In particular, the greater the proportion of time after tapping in a cycle, the greater the power saving ratio.

Taking injection molding machines as an example, most injection molding machines use hydraulic transmission and electro-hydraulic proportional control technology. The hydraulic system of the equipment is composed of positive displacement hydraulic oil pumps (vane pumps, plunger pumps, piston pumps, etc.) and related oil circuits and accessories. The flow rate of the hydraulic oil pump is directly proportional to the rotational speed. Under the power supply condition of power frequency 50Hz, the motor of the oil pump operates at the rated rotational speed. The power consumption of the hydraulic system accounts for about 80% of the total power consumption of the injection molding machine. The injection molding machine is in different sections such as mold clamping, mold locking, injection, pressure holding and cooling, mold opening and ejection, etc. The oil pressure and oil volume required by the injection molding machine vary greatly, especially during the cooling and holding time, the flow rate is almost zero. The oil pump motor runs at rated speed, the oil pressure rises, and the high-pressure oil overflows through the relief valve, causing a large amount of waste of electrical energy.

Use PLC and other control equipment to detect the current working status of the injection molding machine. When the injection molding machine is in different states such as mold clamping, injection, pressure holding, sol, cooling, mold opening, thimble, and standby, use a frequency converter to adjust the speed of the oil pump. Make the oil supply of the oil pump match the oil demand of the injection molding machine, so as to avoid the phenomenon of high-pressure overflow and reduce the power consumption of the oil pump. Figure 24.5 is the schematic diagram of the power saving control system of the injection molding machine.

In Fig. 24.5, PLC is used to detect the running state of the injection molding machine. According to the different sections of the injection molding machine, such as mold clamping, mold locking, injection, pressure holding and cooling, and mold opening, the analog output of PLC controls the output frequency of the inverter. Use the frequency converter to adjust the speed of the oil pump. When the injection molding machine is at a low oil consumption, reduce the speed of the oil pump to

Fig. 24.5 Energy saving of injection molding machine oil pump

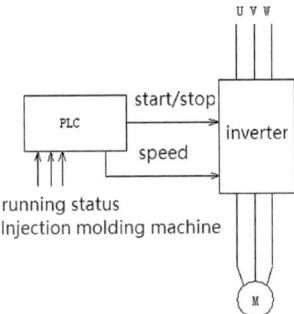

Fig. 24.6 Two ways to reduce voltage

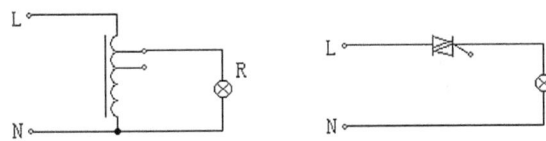

avoid the problems of high oil pressure and too much energy consumption, so as to achieve energy-saving operation.

24.4 Lighting Step-Down and Power Saving

Lighting electricity consumption accounts for about 10% of the country's total electricity consumption. It can be said that since human beings entered the era of electrification, lighting has been everywhere. How to achieve energy-saving lighting is of great significance.

Energy saving in the general sense refers to the comparison on the basis of completing the same work. According to the same point of view, energy saving work in the field of lighting needs to maintain the illuminance of lighting fixtures to save energy. In fact, due to human eyes, the response to the light intensity is related to the surrounding environment and the length of time in the dark, so it may not be the best to mechanically follow the method of maintaining the constant illuminance.

According to a large number of experimental statistical results, the acuity of human eyes to illumination is not linear, but logarithmic. According to the logarithmic theory of vision and light intensity, if the light intensity decreases by 10%, vision decreases by 1%. When it is 10%, the illuminance of commonly used electric light sources, such as fluorescent lamps, is only reduced by about 7%. The slight change in illuminance caused by a small reduction in light input power can hardly be felt by the human eye, but it is beneficial to prolonging the life of lamps, reducing maintenance costs and realizing energy-saving operation. But there is a positive meaning.

In the lighting power supply system, in order to avoid the terminal voltage being too low due to line loss and peak power consumption, it is often transmitted at a higher voltage. The design of street lamps also takes more consideration of the voltage and brightness of the lamps at the end of the line. After midnight, there are few pedestrians on the road, and because of the low power consumption, the grid voltage rises, resulting in an inversion of illuminance and demand, which not only shortens the life of the lamps, but also greatly increases power consumption.

In order to reduce this waste, we can use autotransformer step-down, series variable inductance coil step-down, thyristor chopper step-down or inverter step-down to reduce the supply voltage of street lamps to achieve energy saving. The magnitude of the voltage reduction depends on the lower limit of illumination allowed by the lighting and the fluctuation range of the circuit voltage. According to the different night time periods and the voltage level, the voltage of the lighting fixtures is properly reduced without affecting the lighting requirements of passers-by to achieve Power-saving operation. Figure 24.6 shows two ways to step down voltage by using transformer taps and stepping down by using thyristor.

For the thyristor step-down method and the method of using the inverter to step down, harmonics will be generated, which will pollute the power grid. The advantages are small size, simple control, and the method of stepping down by using transformer taps or variable inductance coils has no harmonics.

For some street lamps, a higher working voltage is only required when starting up, and the power supply voltage can be appropriately reduced after being ignited, so as to maintain its normal illuminance, effectively save electricity, and prolong the life of the lamp.

Taking AC220V powered lighting fixtures as an example, we can control them in this way. At the beginning, we can soft start with 90% of the voltage, and then slowly increase to 100% of the rated lighting voltage to preheat the lamps, which can also reduce the voltage to the lamps. Then reduce the lamp voltage to 95% for normal power supply. According to the flow of pedestrians on the local roads, determine a midnight time (such as 23:00). After this time, reduce the voltage to 90% of the rated voltage, and it will dawn in the early morning for a period of time before (such as the increase of pedestrians on the road starting at 5 o'clock), the voltage will be raised to 95% for power supply until it finally goes out. If there is a period when the grid voltage is too high in the middle process, we can increase the step-down range, and the power saving ratio will be greater. Of course, this is just an example. In practice, readers can implement other better power-saving control strategies according to the specific conditions of their own enterprises.

Energy-saving methods such as turning off lights when people go out are beyond the scope of this book.

24.5   Waste Heat Recovery

In steel, chemical, cement, textile, glass, ceramics, boilers, power generation and other industrial fields, a large amount of waste heat is wasted, including high-temperature product and slag waste heat, high-temperature waste gas waste heat, chemical reaction waste heat, waste steam and wastewater waste heat, cooling waste heat from media, waste heat from combustible waste gas, liquid waste, etc. The total waste heat resources of various industries account for about 17 to 67% of their total fuel consumption, and about 60% of the total waste heat resources can be recycled.

In the metallurgical industry, the waste heat that can be used includes: the waste heat in the billet heating furnace, the waste heat in the rolling steel continuous heating and soaking furnace. The waste heat in the wire rod annealing furnace, the waste heat in the sintering machine.

In the chemical industry, it can be used the waste heat used includes: the waste heat of synthetic ammonia blowing gas combustion, the waste heat of flue gas from the first (second stage) furnace of synthetic ammonia, and the waste heat of upstream and downstream gas of synthetic ammonia.

In the building materials industry, the waste heat that can be used includes: waste heat in cement kilns, the waste heat in various ceramic down-burning furnaces and tunnel kilns, the waste heat in glass kilns. The waste heat in kaolin spray drying hot blast furnace.

In petrochemical industry, the waste heat that can be used includes: waste heat in various heating furnaces, waste heat in hydrocarbon pyrolysis furnace (working temperature is about $750 \sim 900°C$), waste heat in catalysis, waste heat in ethylbenzene dehydrogenation reactor, and waste heat in cyclohexanol dehydrogenation chemical reactor recycling.

In the textile printing and dyeing industry, the waste heat that can be used includes: boiler flue gas waste heat, setting machine waste heat, printing and dyeing hot sewage waste heat.

In the sulfuric acid industry, the waste heat that can be used is: SO_2 high temperature furnace gas from boiling Medium waste heat, waste heat in the boiling layer of the boiling roaster for sulfuric acid production, waste heat in hydrochloric acid and nitric acid furnaces.

There are many ways to recover waste heat, the efficiency of comprehensive utilization or direct utilization is high, and the efficiency of indirect utilization is low. The sequence of utilization of steam waste heat is: power heating, power generation and heating, production process use, direct replacement of motor-driven equipment, use of steam turbine to generate electricity, and domestic use. The utilization sequence of hot water waste heat is: for production process utilization, return to boiler utilization, and domestic use. The utilization sequence of air waste heat is: production utilization, HVAC utilization, power utilization, power generation utilization.

Figure 24.7 is the appearance of a power plant flue gas waste heat recovery and utilization equipment.

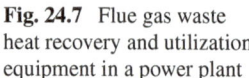
Fig. 24.7 Flue gas waste heat recovery and utilization equipment in a power plant

24.6 Solar Photovoltaic Power Generation Technology

Semiconductor Photovoltaic Effect: When the light shines on the solar cell, photons with sufficient energy excite electrons from covalent bonds in P-type silicon and N-type silicon, generating electron–hole pairs, and the electrons move to the positively charged N region, the holes move to the negatively charged P region, the charge is separated, and a voltage is generated between the P region and the N region. For solar silicon cells, the value of the open circuit voltage is about 0.5 to 0.6V.

Solar cells use the photovoltaic effect of semiconductors to directly convert light energy into electrical energy: silicon atoms have 4 electrons. If phosphorus atoms with 5 electrons are mixed into pure silicon, it becomes a negatively charged N-type semiconductor. Boron atoms with 3 electrons are doped in silicon to form a positively charged P-type semiconductor. When they are combined together, a potential difference will be formed on the contact surface, and sunlight will irradiate the P–N junction, and holes will flow from the N pole region to the P–N junction. The P pole area moves, and the electrons move from the P pole area to the N pole area to form a current, which becomes a solar cell. The more light-energy absorbed by the surface of the solar cell, the greater the current formed in the solar cell. A large number of solar cells are packaged in series to form a large-area solar cell module.

The shape of the solar cell is shown in Fig. 24.8.

Off-grid photovoltaic power station: This system is composed of solar cell square array, system controller, battery pack, DC/AC inverter, etc. It is not connected to the public power grid and is mainly used in areas without public power grids, such as pastoral areas and remote People in mountain villages, plateaus, islands, and deserts provide electricity for lighting, television, broadcasting, and communications, and provide power for communication relay stations, weather stations, border posts, navigation marks, and highways.

Grid-connected photovoltaic power station: The power station is composed of a solar cell array, a system controller, and a grid-connected inverter.

Figure 24.9 is a photo of a solar photovoltaic power plant.

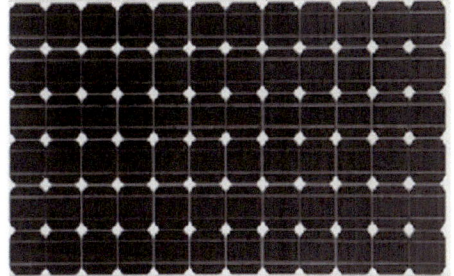

Fig. 24.8 Solar cell

Fig. 24.9 A solar
photovoltaic power station

24.7 Wind Power Technology

Wind power generation is to convert wind energy into mechanical work, and then generate electric energy. The principle is to use wind power to drive the windmill blades to rotate and drive the generator to generate electricity.

According to the orientation of the fixed axis of the blade, there are two types of wind turbines: horizontal axis type and vertical axis type. Horizontal axis wind turbines are currently the mainstream in the world.

The diameter of the blades of wind turbines can reach 216 m, and the power of a single machine can reach 11MW. There are lightning protection strips in the blades. When the blades are struck by lightning, the lightning protection strips will lead the lightning current into the ground.

Like the pump and fan equipment, the power generated by the wind turbine is proportional to the cube of the wind speed.

Because wind power does not use fuel, and does not produce radiation or air pollution, it is recognized as a green energy source.

The internal structure of the horizontal axis wind turbine is shown in Fig. 24.10, and a wind farm is shown in Fig. 24.11.

Fig. 24.10 Internal structure of horizontal axis wind turbine

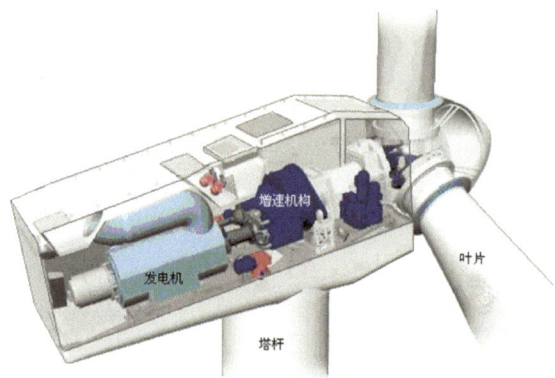

Fig. 24.11 A wind farm

Chapter 25
Energy Efficiency Optimization of Multi-Engine Launch Vehicles

Space exploration is in the ascendant. In order to carry more things and fly farther and higher, people are beginning to use multi-engine high-thrust rockets (Fig. 25.1). How to optimize the cooperation of these engines to make the overall energy efficiency of the rocket higher and lower cost are becoming increasingly important.

Generally, the energy efficiency of electric heating, chemical reactions and some combustion processes is very high, because most of the energy loss in nature is ultimately lost in the form of heat dissipation. However, in the process of doing work, there is inevitably a conduction process, that is, there is heat energy loss. Therefore, the actual operating efficiency is not 100%. Rocket engines are no exception.

25.1 The Energy Efficiency of a Rocket Engine

For the energy efficiency curve of a rocket engine, for the convenience of measurement, we define that the abscissa is the mass of fuel burned per unit time of the i-th engine, m_i, and the ordinate is the energy efficiency of the i-th engine, η_i, η_i is replaced by kF_i/m_i, k is a constant, F_i represents the thrust of the i-th engine during unit time, the curve is as shown in Fig. 25.2.

Assume the energy efficiency functions $\eta_1(m)$ and $\eta_i(m)$ of the first and the i-th engine is as shown in Fig. 25.3.

If the following equation holds:

$$\eta_i(m_i) = \eta_1\left(\frac{m_1}{\beta_i}\right) \tag{25.1}$$

We call the first engine and the i-th engine as the energy efficiency similarity engines.

© The Author(s) 2024
F. Yao and Y. Yao, *Efficient Energy-Saving Control and Optimization for Multi-Unit Systems*, https://doi.org/10.1007/978-981-97-4492-3_25

Fig. 25.1 Multi-engine rocket

Fig. 25.2 Energy efficiency curve of rocket engine

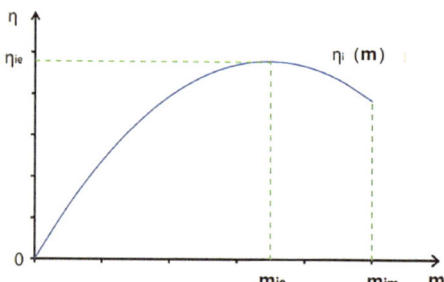

Fig. 25.3 The energy efficiency curves of the first and the i-th engine

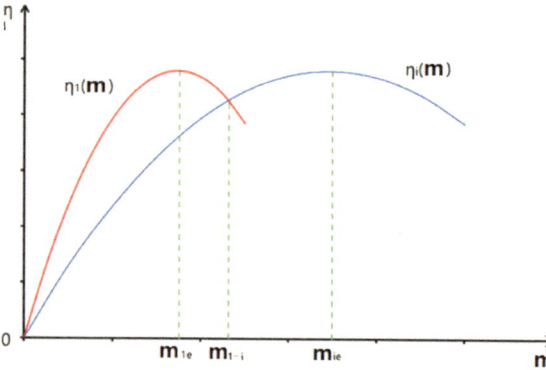

25.2 Analysis of Launch Process of Launch Vehicle

The total rocket thrust F is

$$F = \sum_{i=1}^{n} F_i \tag{25.2}$$

The energy efficiency optimization problem of multi-engine rocket is

$$\text{Max } S = \text{Max} \int_{t=0}^{\infty} V dt \tag{25.3}$$

In Eq. (25.3), S is the total distance traveled by the rocket, and V is the speed of the rocket at time t.

In the vacuum resistance-free state, the kinetic energy and pressure energy output by the engine become the potential energy of the rocket, causing the rocket to rise ΔH, with the following relationship

$$\sum_{i=1}^{n} \left(\frac{1}{2} m_i V_i^2 + \frac{P_i}{\gamma g_h} \right) = \left(M - \sum_{i=1}^{n} m_i \right) g_h \Delta H \tag{25.4}$$

g_h is the acceleration of gravity at different altitudes, n is the number of working engines, V_i is the speed of fuel ejection by the i-th engine, P_i is the pressure of the gas ejected by the i-th engine, and M is the rocket total mass at the previous time t.

Taking resistance such as wind resistance into account, according to Newton's second law, we have

$$\sum_{i=1}^{n} F_i - \left(M - \sum_{i=1}^{n} m_i \right) g_h - k_1 V^2 = \left(M - \sum_{i=1}^{n} m_i \right) \frac{dV}{dt} \tag{25.5}$$

In Eq. (25.5), k_1 is the wind resistance coefficient, which changes with the thinness of the air or the altitude of the rocket. The expression for F becomes

$$F = \sum_{i=1}^{n} F_i = \left(M - \sum_{i=1}^{n} m_i \right) g_h + k_1 V^2 + \left(M - \sum_{i=1}^{n} m_i \right) \frac{dV}{dt} \tag{25.6}$$

In a short period of time t, g_h and k_1 are approximately constant. There are two variables, V and M_1. V is the current speed of the rocket, and M_1 is the current total mass of the rocket.

$$M_1 = \left(M - \sum_{i=1}^{n} m_i \right) \tag{25.7}$$

The total mass of the rocket has been decreasing and the wind resistance has been increasing. Derive V and M_1 to get the optimal thrust F.

For each F, F_i is arranged according to the principle of maximum energy efficiency, that is, the principle of minimum fuel consumption, and the corresponding fuel combustion quality m_i is calculated. Then we have

$$F = \sum_{i=1}^{n} F_i = k_2 \sum_{i=1}^{n} m_i \eta_i (m_i) \tag{25.8}$$

$$m_t = \sum_{i=1}^{n} m_i$$

where m_t is the total mass ejected per unit time by the rocket.

25.3   Optimal Scheduling of Multiple Engines in a Rocket

1. If the n engines are identical, the energy expression of the thrust becomes

$$F = k_2 m_t \sum_{i=1}^{n} \theta_i \eta (\theta_i m_t) \tag{25.9}$$

where θ_i is

$$\theta_i = \frac{m_i}{m_t} \tag{25.10}$$

$$\sum_{i=1}^{n} \theta_i = 1$$

Consider the maximization problem of the total thrust F

$$\max F \tag{25.11}$$

$$s.t. \ \theta_i > 0, \ i = 1, 2, ...n$$

$$\sum_{i=1}^{n} \theta_i = 1$$

$$m_t = cons \tan t$$

This problem can also be written as

$$\max k_2 m_t \sum_{i=1}^{n} \theta_i \eta(\theta_i m_t) \tag{25.12}$$

$$s.t. \ \theta_i > 0, i = 1, 2, ...n$$

$$\sum_{i=1}^{n} \theta_i = 1$$

$$m_t = cons \tan t$$

We consider three cases:

(1) $n = 2$

The rocket has two variables and has

$$\theta_1 + \theta_2 = 1$$
$$\theta_1 > 0$$
$$\theta_2 > 0 \tag{25.13}$$

The objective function F can be expressed as

$$F = k_2 m_t (\theta_1 \eta(\theta_1 m_t) + \theta_2 \eta(\theta_2 m_t)) = k_2 m_t (\theta_1 \eta(\theta_1 m_t) + (1 - \theta_1)\eta((1 - \theta_1)m_t)) \tag{25.14}$$

The optimization condition is

$$F'(\theta_1) = 0 \tag{25.15}$$

It is easy to see that

$$\theta_1 = \frac{1}{2} \tag{25.16}$$

for the optimization point. Then we have

$$\theta_2 = 1 - \theta_1 = \theta_1 = \frac{1}{2} = \frac{1}{n} \tag{25.17}$$

That is, the optimal control method is to keep

$$m_1 = m_2 = \frac{m_t}{2} = \frac{m_t}{n} \tag{25.18}$$

The total F is

$$F = k_2 m_t \eta\left(\frac{m_t}{2}\right) = k_2 m_t \eta\left(\frac{m_t}{n}\right) \tag{25.19}$$

Since the shape of the overall efficiency curve of the rocket is the same as that of a single engine, so the second derivative of F is also less than zero

$$F''(\theta_1) < 0 \qquad\qquad (25.20)$$

F is the unique maximum value.

$$\text{maxF} = k_2 m_t \eta\left(\frac{m_t}{2}\right) \qquad\qquad (25.21)$$

The overall energy efficiency η_t of the rocket is the only maximum value.

$$max\eta_t = \eta\left(\frac{m_t}{n}\right) \qquad\qquad (25.22)$$

(2) n = 3

The rocket has three variables, based on known conditions, we have

$$\theta_1 + \theta_2 + \theta_3 = 1 \qquad\qquad (25.23)$$

$$\theta_1 > 0$$
$$\theta_2 > 0$$
$$\theta_3 > 0$$

The F expression becomes

$$F = k_2 m_t (\theta_1 \eta(\theta_1 m_t) + \theta_2 \eta((\theta_2)m_t) + \theta_3 \eta((\theta_3)m_t)) \qquad\qquad (25.24)$$

Assuming that θ_3 is fixed and an optimization point, only θ_1 and θ_2 are variables, we have

$$\theta_1 + \theta_2 = 1 - \theta_3 = constant \qquad\qquad (25.25)$$

Based on the conclusion of n = 2 above, there are

$$\theta_1 = \theta_2 \qquad\qquad (25.26)$$

is the optimal point.
Assuming that θ_2 is fixed and is an optimization point, only θ_1 and θ_3 are variables, we have

$$\theta_1 + \theta_3 = 1 - \theta_2 = constant \qquad\qquad (25.27)$$

According to the conclusion of n = 2 above, there are

$$\theta_1 = \theta_3 \tag{25.28}$$

is the optimal point.

Similarly, assuming that θ_1 is fixed and is an optimization point, only θ_2 and θ_3 are variables, we have

$$\theta_2 + \theta_3 = 1 - \theta_1 = constant \tag{25.29}$$

According to the conclusion of $n = 2$ above, we have

$$\theta_2 = \theta_3 \tag{25.30}$$

to be the optimal point.

So, we have the optimal point

$$\theta_1 = \theta_2 = \theta_3 = \frac{1}{3} = \frac{1}{n} \tag{25.31}$$

That is, the optimal control method is to keep

$$m_1 = m_2 = m_3 = \frac{m_t}{3} = \frac{m_t}{n} \tag{25.32}$$

The maximum value of the total F is

$$maxF = k_2 m_t \eta\left(\frac{m_t}{3}\right) = k_2 m_t \eta\left(\frac{m_t}{n}\right) \tag{25.33}$$

The maximum value of the overall energy efficiency η_t is

$$\max \eta_t = \eta\left(\frac{m_t}{3}\right) = \eta\left(\frac{m_t}{n}\right) \tag{25.34}$$

(3) $n = k$

The rocket has k variables, the above conclusion can be extended to the case of $n = k$, the optimal point is

$$\theta_1 = \theta_2 = ... = \theta_k = \frac{1}{k} \tag{25.35}$$

That is, the optimal control method is to keep

$$m_1 = m_2 = ... = m_k = \frac{m_t}{k} \tag{25.36}$$

The maximum value of F is

$$maxF = k_2 m_t \eta \left(\frac{m_t}{k} \right) \tag{25.37}$$

The maximum value of the overall energy efficiency ηt is

$$\max \eta_t = \left(\frac{m_t}{k} \right) \tag{25.38}$$

2. If the n engines are the energy efficiency similarity engines, the energy expression of the total thrust becomes

$$F = F = k_2 m_t \sum_{i=1}^{n} \theta_i \eta_i (\theta_i m_t) \tag{25.39}$$

where

$$\theta_i = \frac{m_i}{m_t}$$
$$\sum_{i=1}^{n} \theta_i = 1 \tag{25.40}$$

Consider the maximization problem of the total thrust F

$$\max F \tag{25.41}$$

$$s.t. \ \theta_i > 0, i = 1, 2, ...n$$
$$\sum_{i=1}^{n} \theta_i = 1$$
$$m_t = cons \tan t$$

This problem can also be written as

$$\max k_2 m_t \sum_{i=1}^{n} \theta_i \eta_i (\theta_i m_t) \tag{25.42}$$

$$s.t. \ \theta_i > 0, i = 1, 2, ...n$$
$$\sum_{i=1}^{n} \theta_i = 1$$
$$m_t = cons \tan t$$

We consider three cases:

(1) n = 2

The rocket has two variables and has

$$\theta_1 + \theta_2 = 1 \tag{25.43}$$

$$\theta_1 > 0$$
$$\theta_2 > 0$$

The objective function F can be expressed as

$$F = k_2 m_t (\theta_1 \eta_1 (\theta_1 m_t) + \theta_2 \eta_2 (\theta_2 m_t)) \tag{25.44}$$

The optimization condition is

$$F'(\theta_1) = 0 \tag{25.45}$$

It is easy to see that

$$\theta_1 = \frac{1}{1 + \beta_2} = \frac{\beta_1}{\beta_1 + \beta_2} \tag{25.46}$$

for the optimization point. Then we have

$$\theta_2 = 1 - \theta_1 = \frac{\beta_2}{1 + \beta_2} = \frac{\beta_2}{\beta_1 + \beta_2} \tag{25.47}$$

That is, the optimal control method is to keep

$$m_1 = \frac{1}{1+\beta_2} m_t$$
$$m_2 = \frac{\beta_2}{1+\beta_2} m_t \tag{25.48}$$

The total thrust F is

$$F = k_2 m_t \eta_1 \left(\frac{1}{1 + \beta_2} m_t \right) \tag{25.49}$$

Since the shape of the overall efficiency curve of the rocket is the same as that of a single engine, so the second derivative of the F is also less than zero

$$F''(\theta_1) < 0 \tag{25.50}$$

F is the only maximum value.

$$\max F = k_2 m_t \eta_1 \left(\frac{1}{1 + \beta_2} m_t \right) \tag{25.51}$$

The overall energy efficiency η_t of the rocket is the only maximum value.

$$maxn_t = \eta_1 \left(\frac{1}{1 + \beta_2} m_t \right)$$ (25.52)

(2) n = 3

The rocket has three variables, based on known conditions, we have

$$\theta_1 + \theta_2 + \theta_3 = 1$$ (25.53)

$$\theta_1 > 0$$
$$\theta_2 > 0$$
$$\theta_3 > 0$$

The F expression becomes

$$F = F = k_2 m_t (\theta_1 \eta_1 (\theta_1 m_t) + \theta_2 \eta_2 (\theta_2 m_t) + \theta_3 \eta_3 (\theta_3 m_t))$$ (25.54)

Assuming that θ_3 is fixed and is an optimization point, only θ_1 and θ_2 are variables, we have

$$\theta_1 + \theta_2 = 1 - \theta_3 = constant$$ (25.55)

Based on the conclusion of n = 2 above, we have

$$\theta_1 = \frac{1}{1+\beta_2}$$
$$\theta_2 = \frac{\beta_2}{1+\beta_2}$$ (25.56)

is the optimal point.

Assuming that θ_2 is fixed and is an optimization point, only θ_1 and θ_3 are variables, there are

$$\theta_1 + \theta_3 = 1 - \theta_2 = constant$$ (25.57)

According to the conclusion of n = 2 above, there are

$$\theta_1 = \frac{1}{1+\beta_3}$$
$$\theta_3 = \frac{\beta_3}{1+\beta_3}$$ (25.58)

is the optimal point.

Similarly, assuming that θ_1 is fixed and is an optimization point, only θ_2 and θ_3 are variables, we have

$$\theta_2 + \theta_3 = 1 - \theta_3 = constant \tag{25.59}$$

According to the conclusion of n = 2 above, there are

$$\theta_2 = \frac{\beta_2}{\beta_2 + \beta_3}$$
$$\theta_3 = \frac{\beta_3}{\beta_2 + \beta_3} \tag{25.60}$$

is the optimal point.

So, we have the optimal point, and the optimal control method is to keep

$$\theta_1 = \frac{\beta_1}{\beta_1 + \beta_2 + \beta_3}$$
$$\theta_2 = \frac{\beta_2}{\beta_1 + \beta_2 + \beta_3}$$
$$\theta_3 = \frac{\beta_3}{\beta_1 + \beta_2 + \beta_3} \tag{25.61}$$

The maximum total trust F is

$$\max F = k_2 m_t \eta_1 \left(\frac{1}{1 + \beta_2 + \beta_3} m_t \right) \tag{25.62}$$

The maximum overall energy efficiency η_t is

$$\max \eta_t = \eta_1 \left(\frac{P_0}{1 + \beta_2 + \beta_3} \right) \tag{25.63}$$

(3) n = k

The rocket has k variables, the above conclusion can be extended to the case of n = k, the optimal point and the optimal control method is to keep

$$\theta_i = \frac{\beta_i}{\sum_{l=1}^{k} \beta_l} \tag{25.64}$$

The maximum total trust F is

$$\max F = k_2 m_t \eta_1 \left(\frac{1}{\sum_{l=1}^{k} \beta_l} m_t \right) \tag{25.65}$$

The maximum overall energy efficiency η_t is

$$\max \eta_t = \eta_1 \left(\frac{1}{\sum\limits_{l=1}^{k} \beta_l} - m_t \right) \tag{25.66}$$

25.4 Optimal Number of Operational Engines in a Rocket

1. A rocket has m identical engines, if the following equations are satisfied

$$\eta\left(\frac{m_t}{n}\right) \geq \eta\left(\frac{m_t}{n-1}\right) \tag{25.67}$$

$$\eta\left(\frac{m_t}{n}\right) \geq \eta\left(\frac{m_t}{n+1}\right)$$

$$n \leq m$$

Then the number n is the number of engines with optimal operation.

2. A rocket has m energy efficiency similarity engines, if the following equations are satisfied

$$\eta_1 \left(\frac{P_0}{\sum\limits_{l=1}^{n} \beta_l} \right) \geq \eta_1 \left(\frac{P_0}{\sum\limits_{l=1}^{n1} \beta_l} \right) \tag{25.68}$$

$$n \leq m$$
$$n1 \leq m$$

n1 is any combination other than the optimal combination of n units this time, and also include other combinations of n units. The number n is the number of engines with optimal operation.

25.5 Optimal Switching Rule for Multiple Engines in a Rocket

1. A rocket has m identical engines, and the number n is the number of engines currently in optimal operation. If m_t increases to m_{t1}, the following relationship holds:

$$\eta\left(\frac{m_{t1}}{n}\right) = \eta\left(\frac{m_{t1}}{n+1}\right) \tag{25.69}$$

$$n \leq m$$

Then m_{t1} is the optimal switching point between the operation of n engines and the operation of $n + 1$ engine. When the total required m_t is greater than m_{t1}, the optimal number of engines in operation is switched from n to $n + 1$. If m_t increases until $m_t/n = m_{1m}$, there isn't the point m_{t1}, then $m_t/n = m_{1m}$ is the switching point from n to $n + 1$.

If m_t is reduced to m_{t2}, the following relationship is established

$$\eta\left(\frac{m_{t2}}{n}\right) = \eta\left(\frac{m_{t2}}{n-1}\right) \tag{25.70}$$

$$n \leq m$$

Then m_{t2} is the optimal switching point between the operation of n engines and the operation of n-1 engines. When the total required power m_t is less than m_{t2}, the optimal number of engines in operation is switched from n to n-1. If m_t reduces until $m_t/(n-1) = m_{1m}$, there isn't the point m_{t2}, then $m_t/(n-1) = m_{1m}$ is the switching point from n to n-1.

The analysis process is shown in Fig. 25.4.

2. A rocket has m energy efficiency similarity engines, and the number n is the number of engines currently in optimal operation. If m_t increases m_{t1}, the following relation holds:

Fig. 25.4 Energy efficiency comparison curve of the rocket with the identical engines

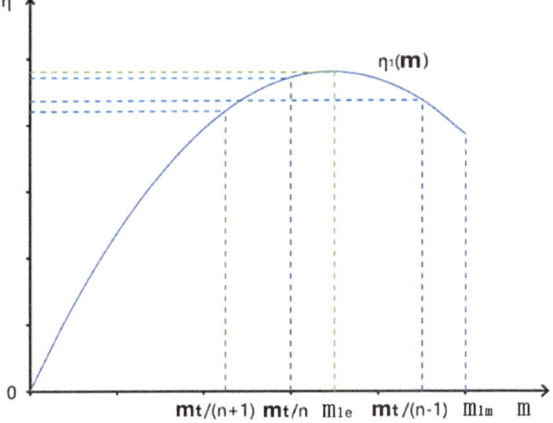

$$\eta_1\left(\frac{m_{t1}}{\sum_{l=1}^{n}\beta_l}\right) = \eta_1\left(\frac{m_{t1}}{\sum_{l=1}^{k1}\beta_l}\right) \qquad (25.71)$$

$$n \leq m$$

Then m_{t1} is the optimal switching point between the operation of n engines and the operation of k1 engines. When the total required power m_t is greater than m_{t1}, the optimal number of engines in operation is switched from n to k1. If m_t increases until $m_t/\sum_{l=1}^{n}\beta_l = m_{1m}$, there isn't the point m_{t1}, then $m_t/\sum_{l=1}^{n}\beta_l = m_{1m}$ is the switching point from n to k1.

If m_t is reduced to m_{t2}, the following relation is established

$$\eta_1\left(\frac{m_{t2}}{\sum_{l=1}^{n}\beta_l}\right) = \eta_1\left(\frac{m_{t2}}{\sum_{l=1}^{k2}\beta_l}\right) \qquad (25.72)$$

$$n \leq m$$

Then m_{t2} is the optimal switching point between the operation of n engines and the operation of k2 engines. When the total required power m_t is less than m_{t2}, the optimal number of engines in operation is switched from n to k2. If m_t reduces until $m_t/\sum_{l=1}^{k2}\beta_l = m_{1m}$, there isn't the point m_{t2}, then $m_t/\sum_{l=1}^{k2}\beta_l = m_{1m}$ is the switching point from n to k2.

The analysis process is shown in Fig. 25.5.

k1 and k2 are any combination other than the optimal combination of n units this time, and also include other combinations of n units. Point $m_t/\sum_{l=1}^{k1}\beta_l$ is the point closest to $m_t/\sum_{l=1}^{n}\beta_l$ to the left of point $m_t/\sum_{l=1}^{n}\beta_l$. Point $m_t/\sum_{l=1}^{k2}\beta_l$ is the point closest to $m_t/\sum_{l=1}^{n}\beta_l$ to the right of point $m_t/\sum_{l=1}^{n}\beta_l$. . If all engines are identical, we have $k1 = n + 1$ and $k2 = n-1$.

Fig. 25.5 Energy efficiency comparison curve of the rocket with the similar efficiency engines

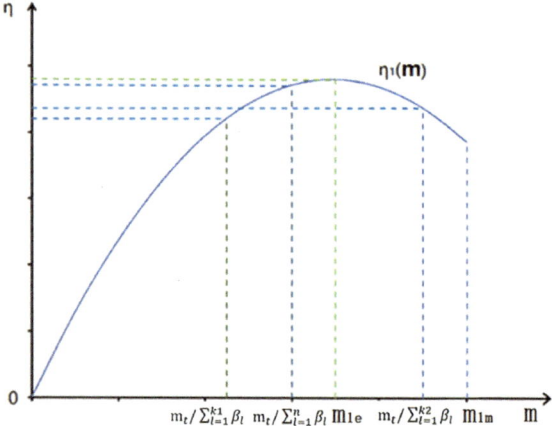

Chapter 26
Energy Efficiency Optimization of Human (or Biological) Team

Why do special forces need to select soldiers with the same physical strength? Why do two adults of the same physique put the heavy object in the middle when lifting something? Why is it that when an adult and a child are lifting something, the weight should be placed closer to the adult? In fact, behind these practices lies the concept of energy efficiency optimization.

For the energy efficiency optimization of humans (or organisms), we can define it as that the physical arrangement method that continues to do work over a long period of time and ultimately maximizes the total amount of work, or minimizes the total physical ability. This is called human (or organisms) energy efficiency optimization Fig. 26.1.

Let's take humans as an example to illustrate below, assuming that the situation of other living creatures is similar.

26.1 Human Energy Efficiency Curve

The human energy efficiency curve can be defined like this. The abscissa is the amount of work done per unit time W/t, labor intensity, which is similar to the output power P of a machine. We use P to represent the labor intensity of human work. The ordinate is the energy efficiency η, which represents the ability to complete the total amount of labor within a period of longer time. The energy efficiency curve of the i-th person is shown in Fig. 26.2.

Assume that the maximum energy efficiency of the first person and the i-th person are equal, but the amount of work done per unit time is different, which is equivalent to the different power of the machine, as shown in Fig. 26.3.

If the energy efficiency curves of person 1 and person i have the following relationship

© The Author(s) 2024

F. Yao and Y. Yao, *Efficient Energy-Saving Control and Optimization for Multi-Unit Systems*, https://doi.org/10.1007/978-981-97-4492-3_26

Fig. 26.1 Physical distribution

Fig. 26.2 Energy efficiency
curve of person i

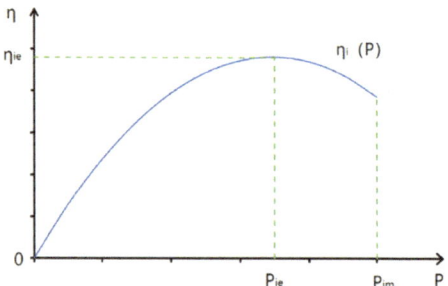

Fig. 26.3 Energy efficiency
curves of the first person and
the i-th person

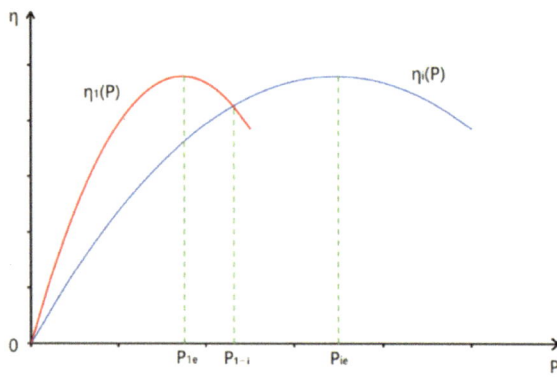

$$\eta_i(P_i) = \eta_1\left(\frac{P_1}{\beta_i}\right) \tag{26.1}$$

where β_i is a constant, we call them similarly efficient people, and $\beta_1 = 1$.
   Assume that there are n people in a team to complete one task.

The total work to be done in a human team is P_0 that is similar to the total energy output in a machine. Assuming P_0 is a fixed value. The work completed by the i-th person is P_i, which is similar to the energy output. The physical energy input of the i-th person is W_i. The total physical energy consumed by all people in the team is P_t that is similar to the total energy input in a machine.

The expression of maximizing P_0 is

$$maxP_0 = max \sum_{i=1}^{n} W_i \eta_i(W_i) \tag{26.2}$$

$$P_t = \sum_{i=1}^{n} W_i$$

The expression of minimizing P_t is

$$minP_t = min \sum_{i=1}^{n} \frac{P_i}{\eta_i(P_i)} \tag{26.3}$$

$$P_0 = \sum_{i=1}^{n} P_i.$$

Note that the arguments of the efficiency functions in the two expressions are different and they are not the identical function.

These are two different expressions of the same problem. We only need to analyze the subsequent minimization expression.

26.2 Energy Efficiency Optimal Control in a Human Team

1. If the n people are identical, P_t becomes

$$P_t = P_0 \sum_{i=1}^{n} \frac{\theta_i}{\eta(\theta_i P_0)} \tag{26.4}$$

where

$$\theta_i = \frac{P_i}{P_0} \tag{26.5}$$

$$\sum_{i=1}^{n} \theta_i = 1.$$

Consider the minimization problem of P_t

$$minP_t \tag{26.6}$$

$$s.t.\theta_i > 0, i = 1, 2, ...n$$

$$\sum_{i=1}^{n} \theta_i = 1$$

$$P_0 = cons\tan t.$$

This problem can also be written as

$$minP_0 \sum_{i=1}^{n} \frac{\theta_i}{\eta(\theta_i P_0)} \tag{26.7}$$

$$s.t.\theta_i > 0, i = 1, 2, ...n$$

$$\sum_{i=1}^{n} \theta_i = 1$$

$$P_0 = cons\tan t.$$

We consider three cases:

(1) $n = 2$

The team has two variables and has

$$\theta_1 + \theta_2 = 1 \tag{26.8}$$

$$\theta_1 > 0$$
$$\theta_2 > 0$$

The objective function P_t can be expressed as

$$P_t = P_0\left(\frac{\theta_1}{\eta(\theta_1 P_0)} + \frac{\theta_2}{\eta(\theta_2 P_0)}\right) = P_0\left(\frac{\theta_1}{\eta(\theta_1 P_0)} + \frac{1 - \theta_1}{\eta((1 - \theta_1)P_0)}\right). \tag{26.9}$$

The optimization condition is

$$P_t'(\theta_1) = 0 \tag{26.10}$$

easy to draw

$$\theta_1 = \frac{1}{2} \tag{26.11}$$

for the optimization point. Then we have

$$\theta_2 = 1 - \theta_1 = \theta_1 = \frac{1}{2} = \frac{1}{n}. \tag{26.12}$$

That is, the optimal control method is to keep

$$P_1 = P_2 = \frac{P_0}{2} = \frac{P_0}{n}. \tag{26.13}$$

The P_t is

$$P_t = \frac{P_0}{\eta\left(\frac{P_0}{2}\right)} = \frac{P_0}{\eta\left(\frac{P_0}{n}\right)} \tag{26.14}$$

Since the shape of the overall efficiency curve of the team is the same as that of one person, so the second derivative of the P_t is also greater than zero

$$P_t^{''}(\theta_1) > 0. \tag{26.15}$$

P_t is the only minimum value.

$$\min P_t = \frac{P_0}{\eta\left(\frac{P_0}{n}\right)} \tag{26.16}$$

The overall energy efficiency η_t of the human team is the only maximum value.

$$max\eta_t = \eta\left(\frac{P_0}{n}\right) \tag{26.17}$$

(2) n = 3

The team has three variables, based on known conditions, we have

$$\theta_1 + \theta_2 + \theta_3 = 1 \tag{26.18}$$

$$\theta_1 > 0$$
$$\theta_2 > 0.$$
$$\theta_3 > 0$$

The P_t expression becomes

$$P_t = P_0 \left(\frac{\theta_1}{\eta(\theta_1 P_0)} + \frac{\theta_2}{\eta(\theta_2 P_0)} + \frac{\theta_3}{\eta((\theta_3 P_0)} \right). \tag{26.19}$$

Assuming that θ_3 is fixed and an optimization point, only θ_1 and θ_2 are variables, we have

$$\theta_1 + \theta_2 = 1 - \theta_3 = constant. \tag{26.20}$$

Based on the conclusion of n = 2 above, we have

$$\theta_1 = \theta_2 \tag{26.21}$$

is the optimal point.

Assuming that θ_2 is fixed and is an optimization point, only θ_1 and θ_3 are variables, we have

$$\theta_1 + \theta_3 = 1 - \theta_2 = constant. \tag{26.22}$$

According to the conclusion of n = 2 above, we have

$$\theta_1 = \theta_3 \tag{26.23}$$

is the optimal point.

Similarly, assuming that θ_1 is fixed and is an optimization point, only θ_2 and θ_3 are variables, we have

$$\theta_2 + \theta_3 = 1 - \theta_1 = constant. \tag{26.24}$$

According to the conclusion of n = 2 above, we have

$$\theta_2 = \theta_3 \tag{26.25}$$

is the optimal point.

So, we have the optimal point

$$\theta_1 = \theta_2 = \theta_3 = \frac{1}{3} = \frac{1}{n}. \tag{26.26}$$

That is, the optimal control method is to keep

$$P_1 = P_2 = P_3 = \frac{P_0}{3} = \frac{P_0}{n}. \tag{26.27}$$

The minimum P_t is

$$minP_t = \frac{P_0}{\eta\left(\frac{P_0}{3}\right)} = \frac{P_0}{\eta\left(\frac{P_0}{n}\right)} \qquad (26.28)$$

The maximum overall energy efficiency η_t is

$$max\eta_t = \eta\left(\frac{P_0}{3}\right) = \eta\left(\frac{P_0}{n}\right). \qquad (26.29)$$

(3) $n = k$

The human team has k variables, the above conclusion can be extended to the case of $n = k$, the optimal point is

$$\theta_1 = \theta_2 = \dots = \theta_k = \frac{1}{k} \qquad (26.30)$$

That is, the optimal method is to keep

$$P_1 = P_2 = \dots = P_k = \frac{P_0}{k}. \qquad (26.31)$$

The minimum P_t is

$$minP_t = P_0 \frac{1}{\eta\left(\frac{P_0}{n}\right)}. \qquad (26.32)$$

The maximum overall energy efficiency η_t is

$$max\eta_t = \eta\left(\frac{P_0}{k}\right) = \eta\left(\frac{P_0}{n}\right) \qquad (26.33)$$

2. If n people are the energy efficiency similarity people, the expression of P_t becomes

$$P_t = P_0 \sum_{i=1}^{n} \frac{\theta_i}{\eta_i(\theta_i P_0)} \qquad (26.34)$$

where

$$\theta_i = \frac{P_i}{P_0} \qquad (26.35)$$

$$\sum_{i=1}^{n} \theta_i = 1.$$

Consider the minimization problem of total power consumption

$$minP_t \tag{26.36}$$

$$s.t. \theta_i > 0, i = 1, 2, ...n$$

$$\sum_{i=1}^{n} \theta_i = 1$$

$$P_0 = cons\tan t.$$

This problem can also be written as

$$minP_0 \sum_{i=1}^{n} \frac{\theta_i}{\eta_i(\theta_i P_0)} \tag{26.37}$$

$$s.t. \theta_i > 0, i = 1, 2, ...n$$

$$\sum_{i=1}^{n} \theta_i = 1$$

$$P_0 = cons\tan t.$$

We consider three cases:

(1) n = 2

The human team has two variables and has

$$\theta_1 + \theta_2 = 1 \tag{26.38}$$

$$\theta_1 > 0$$
$$\theta_2 > 0$$

The objective function P_t can be expressed as

$$P_t = P_0\left(\frac{\theta_1}{\eta_1(\theta_1 P_0)} + \frac{\theta_2}{\eta_2(\theta_2 P_0)}\right) = P_0\left(\frac{\theta_1}{\eta_1(\theta_1 P_0)} + \frac{1-\theta_1}{\eta_2((1-\theta_1)P_0)}\right). \tag{26.39}$$

The optimization condition is

$$P_t{}'(\theta_1) = 0. \tag{26.40}$$

It is easy to see that

$$\theta_1 = \frac{1}{1 + \beta_2} = \frac{\beta_1}{\beta_1 + \beta_2} \tag{26.41}$$

for the optimization point. Then we have

$$\theta_2 = 1 - \theta_1 = \frac{\beta_2}{1 + \beta_2} = \frac{\beta_2}{\beta_1 + \beta_2}. \tag{26.42}$$

That is, the optimal control method is to keep

$$\begin{aligned} P_1 &= \frac{1}{1+\beta_2} P_0 \\ P_2 &= \frac{\beta_2}{1+\beta_2} P_0 \end{aligned}. \tag{26.43}$$

The P_t is

$$P_t = \frac{P_0}{\eta_1 \left(\frac{P_0}{1+\beta_2} \right)} \tag{26.44}$$

Since the shape of the overall efficiency curve of the human team is the same as that of one person, so the second derivative of the P_t is also greater than zero

$$P_t^{''}(\theta_1) > 0. \tag{26.45}$$

P_t is the only minimum value with

$$minP_t = \frac{P_0}{\eta_1 \left(\frac{P_0}{1+\beta_2} \right)}. \tag{26.46}$$

The overall energy efficiency η_t of the team is the only maximum value.

$$maxη_t = \eta_1 \left(\frac{P_0}{1 + \beta_2} \right) \tag{26.47}$$

(2) n = 3

The team has three variables, based on known conditions, we have

$$\theta_1 + \theta_2 + \theta_3 = 1 \tag{26.48}$$

$$\begin{aligned} \theta_1 &> 0 \\ \theta_2 &> 0. \\ \theta_3 &> 0 \end{aligned}$$

The P_t expression becomes

$$P_t = P_0 \left(\frac{\theta_1}{\eta_1 (\theta_1 P_0)} + \frac{\theta_2}{\eta_2 (\theta_2 P_0)} + \frac{\theta_3}{\eta_3 ((\theta_3 P_0)} \right). \tag{26.49}$$

Assuming that θ_3 is fixed and is an optimization point, only θ_1 and θ_2 are variables, we have

$$\theta_1 + \theta_2 = 1 - \theta_3 = constant. \tag{26.50}$$

Based on the conclusion of n = 2 above, there are

$$\begin{aligned} \theta_1 &= \frac{1}{1+\beta_2} \\ \theta_2 &= \frac{\beta_2}{1+\beta_2} \end{aligned}. \tag{26.51}$$

is the optimal point.

Assuming that θ_2 is fixed and is an optimization point, only θ_1 and θ_3 are variables, we have

$$\theta_1 + \theta_3 = 1 - \theta_2 = constant. \tag{26.52}$$

According to the conclusion of n = 2 above, there are

$$\begin{aligned} \theta_1 &= \frac{1}{1+\beta_3} \\ \theta_3 &= \frac{\beta_3}{1+\beta_3} \end{aligned} \tag{26.53}$$

is the optimal point.

Similarly, assuming that θ_1 is fixed and is an optimization point, only θ_2 and θ_3 are variables, we have

$$\theta_2 + \theta_3 = 1 - \theta_3 = constant. \tag{26.54}$$

According to the conclusion of n = 2 above, we have

$$\begin{aligned} \theta_2 &= \frac{\beta_2}{\beta_2+\beta_3} \\ \theta_3 &= \frac{\beta_3}{\beta_2+\beta_3} \end{aligned} \tag{26.55}$$

to be the optimal point.

Thus, the optimal control method is to keep

$$\begin{aligned} \theta_1 &= \frac{\beta_1}{\beta_1+\beta_2+\beta_3} \\ \theta_2 &= \frac{\beta_2}{\beta_1+\beta_2+\beta_3} \\ \theta_3 &= \frac{\beta_3}{\beta_1+\beta_2+\beta_3} \end{aligned} \tag{26.56}$$

The minimum P_t is

$$minP_t = \frac{P_0}{\eta_1\left(\frac{P_0}{1+\beta_2+\beta_3}\right)} \qquad (26.57)$$

The maximum overall energy efficiency η_t is

$$max\eta_t = \eta_1\left(\frac{P_0}{1+\beta_2+\beta_3}\right). \qquad (26.58)$$

(3) $n = k$

The human team has k variables, the above conclusion can be extended to the case of $n = k$, the optimal point and the optimal method is to keep

$$\theta_i = \frac{\beta_i}{\sum_{l=1}^{k}\beta_l}. \qquad (26.59)$$

The minimum P_t is

$$minP_t = P_0\frac{1}{\eta_1\left(\frac{P_0}{\sum_{l=1}^{k}\beta_l}\right)} \qquad (26.60)$$

The maximum overall energy efficiency η_t is

$$max\,\eta_t = \eta_1\left(\frac{P_0}{\sum_{l=1}^{k}\beta_l}\right). \qquad (26.61)$$

26.3 Optimal Number of Working People in a Human Team

1. A human team has m identical people, if the following equations are satisfied

$$\eta\left(\frac{P_0}{n}\right) \geq \eta\left(\frac{P_0}{n-1}\right) \qquad (26.62)$$

$$\frac{\eta\left(\frac{P_0}{n}\right) \geq \eta\left(\frac{P_0}{n+1}\right)}{n \leq m}.$$

Then the number n is the optimal number of working people.

2. A human team has m energy efficiency similarity people, if the following equations are satisfied

$$\eta_1 \left(\frac{P_0}{\sum_{l=1}^{n} \beta_l} \right) \geq \eta_1 \left(\frac{P_0}{\sum_{l=1}^{n1} \beta_l} \right) \tag{26.63}$$

$$n \leq m$$
$$n1 \leq m \,.$$

n1 is any combination other than the optimal combination of n working people this time, and also include other combinations of n working people. The number n is the number of people with optimal working.

26.4   Optimal Scheduling for the Number of Working People

As long as people are sent out, they have to eat whether they work or not. In this regard, people are different from machines. Therefore, when a team is sent out to perform a task, as long as the number of people is determined, there is no problem of optimizing the number of people during the process. There is a problem of optimizing the number of people before departure.

1. A human team has m identical people, and the number n is the optimal number of working people. If P_0 increases to P_{01}, the following relation holds:

$$\eta \left(\frac{P_{01}}{n} \right) = \eta \left(\frac{P_{01}}{n+1} \right) \tag{26.64}$$

$$n \leq m$$

Then P_{01} is the optimal scheduling point between n people working and $n + 1$ people working. When the total required power P_0 is greater than P_{01}, the optimal number of working people is adjusted from n to $n + 1$. If P_0 increases until $P_0/n = P_{1m}$, there isn't the point P_{01}, then $P_0/n = P_{1m}$ is the adjusting point from n to $n + 1$.

If P_0 is reduced to P_{02}, the following relation is established

$$\eta \left(\frac{P_{02}}{n} \right) = \eta \left(\frac{P_{02}}{n-1} \right). \tag{26.65}$$

$$n \leq m$$

Then P_{02} is the optimal adjusting point between the n working people and the n-1 working people. When the total required power P_0 is less than P_{02}, the optimal number of working people is adjusted from n to n-1. If P_0 reduces until $P_0/(n-1) = P_{1m}$, there isn't the point P_{02}, then $P_0/(n-1) = P_{1m}$ is the adjusting point from n to n-1.

Fig. 26.4 Energy efficiency comparison curve of the team with identical efficiency people

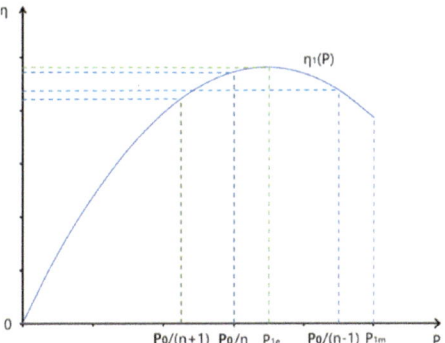

The analysis process is shown in Fig. 26.4.

2. A human team has m energy efficiency similarity people, and the number n is the number of the optimal working people. If P_0 increases to P_{01}, the following relation holds:

$$\eta_1\left(\frac{P_{01}}{\sum_{l=1}^{n}\beta_l}\right) = \eta_1\left(\frac{P_{01}}{\sum_{l=1}^{k_1}\beta_l}\right). \tag{26.66}$$
$$n \leq m$$

Then P_{01} is the optimal adjusting point between the operation of n working people and the operation of k_1 working people. When the total required power P_0 is greater than P_{01}, the optimal number of working people is adjusted from n to k_1. If P_0 increases until $P_0/\sum_{l=1}^{n}\beta_l = P_{1m}$, there isn't the point P_{01}, then $P_0/\sum_{l=1}^{n}\beta_l = P_{1m}$ is the adjusting point from n to k_1.

If P_0 is reduced to P_{02}, the following relation is established

$$\eta_1\left(\frac{P_{02}}{\sum_{l=1}^{n}\beta_l}\right) = \eta_1\left(\frac{P_{02}}{\sum_{l=1}^{k_2}\beta_l}\right). \tag{26.67}$$
$$n \leq m$$

Then P_{02} is the optimal switching point between the operation of n transformers and the operation of k_2 transformers. When the total required power P_0 is less than P_{02}, the optimal number of transformers in operation is switched from n to k_2. If P_0 reduces until $P_0/\sum_{l=1}^{k_2}\beta_l = P_{1m}$, there isn't the point P_{02}, then $P_0/\sum_{l=1}^{k_2}\beta_l = P_{1m}$ is the adjusting point from n to k_2.

The analysis process is shown in Fig. 26.5.

Fig. 26.5 Energy efficiency comparison curve of the team with similarly efficiency people

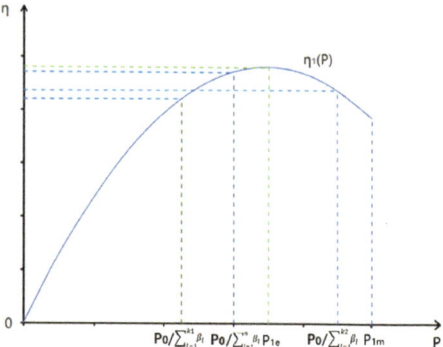

k1 and k2 are any combination other than the optimal combination of n working people this time, and also include other combinations of n working people. Point $P_0/\sum_{l=1}^{k1} \beta_l$ is the point closest to $P_0/\sum_{l=1}^{n} \beta_l$ to the left of point $P_0/\sum_{l=1}^{n} \beta_l$. Point $P_0/\sum_{l=1}^{k2} \beta_l$ is the point closest to $P_0/\sum_{l=1}^{n} \beta_l$ to the right of point $P_0/\sum_{l=1}^{n} \beta_l$. If all people are identical, we have $k1 = n + 1$ and $k2 = n-1$.